PROFESSOR
MAXWELL'S
DUPLICITOUS
DEMON

PROFESSOR MAXWELL'S DUPLICITOUS DEMON

THE LIFE AND SCIENCE OF JAMES CLERK MAXWELL

BRIAN CLEGG

ICON

Published in the UK and USA in 2019 by
Icon Books Ltd, Omnibus Business Centre,
39–41 North Road, London N7 9DP
email: info@iconbooks.com
www.iconbooks.com

Sold in the UK, Europe and Asia
by Faber & Faber Ltd, Bloomsbury House,
74–77 Great Russell Street,
London WC1B 3DA or their agents

Distributed in the UK, Europe and Asia
by Grantham Book Services, Trent Road,
Grantham NG31 7XQ

Distributed in the USA
by Publishers Group West,
1700 Fourth Street, Berkeley, CA 94710

Distributed in Canada by Publishers Group Canada,
76 Stafford Street, Unit 300
Toronto, Ontario M6J 2S1

Distributed in Australia and New Zealand by
Allen & Unwin Pty Ltd, PO Box 8500,
83 Alexander Street, Crows Nest, NSW 2065

Distributed in South Africa by
Jonathan Ball, Office B4, The District,
41 Sir Lowry Road, Woodstock 7925

Distributed in India by Penguin Books India,
7th Floor, Infinity Tower – C, DLF Cyber City,
Gurgaon 122002, Haryana

ISBN: 978-178578-495-8

Typeset in Ten Oldstyle by Marie Doherty

Printed and bound in the UK by
Clays Ltd, Elcograf S.p.A.

Contents

ABOUT THE AUTHOR

Brian Clegg is the author of many books, including most recently *The Graphene Revolution* (Icon, 2018) and *The Reality Frame* (Icon, 2017). His *Dice World* and *A Brief History of Infinity* were both longlisted for the Royal Society Prize for Science Books. Brian has written for numerous publications including *The Wall Street Journal, Nature, BBC Focus, Physics World, The Times, The Observer, Good Housekeeping* and *Playboy*. He is the editor of popularscience.co.uk and blogs at brianclegg.blogspot.com.

www.brianclegg.net

For Gillian, Rebecca and Chelsea

Acknowledgements

As always, thanks to the brilliant team at Icon Books who were involved in producing this book, notably Duncan Heath.

Thanks also to the various experts who have written on James Clerk Maxwell, and to the help from David Forfar and John Arthur of the James Clerk Maxwell Society.

In which the demon is summoned

I appreciate that demons rarely feature in popular science titles. Not even in books on the god particle,* which is somewhat remiss. Yet a demon I am. I was originally summoned into being by the eminently respectable, God-fearing Scottish professor James Clerk Maxwell, and proclaimed to be a demon by his fellow Scot and physicist William Thomson. I was born – as are so many things in your universe – out of the second law of thermodynamics.

This 'law of thermodynamics' business may sound boringly mired in the steam age, and that's certainly how it originated. But the second law determines how the universe works. Strictly speaking, incidentally, the second law is the third law, as an extra one was added in at the top of the list after the first two were proclaimed, but to avoid – or possibly cause – confusion, the late-comer was named the zeroth law. The second law

---

* For those not familiar with this term, it is a nickname for the Higgs boson, which came to public attention when it was discovered using the Large Hadron Collider at CERN in 2012. Amusingly for those of us with demonic tendencies, physicist Leon Lederman wanted to call his book on the search for the particle *The Goddamn Particle*, because the Higgs was such a pain to pin down. The publishers objected that this might be considered too irreverent by the public and resorted to the misleading alternative of *The God Particle*, which really winds up most physicists.

can be phrased in two ways, either of which sounds perfectly innocuous. Yet in those simple statements lie the foundations of reality and the doom of everything.

It's the second law that decides that effect follows inevitably from cause. It's the second law that ensures that books on perpetual motion machines remain on the fiction shelves in the library. Indeed, it's the second law that determines the flow of time in your world (it's far more flexible in mine). If you could prove that the second law could be broken, you would set chaos loose to reign in the world. As a demon, this sounds an attractive proposition – and it's appropriate, as breaking that law is exactly what I was created to do.

How does my charge sheet read? You can either say that the law states that heat passes from a hotter to a colder body, or that entropy – the measure of the disorder in a system – always stays the same or increases. But I was brought into being to challenge this law. Do you think it doesn't matter if some piddling law of physics is broken? This is the law that explains why a dropped glass breaks and never unbreaks. It makes it possible for life to exist on Earth and it predicts the end of the universe. And without it, the many engines that your lives depend on, from cars to computers, would fail. So, don't disrespect the second law.

The early twentieth-century English physicist and science writer Arthur Eddington* said: 'If someone points out to you that your pet theory of the universe is in disagreement with Maxwell's equations [James Clerk M's masterpiece that describe how electromagnetism works] – then so much the worse for Maxwell's equations. If it is found to be contradicted

---

* A man totally lacking in the wondrous beard sported by each of his Scottish counterparts.

by observation – well these experimentalists do bungle things sometimes. But if your theory is found to be against the second law of thermodynamics I can give you no hope; there is nothing for it but to collapse in the deepest humiliation.'

Which raises the curtain for me. My sole purpose in life is to show that the second law of thermodynamics can indeed be broken. I enable heat to travel from a colder to a hotter place. Uncomfortably for a demon, I am able to *reduce* the level of disorder in the world. And if I can truly achieve this, it's not me, but every physicist since Victorian times who must collapse in the deepest humiliation.

I am, as Churchill might have put it, a riddle, wrapped in a mystery, inside an enigma. Whether anyone has been able to find the key to defeat me remains to be seen in the pages to come. But first, we need to discover the young James Clerk Maxwell.

At the risk of sounding like Frankenstein's monster, prepare to meet my creator.

Chapter 1

Not a little uncouth in his manners

There was nothing to suggest the coming of a demon in James Clerk Maxwell's early life. We ought to get that convoluted name untangled first of all. Over the years, those writing about him have never been sure what to call him. Some have resorted to Clerk Maxwell or even an approach he would never have countenanced, the hyphenated Clerk-Maxwell, but his name was not really double-barrelled and 'Maxwell' does the job far better.

Maxwell's father was originally called John Clerk (pronounced to rhyme with 'park'). This family, existing on the boundary between the upper middle class and the aristocracy, had a complex history. One of Maxwell's distant ancestors, another John Clerk, had bought the vast lowland Scottish estate of Penicuik, and with it a baronetcy* back in 1646. His second grandson married an Agnes Maxwell, who brought with her the equally impressive estate of Middlebie. Over the years (and quite a lot of intermarrying of cousins) the name 'Clerk' was

---

* It's the demon here – I'll be handling the footnotes throughout the book. There is something decidedly demonic about footnotes. For those of you not familiar with the English system of titles, a baronet is the only hereditary title that does not make someone a lord – they're a knight. If buying one sounds a little cheesy, bear in mind the whole idea was dreamed up by James I as a fundraiser.

always associated with Penicuik and Maxwell with Middlebie – and when appropriately named cousins came together, they sometimes took the name Clerk Maxwell.

By Maxwell's father's time, Middlebie was only a shadow of its former self, a 'small' 1,500-acre (600-hectare) holding, which is why their estate house ended up a good 30 miles from the town of Middlebie itself. The rest of the estate was sold off to cover some risky speculation in mining and manufacturing by Maxwell's great-grandfather. John Clerk's older brother George was the principal heir, but part of John's inheritance was what was left of the Middlebie estate. This was not an act of generosity on George Clerk's part. The estate was entailed such that Middlebie and Penicuik could not be held together – otherwise, he would likely have held on to the whole thing. Splitting estates was considered bad form. When John Clerk received this new position, he took the traditional laird's name, tacking 'Maxwell' on after Clerk.

Edinburgh and Glenlair

James Clerk Maxwell was born on 13 June 1831, at his parents' home, 14 India Street, Edinburgh – now, appropriately enough, the home of the James Clerk Maxwell Foundation. This was a three-storey townhouse on a cobbled street set back from Queen Street, one of the three parallel roads that form the heart of the city. Maxwell was a late and, in all probability, a spoiled child. His mother, Frances Cay before marriage, had lost her first child Elizabeth as a baby. Frances was almost forty when Maxwell turned up.

Maxwell's father, John, had been a successful advocate (the Scottish equivalent of a barrister), but by the time Maxwell was two, John Clerk Maxwell had settled into his new role of country landowner. The family left the Edinburgh house behind, still owning it but renting it out throughout Maxwell's life. Middlebie had no grand manor, unlike brother George's imposing Palladian-style Penicuik House,* but John and Frances arranged for a relatively humble home, Glenlair, to be built for them on the farmland known as Nether Corsock.

The social distance between the lively city of Edinburgh and the rural isolation of Middlebie was far more than the 80 or so miles between them suggests. Edinburgh was a modern Victorian city, encouraging scientific and literary thought. Middlebie might as well have remained stuck two centuries in the past. And that 80 miles was made to seem greater still by the difficulties of travelling in rural parts of Scotland. The route, via Beattock, took two complete days, needing a stop along the way. The vehicles available were hardly state-of-the-art. In the biography of Maxwell written just three years after his death by Lewis Campbell, a lifelong friend since school who became a professor of classics, and William Garnett, another friend who was an English electrical engineer, it is noted that:

> Carriages in the modern sense were hardly known to the Vale of Urr. A sort of double-gig with a hood was the best apology for a travelling coach, and the most active mode of locomotion was in a kind of rough dog-cart, known in the family speech as a 'hurly'.

It's indicative of John's nature – which seems to have been inherited by his son – that when outbuildings were added to the house

---

* Penicuik House has been a shell since being destroyed by fire in 1899, but it was partly restored around 2014 and is now open to visitors.

in 1841, not only did John plan what was required, he drew up the working plans for the builders to use. Although he was a lawyer, according to Maxwell's early biographers, when not on a case, John 'dabbled between-whiles in scientific experiment'. He even published a paper in *The Edinburgh New Philosophical Journal* on an automated printing device entitled 'Outline of a plan for combining machinery with the mechanical printing-press'. John Clerk Maxwell was exactly the right kind of father to encourage his son to take an interest in the natural world.

Maxwell's first eight years must have seemed idyllic for a well-off child of the period. His parents allowed him remarkable freedom, neither preventing him from mixing with the local farm children, nor beating out of him the thick Galloway accent he picked up from his friends, which surely must have put a strain on their class-driven sensibilities. In fact, they seem to have been unusually unstuffy for a Victorian family.* Theirs was a home where there was little room for formality, but plenty of humour, an approach to life that would later stand Maxwell in good stead.

The estate combined the contrasting terrains of moorland and farmland and ran alongside the curving banks of the River Urr. A small burn feeding the Urr ran at the edge of the meadow beyond the house. By digging out a hollow in the bed of the burn, the Maxwells provided themselves with a swimming pool. Though it would have been freezing cold even at the height of summer, it was no doubt a great attraction for the young Maxwell.

Given their relative wealth, the Maxwells could have readily afforded a tutor for their son. It's telling that when Mary Godwin

---

* Strictly speaking, the Victorian era didn't begin until Maxwell was six, but I am inclined to allow the author some leeway here.

(later Mary Shelley), the author of *Frankenstein*, was young, her family was described as being of a 'very restricted income', yet her brothers were sent to boarding school and she had 'tutors in music and drawing as well as a governess'. The Maxwell family was far better off than the Godwins, but displaying an unusual interest in her child for a wealthy parent of the day, Frances looked after Maxwell's schooling herself. Things would soon change, though. The death of Frances from abdominal cancer in 1839, aged just 47, must have caused the bottom to drop out of the eight-year-old James's world.

While his father, John, had certainly gone along with Frances' wish to devote her time to raising the boy, he either wasn't able or didn't wish to do the same himself. It was one thing to let the young Maxwell play with his local contemporaries, but the nearby schools were very limited in their educational standards and John could not see his son attending one. For a while, he tried out a young man as a tutor, just sixteen when he took on the job. The teenager had neither the talent nor the experience to keep the bright and curious young Maxwell interested and his efforts failed miserably. Maxwell became difficult and would not accept his lead.

The tutor (whom Maxwell later felt it inappropriate to name) was also rough, even by the standards of the period. Maxwell's experience apparently included being 'smitten on the head with a ruler and [having] one's ears pulled 'til they bled'. As his contemporary biographers who knew him well put it, the effects of this harsh treatment remained 'in a certain hesitation of manner and obliquity of reply which Maxwell was long in getting over, if, indeed, he ever quite got over them'.

In this difficult time, Maxwell's release was the chance to roam free on the estate, observing the natural world close-up.

This is something that his father had always encouraged, and Maxwell took a particular interest in the variations in colour he saw in nature. He was especially interested in crystals, which fascinated him in the way that their colours changed as they were put under pressure. His father's friend, Hugh Blackburn, a professor from Glasgow University, added a novel delight, allowing Maxwell to help him launch a series of hot air balloons from the Glenlair estate.

Maxwell had the usual youngster's excitement and interest in everything around him. According to the early biography, among his favourite phrases were 'Show me how it does', and 'What's the go o' that*?' This enthusiastic curiosity about the world around us seems natural in youth – speak to children at primary school and you can't miss the way that they are enthused by science – but many lose that sense of wonder during their secondary school years. Maxwell held on to a childlike fascination for the rest of his life.

The Academy

It was clear, though, that the attempt to use the failing tutor to deal with Maxwell's education was a disaster that could not be sustained; Frances' sister, Jane Cay, who lived in Edinburgh, came to the rescue. She suggested to John that Maxwell could come to live in the city with John's unmarried sister Isabella. Isabella's house was ideally placed to walk to the prestigious Edinburgh Academy – Maxwell could get a decent education and live under the supervision of his aunts during term time, then return to roam free on the Glenlair estate in the holidays. This wasn't, however, a matter of his father dismissing Maxwell

---

* In other words, 'What makes it go?'

solely to his aunts' care – in the winter particularly, John Clerk Maxwell spent regular evenings in Edinburgh with his son.

Glenlair wasn't a grand aristocratic country house – it was effectively a large farmhouse,* though Maxwell would extend it considerably in the 1860s. It was big enough to entertain and to have space for Maxwell's scientific ventures when he was older, but on a scale where it still felt homely. Glenlair would remain an important focal point for Maxwell throughout his life.†

Despite the suggestion that he was rendered hesitant by the bad treatment of his tutor, Maxwell seems not to have been a sensitive child. And it's just as well, given his reception when he was sent to the Edinburgh Academy for the first time at the age of ten. Schoolchildren have never been slow to pick on those who are different, and Maxwell offered them rich opportunities for mockery, especially as the first-year class was full and so he was plunged straight in with older, better-established boys.

It wasn't just his accent, marking him out as provincial, that made the young Maxwell a target for mockery. He arrived at the school dressed in a combination of tweed jacket, frilly-collared shirt and brass-buckled shoes that were guaranteed to make him look like a mongrel throwback from fashion history. Maxwell reported that he returned home on the first day with his tunic

---

* Maxwell was not the first great physicist to be brought up in a house with airs and graces beyond its physical reality. Newton's childhood home, the impressive-sounding Woolsthorpe Manor, was equally nothing more than a large farmhouse. It's interesting to speculate whether the hands-on life of a farm provides an ideal encouragement to take an interest in the world around us.

† Unlike the other houses Maxwell lived in throughout his working life, most of which remain in good condition, Glenlair is mostly a ruin since a fire in the 1920s, though the oldest part of the house remained habitable and was renovated in the 1990s.

reduced to rags, though he appeared to find this more amusing than frightening.

The Academy was a relatively new school, which had been open for just eighteen years when Maxwell first attended. It was set up to compete with the classical education provided by English public schools. As such, it had a focus on giving its pupils independence and hard discipline alongside a rigid curriculum that focused intensely on the classics with perhaps a spot of maths; there was very little science. As the father of the founder of the Scouting movement Robert Baden-Powell commented in 1832: 'Scientific knowledge is rapidly spreading among all classes except the higher, and the consequence must be, that that class will not long remain the higher.'

Having such a limited curriculum seemed to be a mark of pride in the public schools. John Sleath, High Master of the prestigious St Paul's School in London during the early part of the nineteenth century, wrote to his parents: 'At St Paul's School we teach nothing but the classics, nothing but Latin and Greek. If you want your boy to learn anything else you must have him taught at home, and for that purpose we give him three half-holidays a week.'

This was a period when public schools were hardly centres of excellence. For example, the pupils of Rugby School took their masters prisoner at sword-point and were overcome after the reading of the Riot Act resulted in an armed rescue. With very little parental supervision, many schools, even the big names, provided a shoddy education in return for their fees. At Eton, to keep the costs of teaching staff down, boys could be taught in groups that were nearly 200 strong. While conditions were not so extreme at Edinburgh, in Maxwell's early years, classes could have 60 or more pupils.

However, reforms were underway in the school system, with more opportunity to have a 'modern' side as an alternative to the classics, and Edinburgh Academy was arguably more up-to-date in its approach than many of its older English equivalents. Even so, not used to the pressures of school life, having always had the time to think at his own pace, Maxwell came across as slow to learn. A combination of this and his rural accent earned him the nickname Dafty, which stuck even when it became clear that he was extremely academically gifted. Inevitably, though, so far away from his familiar home and the estate, it took Maxwell a while to bed in. A classmate called him 'A locomotive under full steam, but with the wheels not gripping the track.'*

Maxwell was not exactly a loner at school, but simply seemed to carry on as he had before, doing his own thing – if others wanted to join him, that was fine, but he seemed in no hurry to conform. Thankfully, his aunts quickly provided him with more conventional clothing when it was realised that his dress appeared more than a little eccentric. Maxwell certainly seemed comfortable when at home at Isabella's house, 31 Heriot Row, a handsome four-storey grey stone townhouse with a small park out the front. He had a chance to explore both the house's excellent library and what the natural world of Edinburgh had on offer for him to observe. Though the school took boarders, it always had day boys as well, including Maxwell.

With time, Maxwell's limited social contact at school grew. A like-minded student who did not consider it embarrassing to

* This was a distinctly trendy simile from Maxwell's contemporary at the Edinburgh Academy – Maxwell started school in 1842, only twelve years after the world's first steam railway, the Liverpool and Manchester, was opened. Presumably, the railway then had the same fascination for schoolchildren as space travel has more recently.

be academic, Lewis Campbell moved to live near Heriot Row, and soon the two boys spent their journeys to and from school together, developing a strong bond that would last a lifetime. They had now reached an age when the school added mathematics to its limited classical curriculum – something omitted in the first two years – and Maxwell not only found that he excelled at the subject, but that he and Campbell shared a love of maths (and a certain amount of rivalry in their ability to solve mathematical problems).

Once this barrier was broken through, it seemed easier to gain friends who had an interest in science and nature, notably Peter Tait. Another lifelong friend, Tait would himself go on to become one of Scotland's leading physics professors, in his early career even managing to beat Maxwell to take an academic post. At school, Maxwell came second to Tait in mathematics in 1846 (at the time his best subjects were scripture, biography and English verses) but pulled ahead in 1847. When secure in his little group with Tait and Campbell, Maxwell loved the opportunity to puzzle through mathematical and physical challenges, something that inspired him at the age of fourteen to come up with his first academic paper – though strangely his investigations owed as much to the arts as the sciences.

The young mathematician

Maxwell's father regularly took him to meetings of both the Royal Society of Edinburgh and the Royal Scottish Society of Arts (RSSA). It was here that Maxwell became familiar with the work of the local artist David Ramsay Hay. In Hay's philosophy, Maxwell found a point of view that was similar to his own – Hay both delighted in the beauty of nature and wanted to apply scientific measurements to it. Maxwell would later spend much effort on the nature of colour and colour vision – Hay was interested in a mathematical representation of the beauty of colour. But, equally, Hay was fascinated by the mathematics of shape and it was here that Maxwell's paper seems to have drawn its inspiration. Hay would later give a paper at the RSSA on 'Description of a machine for drawing a perfect egg-oval'. Maxwell's youthful paper was on the subject of curves such as ovals that can be drawn using a pencil, a piece of string and pins.

Maxwell's experimental apparatus resembled a primitive version of the popular 1960s toy Spirograph. By placing pins through a sheet of paper into a piece of card and looping a length of string around the pins, it's possible with some care to draw simple geometric shapes. With a single pin, you get a circle. Two pins produce the dual foci of an egg-like ellipse. This much was standard school fare, but Maxwell took the investigation significantly further. He looked at what would happen with the string tied to one or more pins and the pencil, allowing for different numbers of loops around each of the two pins, and worked out an equation that linked the number of loops, the distance between the pins and the length of the string.

Maxwell shared his work with his father, who showed it to his friend James Forbes, Professor of Natural Philosophy* at Edinburgh University. Fascinated by this precocious piece of work, Forbes brought in a mathematician from the university,

---

* Natural Philosophy was the generic term for science until the nineteenth century, as science originally simply meant a topic of knowledge – so, for instance, the favourite subject of us demons, theology, was known as the 'queen of the sciences'. A practitioner of what we would now call science was known as a natural philosopher. As philosophy became an increasingly specific discipline, the label changed to natural sciences, a term still used by some of the older universities.

Philip Kelland, who checked through the literature for prec-
edents.* Although some similar work had been done by the
French scientist and philosopher René Descartes, Kelland
discovered that not only was Maxwell's approach simpler and
easier to understand than Descartes', it was more general than
the results that Descartes had published.

Given the originality of young Maxwell's work, Forbes was
not going to let the effort go by rewarded with nothing more
than a pat on the head. He managed to present Maxwell's paper,
now grandly titled 'Observations on Circumscribed Figures
Having a Plurality of Foci, and Radii of Various Proportions',
at the Royal Society of Edinburgh in April 1846. The fourteen-
year-old Maxwell could not present the paper himself as he was
both too young to do so and not a member, but he had regularly attended Royal Society meetings with his father and was
present to hear his work read. The paper was well received and
cemented Maxwell's growing feeling that his future lay in science and mathematics. It is too long (and, frankly, too boring)
to reproduce here, but here is the opening sentence to get a
feel for the young Maxwell's precocious (and somewhat long-
winded) output:

> Some time ago while considering the analogy of the Circle and
> the Ellipsis – and the common method of drawing the latter fig-
> ure by means of a cord of any given length – fixed by the ends of
> the foci – which rests on the principle, that the sum of the two
> lines drawn from the foci to any point in the circumference is

---

* There were none of your lazy internet searches back then, of course; this
was a matter of physically sorting through books and journals. It's worth thinking that without Maxwell's work, we might not even have the internet today.

a constant quantity, it occurred to me that the *Sum* of the Radii being constant was the essential condition in all circumscribed figures, and that the foci may be of any number and the radii of various proportions.

Maxwell must have been delighted to see as august a body as the Royal Society of Edinburgh begin the description of his work with: 'Mr Clerk Maxwell ingeniously suggests the extension of the common theory of the foci of the conic sections to curves of a higher degree of complication in the following manner:—.' Although Maxwell continued with his general education, he began to read voraciously from the books and papers of the scientific greats, developing a particular affection for the down-to-earth approach of the self-taught English scientist Michael Faraday, who had become a leading light of the Royal Institution in London by the time Maxwell was at school.

Churchman and country squire

We tend these days to make a clear distinction between scientific study and religious beliefs, but Maxwell came from the last generation in the British tradition where there was no feeling of conflict between the two. Like many of the scientific greats before him (including Faraday and, in his own strange way, Newton), Maxwell had a deeply held religious faith. On his breaks from Edinburgh, back home in Glenlair, the family and their servants would join together each day in prayer, and the entire household made the five-mile trek each Sunday to attend the Presbyterian Church of Scotland's Parton Kirk – where his mother was buried, in a grave inside the ruins of the Old Kirk that would eventually also hold his father, Maxwell himself and Maxwell's widow. While he was in Edinburgh, his Aunt Jane made sure this observance was kept up, taking him to attend both Episcopal and Presbyterian churches, cementing a religious faith that remained strong throughout Maxwell's life.

Regular breaks at Glenlair would remain an essential for Maxwell, whether he was a student or professor. It was a total break from the bustle of the city or the rigours of an academic institution. In his obituary for Maxwell, his friend Peter Tait would comment of his schooldays:

> [H]e spent his occasional holidays in reading old ballads, drawing curious diagrams, and making rude* mechanical models. His absorption in such pursuits, totally unintelligible to his schoolfellows (who were then quite innocent of mathematics), of course procured him a not very complimentary nickname ...

Wherever he worked as an adult – he would later be based in Aberdeen, London and Cambridge – Maxwell's summers would be spent on the Glenlair estate. When at home, in almost all respects, Maxwell would be a typical country gentleman of the period – except for his unusual enthusiasm for and delight in nature. Where most of his contemporaries enjoyed nothing better than a mass slaughter in the shooting season, Maxwell never took part in hunting and shooting.

Even though he continued to live at 31 Heriot Row with Aunt Isabella, the influence of Maxwell's family was about to wane, as he transferred from the Academy to Edinburgh University at

---

* This is typical of the humour of Maxwell's circle: 'rude' here means 'rough and ready', but is used to get in a sly reference to Shakespeare, who has Puck speak of 'rude mechanicals' in *A Midsummer Night's Dream*.

the age of sixteen. It might seem after his clear demonstration of mathematical originality that he would have already set his sights on a mathematical or scientific career, but this was a period when professional scientists like Michael Faraday and Faraday's former boss Sir Humphry Davy were in the minority. The word 'scientist' was only coined in 1834 when Maxwell was three, and took a while to settle in. Some alternatives of the period were 'scientician' and 'scientman'. Maxwell is often considered one of the first truly modern scientists.

It was not that landed gentry did not partake in science. It was just that someone of Maxwell's status was far more likely to perform their scientific work as an amusement, a hobby to pass the time – so Maxwell's original intention had been to follow his father in entering the law. However, Edinburgh University was still using the traditional broad approach of the ancient university curricula, so had both mathematical and natural philosophy (science) content in its degree course. It's notable from a letter that Maxwell wrote to Lewis Campbell in November 1847 that the maths and science were the parts that dominated his interest:

As you say, sir, I have no idle time. I look over notes and such like until 9.35, then I go to Coll., and I always go one way and cross streets at the same places; then at 10 comes Kelland [mathematics lecturer, Philip Kelland]. He is telling us about arithmetic, and how the common rules are the best. At 11 there is Forbes [Maxwell's father's friend, the physics professor], who has now finished introduction and properties of bodies, and is beginning Mechanics in earnest. Then at 12, if it is fine, I perambulate the Meadows; if not, I go to the Library and do references. At 1 I go to Logic [with Sir William Hamilton].

The only passing mention of the classics in his letter is to say, 'I intend to read a few Greek and Latin [textbooks] beside'. Classics was a compulsory part of the majority of university courses. There is no mention at all of the law – this would be picked up after his university degree.

Perhaps most importantly, Maxwell had access to the university's limited laboratory equipment (likely to be in an outhouse, as there was no purpose-built lab at Edinburgh in 1847) when he had the time, encouraged by family friend Professor James Forbes. It was here, and in a workroom at Glenlair during the long summer vacation, as much as in the formal training he received at the university in logic and natural philosophy, that Maxwell's unstructured, youthful scientific curiosity was forged into a first-class scientific mind.

The university life

At the time, some of the personal oddities that had got Maxwell mocked at school still remained part of his nature. His early biographers note: 'When he entered the University of Edinburgh, James Clerk Maxwell still occasioned some concern to the more conventional among his friends by the originality and simplicity of his ways. His replies in ordinary conversation were indirect and enigmatical, often uttered with hesitation and in a monotonous key.' While he grew out of this (apart, apparently, from when 'ironically assumed'), his relative frugality, preferring the third-class railway carriage to the first, and a tendency to lose himself in thought while at the dinner table would remain with him for life.

The experimental side of the course at Edinburgh was limited and sometimes verged on the amateurish. Maxwell noted in a letter to his friend Lewis Campbell:

> On Saturday, the natural philosophers ran up Arthur's Seat with the barometer. The Professor [presumably Forbes] set it up at the top and let us pant at it till it ran down with drops. He did not set it straight, and made the hill grow fifty feet; but we got it down again.

The barometer in question was likely to be an inverted tube of mercury, measuring atmospheric pressure which was then used to calculate the height above sea level of the famous rocky outcrop above Edinburgh.

In the same letter, Maxwell makes a first mention of a devil that would be a companion for much of his life – though not the titular demon of this book. He wrote:

> Then a game of the Devil, of whom there is a duality and a quaternity of sticks, so that I can play either conjunctly or severally. I can jump over him and bring him round without leaving go of the sticks, I can also keep him up behind me.

This refers to the game known as 'the devil on two sticks', now more commonly called diabolo, where a double cone, joined point to point, is kept in the air using a string between two rods.

For much of his career, Maxwell would supplement his academic work with experiments in a series of home laboratories that would eventually have been better equipped than a university. It was not until he was involved in setting up the prestigious Cavendish Laboratory in Cambridge (see Chapter 8) that he would have significant access to a professional, university-based workshop. During his time at Edinburgh University, he got together a small lab at Glenlair in a room over the wash-house. In the summer of 1848 (when Maxwell was seventeen), he wrote to Lewis Campbell:

I have regularly set up shop now above the wash-house at the gate, in a garret. I have an old door set on two barrels, and two chairs, one of which is safe, and a skylight above, which will slide up and down.

On the door (or table), there is [*sic*] a lot of bowls, jugs, plates, jam pigs,* etc., containing water, salt, soda, sulphuric acid, blue vitriol,† plumbago ore;‡ also broken glass, iron and copper wire, copper and zinc plate, bees' wax, sealing wax, clay, rosin, charcoal, a lens, a Smee's Galvanic apparatus,§ and a countless variety of little beetles, spiders, and wood lice, which fall into the different liquids and poison themselves … I am making copper seals with the device of a beetle. First, I thought a beetle was a good conductor, so I embedded one in wax (not at all cruel, because I slew him in boiling water in which he never kicked), leaving his back out, but he would not do.

Although Maxwell was busy with his experiments that summer, it didn't stop him from writing highly mathematical papers. He had continued to do this since his first success at the Royal Society of Edinburgh aged fourteen, though often the documents were

---

* Not as exciting as it sounds: nothing more exotic than jam jars.

† The bright blue chemical compound, copper sulfate.

‡ Confusingly, since the name is reminiscent of the Latin for lead (*plumbum*), plumbago is in fact naturally occurring carbon – graphite. The ore was frequently confused for lead ore or galena as both are found as shiny black deposits – hence the way we still refer to the graphite in a pencil as its lead.

§ Surely the highlight of Maxwell's equipment, this was an impressive mahogany-framed six-cell battery retailing at an expensive £3 10 shillings. The advertising for the apparatus claimed that it would 'heat to redness 4 inches of platinum wire, fuse iron wire with facility, and empower a sufficiently strong electro-magnet to sustain many hundredweights'.

just handwritten for the consumption of his friends. However, in 1848 he wrote a paper stretching to 22 long pages called 'On the Theory of Rolling Curves', published the following year in the *Transactions of the Royal Society of Edinburgh*. This combines geometry with some sophisticated algebra and calculus, describing how one curve, rolling along another curve (which is 'fixed to the paper') would produce a third curve.

Quoted in the biography by Campbell and Garnett, Maxwell remarks that his decision to switch from a legal track was made to pursue 'another kind of laws'. Most undergraduates content themselves with the work programme that the university sets, but Maxwell was already at his best when exploring on his own, continuing his early experiments with some remarkably sophisticated developments – something that comes across particularly in his work on stress and polarised light.

A particular light

Maxwell had been introduced to the topic of polarisation – a variation in the direction of oscillation of waves of light, which can be separated by special materials – while still at school. His mother's older brother, John Cay, took Maxwell and Lewis Campbell to visit the optical expert William Nicol, who had found a way to produce polarised light at will.

The concept of polarisation dated back to 1669, when Danish natural philosopher Erasmus Bartholin had been the first to explain the workings of an odd crystal known as Iceland spar. This is a form of calcite – crystalline calcium carbonate. If you put a chunk of the transparent crystal on top of, say, a document, you see not one, but two copies of the writing, shifted with respect to each other. The phenomenon itself had been known for centuries – it has even been suggested that the Vikings may have used 'sunstones' with a piece of Iceland spar in them as a navigating device to estimate distances. But Bartholin's insight was to realise that the crystal split two different forms of light that were both present in ordinary sunlight.

When at the start of the nineteenth century Thomas Young demonstrated that light was a wave that rippled from side to side as it moved forward (known as a lateral or transverse wave), the French physicist Augustin Fresnel realised that this provided an explanation for the special ability of Iceland spar. Light waves from a source such as the Sun would be oriented in all directions – some would be rippling side to side while others oscillated up and down – in fact the waving could take place in any direction at right angles to the direction of the light beam's travel. If the crystal split apart waves rippling in different directions – the direction of the side-to-side ripple being described as its direction of polarisation – then the two images could be the result of the crystal separating rays with two different directions of polarisation.

When Maxwell's uncle John Cay took him and his friend to visit William Nicol, they were shown prisms made from Iceland spar which had the effect of splitting off just one polarisation of light (for a time these optical devices were known as nicols, after their maker). This seems to have inspired Maxwell while he was at Edinburgh University, with polarised light soon becoming the prime focus of his spare-time experiments. It was known that when such light is passed through ordinary glass there is relatively little effect. However, if the same light is shone through unannealed glass, glass that has been heated until it is glowing and then cooled very quickly, the polarised light produces a coloured pattern, caused by the internal stresses in the glass.

Initially, Maxwell experimented with pieces of window glass, heating them to red heat then rapidly cooling them.* In a letter to Lewis Campbell he wrote:

> I cut out triangles, squares, etc., with a diamond, about 8 or 9 of a kind, and take them to the kitchen, and put them on a piece of iron in the fire one by one. When the bit is red hot, I drop it into a plate of iron sparks [filings] to cool, and so on till all are done.

To produce polarised light, he made his own polarisers using a matchbox with pieces of glass set in it to produce reflections (reflected light is partially polarised); he also attempted to make polarisers from crystalline saltpetre (potassium nitrate). Maxwell made watercolour paintings of the brightly coloured patterns that he obtained in his heated and cooled window glass, some of which he sent to William Nicol, who was sufficiently impressed to send Maxwell a pair of his optically precise nicols, producing far better polarised light than Maxwell had been able to obtain with his do-it-yourself matchbox devices.

From an engineering viewpoint, getting an understanding of the stresses inside an object is essential to predict how it will stand up to strain when it is put in use. Maxwell had the insight to see that if, for example, a girder could be made of a transparent material, it would be possible to use polarised light to study the internal stresses as the girder begins to bear a load. Clearly this isn't possible using an actual iron or steel girder – but if a model of it could be constructed in a suitable transparent

---

* Not to be tried at home as the glass may well shatter. As was not unusual at the time, Maxwell seems to have had little concern for personal health and safety.

material, it could be used to discover how stresses form in the structure and change under load, reducing the risk of structural collapse.

Unfortunately, glass doesn't respond well to strain, and the clear plastics and resins that would later be used in this 'photo-elastic' method that Maxwell devised, and which is still used by engineers, weren't available at the time. Instead, with that same make-do-and-mend approach that had seen him attempt to use beetles as part of his electrical toolkit, he got hold of some gelatine from the Glenlair kitchen and used it to make clear jelly shapes. Maxwell was delighted to discover that his jellified models produced exactly the kinds of stress patterns he hoped for as he put them under strain.

The path to Cambridge

When not doing experiments, Maxwell would be working through numerous physical propositions or 'props' as he and his friends called them, often studying the most mundane of objects and trying to deduce something interesting from them. Sometimes these can seem a little bizarre. For example, in a letter written from Glenlair in October 1849 he noted: 'I have got an observation of the latitude just now with a saucer of treacle, but it is very windy.'

As well as more practical work, Maxwell followed up his pin and string mathematical paper and other topics while still an undergraduate. His most outstanding attempt of the period was to derive a mathematical analysis of the stress patterns he had observed using his photoelastic technique. He confirmed these mathematical formulae, covering different basic 3D shapes such as cylinders and beams, as much as he was able with his experimental work. This was a remarkable achievement for someone with his very limited experience, but he was to discover that it wasn't enough to perform careful experiments or to produce mathematics that successfully described them. It was also important that you could communicate your scientific findings effectively. He wrote up his work and asked Professor Forbes to present it to the Royal Society of Edinburgh.

Forbes may have been highly impressed with the younger Maxwell's ventures into mathematics, but this new paper was more directly impinging on his own field, and Maxwell was now nearing adulthood. Forbes did not think much of his writing style in the paper, which was refereed by Maxwell's mathematics lecturer, Philip Kelland. Forbes commented that Professor Kelland 'complains of the great obscurity of several parts owing to the abrupt transitions and want of distinction between what is assumed and what is proved in various passages'.

Professor Forbes went on to say: 'it must be useless to publish a paper for the use of scientific readers generally, the steps of which cannot, in many places, be followed by so expert an algebraist as Prof. Kelland; – if, indeed, they be *steps* at all ...' This kind of criticism could have been deadly for a beginner who took it personally, but it spurred Maxwell into studying the best of the period's scientific writing, analysing the wording and structure to see what made it effective and incorporating what he discovered into his own style. While he never became one of the greats of science communication, after this his papers were usually lucid and well written.

There was something about Maxwell's personality that made him able to adapt well to constructive criticism in this way. He seems to have had the ideal balance of freedom to experiment and try things out, with a network of peers who were prepared to point out his failings and help him overcome them. Like his scientific hero Michael Faraday, Maxwell never gave himself the airs and graces of some of their contemporaries such as Sir Humphry Davy in London or, in later years, Maxwell's regular correspondent William Thomson, who would become Sir William and then Lord Kelvin. Maxwell's religious upbringing, his mixing with the country children on the estate and his down-to-earth humour seem to have protected him from ever having an over-inflated sense of his own importance.

Maxwell's paper on the mathematics of the stresses observed in his photoelastic experiments was originally submitted to Professor Forbes in December 1849. After Forbes and Kelland's feedback, Maxwell redrafted it in the spring of 1850 and the revised version, which had large chunks of the original omitted or reworded, appeared in the *Transactions of the Royal Society of Edinburgh* that year. It was a long piece of work running to 43 pages, which combined some of his own experimental observations with a much wider mathematical analysis.

Although Edinburgh allowed Maxwell considerable freedom in pushing forward his scientific thinking, it was still primarily seen as a track for him to achieve a degree on the way to a career in law. But as he wrote to Lewis Campbell on 22 March 1850:

> I have notions of reading the whole of *Corpus Juris* and Pandects [for his studies of law] in no time at all; but these are getting somewhat dim, as the Cambridge scheme has been howked up from its repose in the regions of abortions, and is as far forward as an inspection of the Cambridge *Calendar* and a communication with Cantabs.*

* I.e. the University of Cambridge, traditionally given the abbreviation 'Cantab' from Universitas Cantabrigiensis, its Latin name.

Maxwell decided that after three years at Edinburgh University, before he had completed his degree, he needed a more thorough scientific and mathematical content to his studies and applied to Peterhouse college, Cambridge, where his friend Peter Tait was already resident. Such a move was not uncommon. Tait had left Edinburgh for Cambridge after just one year and another friend, Allan Stewart, after two years. This change of academic venue required Maxwell's father's support, which seems to have been given wholeheartedly. John Clerk Maxwell travelled down to Cambridge with his son on 18 October 1850, as the young scientist started on the next leg of his academic journey.

In which electricity meets magnetism

With my creator safely set off on his way to Cambridge, this is a good opportunity to give you a little background on a topic that would come to dominate much of Jimmy's* life – electromagnetism. While inevitably from my own viewpoint it's JCM's interest in thermodynamics and the related field of statistical mechanics that makes him of particular value, a truly objective observer[†] would probably have to assess his contributions to electromagnetism as having the greater impact on the world.

Both electricity and magnetism were known as strange natural phenomena long before there was any idea of what might be going on, each considered totally separate from the other.[‡] We can get an idea of where the concept of electricity originated from its name. Both 'electricity' and 'electron' are derived from *electrum*, the Latin for amber (a name itself derived from a similar Greek word).

---

* To me, Maxwell will always be Jimmy or Jim, but the wording offends my editor, so henceforth I shall refer to him as JCM.
† Not a demon, then.
‡ Bizarrely, you humans still teach your young children about electricity and magnetism separately. Sometimes I suspect that demons are responsible for your education system.

Natural electricity

When amber is rubbed, it generates static electricity in the same way that a balloon gets charged up when it is rubbed on your hair. This so-called triboelectric effect involves electrons coming loose as a result of the rubbing process, leaving the object rubbed and what it's rubbed with each having the opposite electrical charge. Electrical attraction means the balloon can then pick up very light objects through an induced charge and it can even produce a tiny spark. Induction comes into the electrical business quite frequently – it simply means that if you bring something that's electrically charged near an object it will tend to repel particles with the same charge in that object, so that the nearest side of the object has an opposite electrical charge, making the object – in this case, perhaps, a small piece of paper – attracted to the original source.

The triboelectric effect is probably also the mechanism responsible on a much larger scale for the charge build-up in the most dramatic natural example of electricity, a bolt of lightning. It's worth spending a little time on lightning to get a feel for the early impression of electricity.

Lightning was surely the first observed electrical phenomenon, though initially its sheer magnitude resulted in it being labelled the work of an irritable god. Thanks to big-budget superhero movies, probably the best-known thunder god these days is the relative latecomer, the Norse god Thor; though earlier, the power of lightning-throwing was thought to be held by Zeus for the Greeks, Jupiter for the Romans, and Indra in the Hindu pantheon. No doubt there have been many more. When gods weren't blamed, lightning, like other strange appearances in the sky such as comets and red moons, was considered to be a warning of dire events to come. In the first century AD, for example, Pliny the Elder said that a thunderstorm was prophetic, direful and cursed.

The superstitious view of lightning is hardly surprising. It's the most dramatic natural phenomenon that the majority of people will experience – certainly the most common of nature's big beasts, with at any one time typically around 1,800 thunderstorms active around the world. And lightning doesn't just look and sound impressive, it has the capability to blast trees and to kill.

One common myth associated with lightning is that it does not strike the same place twice. Some rural areas of Britain used to have a brisk trade in what were known as thunderstones. These were stones that had a hole in the middle, which were bought as a safeguard to place up the chimney of a house. The idea was that the thunderstone had already been struck by lightning, which was thought to have caused the hole, so lightning would be unable to hit the chimney without breaking the 'striking the same place twice' rule.

Unfortunately, there are two problems with this folk remedy. One is that lightning is entirely happy to strike the same place twice and quite often does so. If a location is susceptible to lightning strikes, it's not unusual for it to get several hits in one day. The Empire State Building, for instance, has received as many as fifteen strikes in a single storm. When you think about it, lightning would have to be conscious and directed if it never returned to the same spot.

Perhaps the most impressive example of multiple strikes, though, was not on a building but on the US park ranger Roy Sullivan, who entered the *Guinness Book of Records* as the person who had been hit by lightning the most times – a total of seven strikes, all of which he survived. The other problem with using thunderstones to ward off lightning is that these unusual formations weren't caused by lightning at all – they are the remains of Stone Age hammers where a wooden handle and leather bindings have long ago rotted away.

In general demonic use (and we use them a lot), thunder and lightning are pretty much interchangeable because we now know that thunder is simply the noise produced by the lightning bolt ripping open the air – traditionally they were considered linked but independent events because of the variable time difference between them. This reflects light's immense speed of 299,792,458 metres per second compared with sound's plodding 343 metres per second at sea level.

It's hard to imagine that something as dramatic as a lightning bolt could be down to a similar procedure to rubbing a balloon on your hair – and, to be honest, the means of lightning production isn't anything like certain – but the model that has been generally supported since the 1950s is a triboelectric mechanism. In a thundercloud there are particles of ice and supercooled water droplets as well as graupel (miniature hailstones), all churning and jostling about in the clouds as warmer and cooler air streams collide. This is thought to result in the heavier graupel getting a relative negative charge. The charge becomes separated as the graupel sink towards the bottom of the cloud while the lighter, positively charged particles rise. This conveyor-like process is thought to occur many times, gradually building up a bigger and bigger electrical potential difference.

It has been suggested by some scientists that cosmic rays could also have a role to play in triggering lightning. Cosmic rays are streams of high-energy charged particles that come rocketing towards the Earth, but that are mostly fended off by the planet's magnetic field and the atmosphere. Russian researchers have suggested that cosmic rays could produce a stream of electrons that build up in a chain reaction as the ice particles circulate. However, other scientists consider this mechanism highly doubtful.

From the skies to the laboratory

What we do know for certain, though, is that lightning is nothing more than an electrical discharge like that produced by rubbing a balloon – though admittedly a discharge with a remarkable kick. The eighteenth-century American journalist, diplomat and scientist Benjamin Franklin is famous for demonstrating this by flying a kite in a thunderstorm with a key attached to the kite string. Or rather he certainly described the experiment in 1750 – it's quite possible, though, that he had someone else carry out the risky procedure.

It's very unlikely that Franklin (or whoever he persuaded to take the risk) flew a kite and waited for it to be struck by lightning, as the legendary experiment is often portrayed. Instead, his proposal was to tap into the electrical charge in the thunderclouds to cause a build-up of electricity on the key by induction, with no lightning strike taking place. The charge from the key was then passed to a Leiden jar, an early method of storing electricity, where it could be demonstrated that the electricity from storm clouds was just the same as the tamer-seeming ground-based variety.

There's no doubt that some people have attempted the experiment Franklin described, but it is phenomenally dangerous. Just consider the amount of energy in a typical lightning flash – perhaps half a billion joules. It takes 10 joules to run a 10-watt bulb for a second. The amount of energy in a lightning bolt is more like the output of a mid-sized power station for a second. The electrical current rips through the air, heating it suddenly and causing the rumble of thunder. The temperature in a bolt peaks at around 20,000–30,000°C, as much as five times the surface temperature of the Sun.*

You can't see electricity itself in a lightning bolt. It is atoms that have received energy from the lightning which blast out light as their electrons are boosted in energy then fall back to their usual levels. And what's produced is the full spectrum of electromagnetic radiation, all the way from radio waves up to X-rays and gamma rays. We don't expect electricity to be able to flow through the air because our atmosphere is quite a good insulator. It takes around 30,000 volts to get a spark to jump across just one centimetre at normal levels of humidity (the damper the air, the easier it is for electricity to flow).

It seems reasonable that dampness helps because we are used to water being seen as a good conductor of electricity (which is why it's not a good idea to get electrical equipment wet). Oddly, though, like air, water is actually a reasonably good insulator. Take absolutely pure water and it hardly conducts at all. But it almost always contains the ions of substances that are dissolved in it, and these carry the current. Ions (electrically charged atoms) are also responsible for carrying the electricity through the air in lightning, but to produce the vast streamers of electric discharge in a bolt of lightning still takes a huge amount of electrical power.

Once a significant secondary charge has been induced, something weird happens. There is a relatively weak flow of electricity between the negative storm cloud and its positive

* Or 'nice and cosy' as we demons would say.

target. This flow of electricity ionises the air. Just as in water, a collection of ions in the air conducts electricity much better than a collection of neutral atoms. This weak discharge from the cloud, called a leader, sets up a path for the main burst of lightning, the return stroke, which goes in the opposite direction – in the case of a ground strike, the main stroke runs from the ground up to the cloud rather than in the obvious direction.

Whether or not Benjamin Franklin took a kite out in a thunderstorm, he certainly did invent the lightning rod (also known as a lightning conductor). The idea of this simple device is to provide a metal spike on the highest part of a building, connected via a thick metal conductor to the ground. This spike is most likely to receive a hit and was intended to conduct the electricity away, reducing damage to the structure. In practice more often than not the lightning rod prevents a lightning strike from ever happening. The rod allows any charge being induced around the spike to leak away to the ground, reducing the chances of a leader forming.

Lightning gives a dramatic natural example of electricity in action, but it's difficult to study because it does not take place in a controlled environment. By the eighteenth century, static electricity was being used in dramatic demonstrations such as the 'electric boy', where a youth was suspended from ribbons and charged up by rubbing glass rods with silk. He would then be used to give the audience shocks and to attract light objects. However, it would be the nineteenth century before usable current electricity, where electrical charges flow through wires, became viable. Before we get on to that, we need to take a step back and bring magnetism alongside its electrical cousin.

Magnetic matter

Magnets have been known since ancient times in the form of naturally occurring lodestones. These are chunks of the mineral magnetite – an oxide of iron. Most magnetite has no special properties, but when it carries certain impurities it has the right structure to become a permanent magnet. It's not 100 per cent certain how the lodestones became magnetised in the first place – the most obvious suspect, the Earth's magnetic field, is far too weak. It is suspected, particularly as lodestones only tend to be found near the surface of the Earth, that they were magnetised by lightning strikes (so, in effect, lodestones are the true thunderstones). With the kind of symmetry we demons prefer to avoid – as humans find it far too attractive – once we've united electricity and magnetism, each is capable of producing the other.

The earliest recorded attempt to take a scientific approach to magnetism came from a thirteenth-century French scholar named Peter de Maricourt, though better known as Peter Peregrinus ('Peregrinus' usually referred to a stranger or foreigner – the implication is that he was a wanderer or pilgrim who didn't stay long at a single institution). We know little of Peter himself, though there was a village called Mehariscourt, near the abbey of Corbie in Picardy, the French region that he is associated with in legend. The thirteenth-century English natural philosopher and friar Roger Bacon said of Peter, whom he had met in Paris:

He gains knowledge of matters of nature, medicine and alchemy through experiment, and all that is in the heaven and in the earth beneath. He is ashamed if any old woman or soldier or countryman knows about something he does not know about the country. So he has pried into the work of metal-founders and what is wrought in gold and silver and other metals and all minerals ...

So without him philosophy cannot be completed, nor yet handled with advantage and certainty. But just as he is too worthy for any reward, so he does not seek one for himself. For if he wished to dwell with kings and princes, he could easily find those who would honour and enrich him. Or if in Paris he were to display his knowledge through the works of wisdom, the whole world would follow him. But because in both these ways he would be hindered from the bulk of the experiments in which lie his chief delight, he neglects all honours and riches, especially since he will be able, when he wishes, to reach riches by his own wisdom.

In 1269 Peter completed his *Epistola de Magnete*, a detailed letter which describes the two differing poles of a magnet, attraction and repulsion, how to magnetise iron with a lodestone, and the Earth's magnetism. He also gives some details on constructing compasses, both by the then common approach of floating a magnet on water and by using the more advanced approach of mounting a thin magnet on a pivot. Peter's work describes practical applications rather than providing any theoretical basis, but it was the definitive document on the subject through to the start of the seventeenth century. Then, Peter's work was eclipsed by that of the English natural philosopher William Gilbert.

Gilbert's book, *De Magnete*, went into far more detail than Peter had been able to in his letter, and gave far greater consideration to how the Earth itself could act as a huge magnet. To explore the way that such a magnet would behave, Gilbert constructed spherical metal magnets known as terrella. These helped him understand the property of dip, where a compass needle does not point horizontally due to its position on the Earth's surface.

Inevitably Gilbert did not get everything right. His biggest error was suggesting that gravity was the same as magnetism, though the similarity of principle was a good observation. But his book, which also covered some aspects of static electricity, restarted the interest in thinking about the nature of magnetism over and above its usefulness for constructing compasses.

The birth of electromagnetism

We're now well placed to move on to the discovery of current electricity – electricity that flows from place to place – thanks to Italian physicist Alessandro Volta's 1799 electrical pile (a collection of which was called a battery), and the Danish Hans Christian Oersted's 1820 discovery that electrical currents produce a magnetic effect.

These were precursors to the remarkable work on electricity and magnetism that Michael Faraday would do. Faraday realised that the two were strongly interconnected and made popular the term 'electromagnetism', devised for their combined study by Oersted (though Faraday wrote it, as would be common initially in English, as electro-magnetism). Faraday's discoveries proved essential for the work that JCM would carry out.

Because of Faraday's importance to the development of JCM's thinking, his role needs exploring further. Faraday's family had come to London from Westmorland, in the English Lake District, before he was born in search of work for his blacksmith father. In 1805, at the age of fourteen, Faraday was apprenticed to bookbinder George Riebau, a refugee from the French revolution. Faraday spent all his spare time in the shop, teaching himself from the books that were in to be bound. These heavy volumes, along with the lectures provided by a self-improvement group, the City Philosophical Society, gave Faraday a single-minded intent to break into the closed world of science.

One of Riebau's clients, a Mr Dance, got Faraday a temporary job assisting the Royal Institution's star scientist, Humphry Davy, while Davy's usual assistant was injured. The Institution was a relatively new organisation, established in 1799 by leading British natural philosophers as a means to spread the word about science to the wider population and to provide facilities for undertaking research. Working with Davy was a dream opportunity for Faraday, but he was soon sent back to the bookbinders. Faraday kept up a steady barrage of applications for jobs in scientific establishments. Eventually, in 1813, he got the permanent post of lab assistant at the Royal Institution.

By 1821, with a promotion under his belt, the young scientist was making steady, if unremarkable progress. He was asked to write an article summarising the latest position on electromagnetism, the emerging field of the interaction between electricity and magnetism. To help do so, Faraday replicated the experiments he had read about. As he passed electricity down a wire running alongside a fixed magnet, he saw something that first puzzled him, then challenged his imagination. When the current flowed, the wire moved, circling round the magnet. As far as he could tell, this was a new discovery, and he realised that it needed publication. He wrote it up with little consultation with his peers and was promptly accused of plagiarism.

The man Faraday was supposed to have stolen the idea from was William Wollaston, one of the scientists whose work Faraday had included in his review. Wollaston had started out as a doctor, but had given up medicine as his eyesight was failing. He had decided (with limited evidence) that electricity spiralled its way along wires like a corkscrew. Wollaston had asked his friend Sir Humphry Davy to search for any evidence of this motion. Davy was unable to do so. However, despite a very limited connection, other than electricity and rotation, between Wollaston's theory and Faraday's experiment, Wollaston was convinced that Faraday had stolen his idea. This appalled Faraday. He asked his mentor Davy to help, but to no avail.

While Davy was willing to accept that Faraday did good work, they were miles apart in social status. Davy was a society darling who regularly met with royalty. Early in his career, Faraday had accompanied Davy and the new Mrs Davy on a trip around European scientific establishments. Rather than treat Faraday as an equal, the Davys expected him to take on the role of valet as well as scientific assistant. As far as Davy was concerned, Wollaston, a professional man, was far more of a social equal. He took Wollaston's side. This was the end of anything more than a purely professional relationship between Faraday and Davy.

Luckily, Davy's influence was not sufficient to persuade others that Faraday had copied Wollaston's ideas. And the steady rotation of the wire around the magnet was more than a pretty demonstration: it formed the basis of the electric motor. Faraday had moved out from under Davy's shadow. Two years later Faraday would be elected to a fellowship of the prestigious Royal Society with only one vote against him. That of Sir Humphry Davy.

It would be ten years before Faraday returned to electricity and magnetism. The pain of the accusations and Davy's betrayal bit deep. He turned his attention to chemistry and took on the administrative job of Director of the Royal Institution Laboratory, establishing the Friday Night Discourses and a series of Christmas events for children. But Faraday could not resist the challenge of electromagnetism for ever. By 1831 there was evidence that electricity flowing through a wire could generate a current in another, unconnected wire, somehow communicating across space.

This near-magical proposition – induction – was enough to restart Faraday's enthusiasm for electromagnetism. He constructed a pair of wire coils, wrapping each long, insulated piece of wire around the straight sides of an elongated hoop of iron. He expected to see a steady flow of electricity in the second coil, somehow leaking through to it via the metal hoop, when he powered up the first. Instead there was only a short flow of current in the second coil when the first coil was turned on or off, which soon disappeared.

How, then, was the first coil of wire managing to produce an effect in the second at a distance? As we have seen, Faraday's first investigation had included the way that a coil of electric wire could produce magnetism. And there was no doubt that magnets worked at a distance – a compass showed that to be the case. When the current flowed in the first wire, then, it would act as a magnet. Faraday realised that if the changing level of magnetism generated a new current in the second wire, rather than electricity leaking across the metal core, it would follow that only a short burst of current would be induced in that second wire. He was soon able to demonstrate the generation of electricity by moving a permanent magnet through a coil, devising a basic electrical generator.

The linkage Faraday demonstrated between electricity and magnetism was difficult for scientists to explain. It was known that when iron filings were sprinkled on a sheet of paper and held above a magnet, the tiny pieces of metal would pull together into curved lines that provided a map of the magnet's invisible power. With his limited mathematical ability, Faraday was unable to provide equations to describe the effect; instead he imagined the results he had observed in terms of these lines, which he called 'lines of force'. If he moved a wire near the magnet, the wire was repeatedly cutting through the lines of force, one after another. Each interaction with these imagined lines in apparently empty space generated a flow of electrical current through the wire.

With this picture in mind, Faraday could reconstruct why the electrical induction occurred the way it did. Before the first electrical coil was switched on, the lines of force did not emerge from the coil. But when he started the current flowing, turning the coil into a magnet, the lines moved out into place like the ribs of an opening umbrella. As the lines of force moved out, they cut through the wire of the second coil, one after the other. The way the current pulsed briefly when the magnet was switched on meant that the lines of force did not appear instantly in place when he switched on the magnetic coil; they gradually moved out into position, otherwise the second coil wouldn't interact with the lines and generate a current. Something was travelling through the air, an invisible magnetic phenomenon.

A matter of speculation

Faraday's lines of force were visionary, but initially he was wary about revealing their full implications. He could not have forgotten what had happened when Davy had abandoned him. Instead of publishing all his results, he hid his most controversial ideas away in a sealed envelope, dated 12 March 1832, intending it to be opened after his death. And this document went one step further, giving a hint of something JCM would eventually make his own. In Faraday's mental model, the lines of force moved outwards from the electromagnet when he switched it on. But what exactly did he think was moving? Faraday wrote:

> I am inclined to compare the diffusion of magnetic forces from a magnetic pole, to the vibrations upon the surface of disturbed water, or those of air in the phenomena of sound: i.e. I am inclined to think the vibratory theory will apply to these phenomena, as it does to sound, and most probably to light.

This inspired linkage of magnetic vibrations – waves – and the nature of light stayed in the safe inside its sealed envelope until just before nine o'clock on the evening of Friday, 10 April 1846, when legend has it Charles Wheatstone, due to give a lecture on his electro-magnetic chronoscope,* panicked and ran out. His friend, Faraday, is said to have presented Wheatstone's brief talk, then, without time for preparation, is supposed to have given the most inspired lecture of his career: a first insight into the inseparable nature of light, electricity and magnetism.

In reality, the Royal Institution's records show that on that evening Faraday substituted for another scientist, James Napier, who had given a week's notice of his absence. It is certainly true, though, that Faraday spoke about Wheatstone's delightfully named but wholly forgettable electro-magnetic chronoscope. And then, when his colleague's notes ran out, Faraday began to improvise.

---

* Not as impressive as it sounds: just an electrically controlled clock.

He described light as a vibration, rippling through the invisible magnetic force lines that filled space. This was a remarkable insight to describe in that lecture theatre in 1846. The Royal Institution had moved to its newly-built home on Albemarle Street in London's fashionable Mayfair. Faraday was positioned at the same polished wooden demonstrator's bench that still stands in the imposing semi-circular theatre. This was before electric lighting, when the only sources of night-time illumination were oil lamps and candles and the flare of gas lights. To Faraday's audience, electricity and magnetism were novelties, the inexplicable connection enabling machinery like the chronoscope. Faraday's leap of genius, connecting the ethereal phenomenon of light with magnets and electrical coils, was inspired.

Faraday would later say that he 'threw out as matter for speculation, the vague impressions of my mind'. But those impressions were the result of long thought and the outcome was remarkable. Faraday told his audience:

> The views which I am so bold as to put forth consider, therefore, radiation as a high species of vibration in the lines of force which are known to connect particles, and also masses of matter, together. It endeavours to dismiss the aether,* but not the vibrations.

As we have seen, unlike JCM, Faraday was no mathematician, yet he had produced a visionary idea of the nature of electricity

---

* The aether or ether was the imagined medium filling all space in which the waves of light were thought to be vibrations. Faraday thought that with fields in place, there was no need for its existence.

and magnetism. He believed they produced a sphere of influence around themselves which he would call a 'field'. Led in part by the way that iron filings line up to connect the two poles of a magnet, Faraday thought of his fields being made up of the lines of force that emanated from magnetic poles or electrical charges. When another electrical or magnetic object broke these lines of force they felt an influence. The field lines were envisaged in a way that tied in with many of the behaviours of electricity and magnetism. For example, when lines of force were compressed together, they repelled each other. And Faraday had also set the seeds of understanding the nature of light by making the remarkable leap from the familiar effects of electricity and magnetism to the concept of waves in a field of force.

All of this would be essential background when JCM began to consider the matter. Although Faraday is rightly celebrated for his contribution to the practical side of devising electric motors and generators, the approach he took theoretically was even more fundamental to the way that physics would develop. His concept of fields, considered by some of his contemporaries as hand-waving, vague nothings, would become the standard way that physicists looked at the world, and remains so to this day, often replacing more familiar models of waves or particles.

Those more mathematical contemporaries who criticised Faraday's fields pointed out that their own approach, which considered electrical charges and magnetic poles as points that influenced other objects at a distance, obeying an inverse square law like gravity, produced numerical results that could be matched to experiment. However, the fields had a huge conceptual advantage. The point-based mathematical models depended on a mysterious influence at a distance. Just as

Newton's gravitational theory did not explain how gravity worked across empty space (this would take Einstein's work), so electricity and magnetism could only be explained as causing strange actions at a distance. It was this that Newton's contemporaries mocked in his approach to gravity as being 'occult'.

Fields, on the other hand, did away with the need for action at a distance. An electrical charge, for example, acted on the field where it was situated, not on another charge at some distant place. That action then rippled through the field lines to reach the other location. But before the field concept could be considered an effective approach for physics, JCM would have to take Faraday's qualitative concept and turn it into a mathematical structure that would enable the workings of electromagnetism to be understood and harnessed.

That, though, is still a long way in the future in our story. With the basic ideas of electromagnetism firmly under our belts, let's rejoin the young Maxwell and his pater on the road to Cambridge.

Chapter 2

A most original young man

On the way to Cambridge in 1850, Maxwell and his father had stopped off at two great cathedrals – Peterborough and Ely* – so the relatively small Peterhouse, to the south of the city centre where King's Parade becomes Trumpington Street, may have seemed far less of an architectural marvel. Yet the college should have been an ideal match for Maxwell on his first significant venture outside Scotland. Having his friends Tait and Stewart there to ease the way would have made the transition to Cambridge's social foibles – rather more sophisticated than Edinburgh at the time – easier to make. Because Peterhouse was

---

* John Clerk Maxwell was a great enthusiast for the importance of visiting places as a means of self-improvement. When Maxwell was getting near to his final exams at Cambridge and suggested a few days' break over the vacation to visit a friend in Birmingham, John wrote to him with a list of suggested venues his son could take in. These included 'armourers, gunmaking and gunproving—swordmaking and proving—*Papier-mâchée* and japanning—silver-plating by cementation and rolling—ditto, electrotype—Elkington's works—Brazier's works, by founding and by striking out in dies—turning—spinning teapot bodies in white metal, etc.—making buttons of sorts, steel pens, needles, pins, and any sorts of small articles which are curiously done by subdivision of labour and by ingenious tools— glass of sorts is among the works of the place, and all kinds of foundry works—engine-making—tools and instruments—optical and [philosophical], both coarse and fine.' Ever the attentive son, Maxwell began with the glassworks.

one of the smaller colleges, it would have felt a less pressured environment than, say, Trinity or King's.

Maxwell was given rooms in college with plenty of natural light, something that suited his inclination to experiment – he had a sizeable collection from his home laboratory sent after him on the journey down to Cambridge.* The availability of good space at Peterhouse seems to have been one of his reasons for choosing the college. Mrs Morrison, the mother of Maxwell's friend Lewis Campbell, noted in her journal before Maxwell had chosen his college that Maxwell 'came in full of Forbes' recommendation of Trinity College above all others at Cambridge and that Peterhouse was less expensive than Caius; that the latter is too full to admit of rooms, and freshmen are obliged to lodge out.' Yet, within weeks of arriving in Cambridge, Maxwell was looking to move on.

Stepping up to Trinity

After a term at Peterhouse, Maxwell transferred to Trinity College – a relatively unusual step to take now, as teaching is largely provided by the university rather than the college. However, in Maxwell's day, the college you were a member of contributed significantly to the quality of the education you received. Far more direct teaching was provided by the college than is the case now, and some aspects of what was provided at Peterhouse were limited (despite having a good record in mathematics).

What's more, private tutors were then prevalent and could have a big influence on a student's success; it seemed that

---

* It was, and to some extent still is the fashion to refer to going 'up' to Cambridge from any point of the compass, but from Maxwell's youthful viewpoint where all of England was in the south, it surely was 'down'.

Maxwell wasn't getting on well with his tutor at Peterhouse. Maxwell's father had also found out that his son had little chance of remaining at Peterhouse after graduating. There would probably be only one fellowship available for Maxwell's year, in which it was clear he had several challengers. At the much larger Trinity College, the alma mater and firm favourite of family friend Professor Forbes, Maxwell would have a better chance of gaining a fellowship and staying on.

It's probably no coincidence that Maxwell was supported in the move to Trinity by James Forbes, who told the master of the college that 'he [Maxwell] is not a little uncouth in his manners, but withal one of the most original young men I have ever met with'. Not entirely surprisingly, given his background, at this stage in his life Maxwell was a strange mixture of wide-ranging information and a rambling lack of structure. Peter Tait commented that Maxwell had 'a mass of knowledge which was really immense for so young a man, but in a state of disorder appalling to his methodical private tutor'.

Although Maxwell threw himself into the academic world and thrived at Cambridge, seeming to benefit particularly from the lively social life and intellectual cut-and-thrust that he experienced in Trinity College, he was determined to pack as much into his time there as he could. He had always been enthusiastic for exercise, but with his daily timetable already filled, for a period he devised a way to keep fit despite the restrictions that required students to remain in college at night time. A student with rooms on the same staircase as Maxwell remembered:

> From 2 to 2:30 a.m. he took exercise by running along the upper corridor, down the stairs, along the lower corridor, then up the stairs, and so on until the inhabitants of the rooms

along his track got up and lay *perdus** behind their sporting-doors[†] to have shots at him with boots, hair brushes, etc., as he passed.

This timing was part of a set of experiments Maxwell undertook to determine the best possible hours for sleep and work – everything about life seems to have provided him with an opportunity for experiment. In 1851, for example, he noted in a letter that he had tried out sleeping after hall (the evening meal in the ornate college hall):

> I went to bed from 5 to 9.30, and read very hard from 10 to 2, and slept again from 2.30 to 7. I intend some time to try for a week together sleeping from 5 to 1, and reading the rest of the morning. This is practical scepticism with respect to early rising.

The results of his experiments were not published, but he seems finally to have settled for more sociable hours.

Becoming an Apostle

Clear evidence that Maxwell's air of a country bumpkin was wearing off came in his election to the Select Essay Club, better known by the nickname 'the Apostles', an elite intellectual club which, as the name suggests, originally had twelve members.

---

* Hidden.
† Cambridge rooms had (and older ones still do have) an odd contrivance of inner and outer doors with only an inch or so between them. The inner door was the 'oak' and the outer door the 'sporting door', closed to indicate that the occupant did not want to be disturbed – having a shut sporting door was referred to as 'sporting the oak'.

It was a university-wide group, but drew largely on a group of wealthy colleges, including Trinity. The club would number the likes of Alfred Tennyson, Bertrand Russell and John Maynard Keynes among its members – and still exists. Unlike the infamous Oxford Bullingdon Club, the Apostles meetings were fixed around tea rather than alcohol-fuelled dining, and it had a far more intellectual basis, though it still suffered from some of the affectations of a secret society. It is highly unlikely that Maxwell would have been considered for such a club if he still came across as 'uncouth in his manners'.

It was perhaps due to a more sophisticated circle of friends that during his time at Cambridge, Maxwell made his only known foray into spiritualism, then at the height of its public interest. His friend Lewis Campbell notes that even the 'occult' sciences, in the then fashionable shapes of electro-biology and table-turning, received a share of Maxwell's 'ironical attention'. This was despite a dark warning from Maxwell's father who noted that an acquaintance had known 'two cases of nervous people whose minds were quite disordered' by electro-biology, and warned:

> I hope it is not in fashion at Cambridge, and at any rate that you do not meddle with it. If it does anything, it is more likely to be harm than good; and if harm ensues, the evil might be irreparable, so let me hear that you have dismissed it.

'Electro-biology' sounds like an early attempt at understanding the electrical aspect of the brain and nervous system, but was in reality an alternative term for animal magnetism. Both labels were attempts to give mesmerism (also known as hypnotism) a more scientific-sounding context. Electro-biology was particularly favoured as a term among stage hypnotists. Table-turning would have been considered even worse by Maxwell's devout father. It was a form of séance where the table was lifted and moved as the participants sat around it, supposedly powered by spirit intervention, but more likely to be due to the practitioner's use of his or her feet or hands.

It's quite possible that Maxwell was put off further exploration by the fate of Michael Faraday, who experimented with table-turning and showed that the movement of the table was due to the pressure of participants' fingers on it, quite possibly unconsciously making it move as they hoped. Faraday was then deluged with letters asking if he could explain various other phenomena as if, Maxwell noted, 'Faraday had made a proclamation of Omniscience. Such is the fate of men who make real experiments in the popular occult sciences ... Our anti-scientific men here triumph over Faraday.' It's likely Maxwell did not want to suffer a similar fate.

It was around this time that Lewis Campbell wrote a pen sketch of Maxwell that is interesting to set against the rather stern-looking portraits of the day. Campbell describes Maxwell as follows:

> His dark brown eye seems to have deepened, some parts of the iris being almost black ... His hair and incipient beard were raven black, with a crisp strength in each particular hair, that gave him more the look of a Nazarite than of a nineteenth century youth. His dress was plain and neat, only remarkable for the absence of anything adventitious (starch, loose collar, studs, etc.), and an 'aesthetic' taste might have perceived in its sober hues the effect of his marvellous eye for harmony of colour.

Cats and rhymes

One thing that didn't change for Maxwell while at Cambridge was his enthusiasm for animals. Many Cambridge colleges banned the keeping of dogs (and still do – the dog belonging to the current Master of Selwyn College is officially designated the college cat), which Maxwell would have found a wrench, but he made up for it by making friends with the cats that were employed to keep mice down in college. Being Maxwell, this led to an experiment which gained him a questionable reputation at the time, as he explained looking back in a letter to his wife Katherine written in 1870 when he was visiting Trinity College as an examiner for mathematics:

> There is a tradition in Trinity that when I was here I discovered a method of throwing a cat so as not to light on its feet, and that I used to throw cats out of windows. I had to explain that the proper object of the research was to find how quick the cat would turn round, and that the proper method was to let the cat drop on a table or bed from about two inches, and that even then the cat lights on her feet.

Maxwell also kept up a habit of writing verse while at Cambridge. This would be a pastime he indulged in throughout his life, covering the whole gamut from translations of classical poetry, through serious odes, love poems to his wife when later married, and ventures into comic verse. Sometimes his work would touch on very specific aspects of his work and experience, such as his Cambridge piece 'Lines written under the conviction that it is not wise to read Mathematics in November after one's fire is out'. The opening verses of another poem from his youthful Cambridge days, entitled 'A Vision (of a Wrangler, of a

University, of Pedantry and of Philosophy)' illustrates Maxwell in full comic mode:

> Deep St Mary's bell had sounded,
> And the twelve notes gently rounded
> Endless chimneys that surrounded
> My abode in Trinity.
> (Letter G, Old Court, South Attics)
> I shut up my mathematics,
> That confounded hydrostatics –
> Sink it in the deepest sea.
>
> In the grate the flickering embers
> Served to show how dull November's
> Fogs had stamped my torpid members,
> Like a plucked and skinny goose.
> And as I prepared for bed, I
> Asked myself with voice unsteady,
> If of all the stuff I read, I
> Ever made the slightest use.

The Wranglers

At the time, all Cambridge students were required to take a series of mathematics papers in their finals. The basic exams, which could be passed by learning the fundamentals of the course, were relatively straightforward. But honours students took papers providing them with an extended set of problems over four days. The questions were deliberately obscure, requiring a different kind of thinking and taxing the students' reasoning skills to the limit. The scale of this undertaking can be seen from the details of the 1854

Tripos,* where honours students took sixteen papers, comprising 44.5 hours of examination, working through 211 questions, while the best of the best went on to spend three more days on the 63 additional questions of the Smith's Prize papers.

Those who completed the whole of this gruelling mathematical challenge and came through with first-class honours were given the title 'Wrangler', their position in the listing as hard-fought as the 'Head of the River' in the rowing races. Achieving top position as 'Senior Wrangler' was widely regarded as the ultimate academic achievement in Britain and was feted far beyond Cambridge.

Maxwell achieved the position of Second Wrangler,† and in the separate, even harder mathematical exams for the Smith's Prize was declared joint winner with Senior Wrangler Edward Routh, who would go on to be influential in the mathematics of moving bodies and would develop the beginnings of what became control systems theory.

It's arguable that Maxwell's mix of an Edinburgh and Cambridge education provided the perfect combination to break out of the approach to physics that was deeply embedded in academia at the time. Had he stayed in the Scottish system he would

* 'Tripos', pronounced 'tri-poss' rather than 'tri-pose', is the unique Cambridge system describing the exams necessary to gain a degree in a particular discipline. The word is said to come from the name of the three-legged stool that was used during oral examination, though there is limited evidence for this. Despite the 's' on the end, it's singular.

† Another Second Wrangler was Maxwell's somewhat older friend, the University of Glasgow physics professor William Thomson. Thomson had apparently been so obviously the favourite for Senior Wrangler that he didn't bother to check the result, instead sending a college servant along to the Senate House where the results were posted to see whom he should commiserate with for being Second Wrangler. The servant returned to say, 'You, sir.'

have been absorbed into a tradition that combined experiment with natural philosophy – but Cambridge gave him the extra abilities to take a more detailed mathematical approach, and the capacity to work with the rigour needed to move on to the next stage of the development of physics. Maxwell developed a working method that was a hybrid of the two traditions.

Maxwell's success in the mathematics exams and the Smith's Prize was exactly what he needed to ensure a continued place at the university, winning him the position of bachelor-scholar at Trinity with the near certainty of a fellowship to follow soon in the future. Now he had more time for his own projects. Continuing his fascination with light from his investigations of polarisations and the stress experiments, Maxwell found a new interest in the way that human beings perceive different colours.

Colour vision

At the time, doctors could do little more than peer at an open eye and hope to see into it, but Maxwell constructed one of the first ophthalmoscopes, in effect a microscope for examining the inner workings of the eye. With it, he subjected human and particularly dogs' eyes to lengthy study, and was able to reveal the network of blood vessels on the inner surface. He wrote to his aunt, Miss Cay, in Edinburgh in the spring of 1854:

> I have made an instrument for seeing into the eye through the pupil. The difficulty is to throw the light in at that small hole and look at it at the same time; but that difficulty is overcome, and I can see a large part of the back of the eye quite distinctly with the image of the candle on it. People find no inconvenience in being examined, and I have got dogs to sit quite still and keep their eyes steady. Dogs' eyes are very beautiful behind, a copper-coloured

ground, with glorious bright patches and networks of blue, yellow and green, with blood-vessels great and small.

This work with his ophthalmoscope appealed to his continued interest in the detail of natural phenomena, but it gave no obvious clues as to the mechanism within the eye that enables us to distinguish between colours.

In working this out, Maxwell had two lines of enquiry. The better-established understanding at the time came from artists, who for centuries had been mixing different pigments to produce a palette of colours. The painters considered that red, yellow and blue were the 'primary' colours, able to produce any of the other colours when mixed together. The versatile English doctor and physicist Thomas Young had suggested that the eye worked in a similar fashion to the artist's palette, but in reverse. Different areas of the retina, Young believed, were sensitive to red, yellow and blue – or to be more precise, he suggested that each 'sensitive filament' of the optic nerve was split into three portions, one dealing with each of the primary colours, hence being able to construct the colour range that we see.

Maxwell was also aware of a strand of more physics-related experiments on light and colour from the natural sciences, which stretched back to Isaac Newton. It was while Newton was himself at Trinity College that he had performed his famous experiments. Piercing the blinds of his room, he had let a narrow beam of sunlight through and split it into the rainbow colours using a prism that he bought at a local fair.*

---

* University members were forbidden from attending the fair and would have been at risk of being arrested by the university's proctors, so the fair was cunningly sited just outside the limits of their authority, on Stourbridge Common.

It was Newton who dreamed up red, orange, yellow, green, blue, indigo and violet as the colours of the rainbow. In reality, seven appears a strange number to have selected. There are far more colours if you examine a rainbow under the microscope, but to the naked eye there only appear to be six broad bands, merging Newton's three variants of blue into two. It's thought Newton went for the number seven so there would be the same number of rainbow colours as musical notes, an appeal to the harmony of nature. The ability of a prism to produce a rainbow was well known – that's why they were on sale at the fair as toys – but the prevailing theory at the time for why this happened was that the incoming white light was being coloured by impurities in the glass.

Newton's genius was to separate off individual colours from a prism's output and send them through a second prism. No individual colour was changed by passing through a second piece of glass – suggesting that the prism did not add the colour, but merely separated out the colours that were already present in the white light of the Sun. Newton confirmed this by focusing the rainbow colours with a lens and producing white light again. If these colours were all in white light, but when that light fell on, say, a red apple we see only red, it seemed reasonable that the apple was absorbing many of the colours in the spectrum while reflecting only the red.

Some colours that we can see, though, simply aren't present in a rainbow spectrum of light. Think of brown or magenta (cerise in fashion terms), for example. Such oddities could only be produced by mixing other colours – but how this was to be done caused confusion. Over time, Newton's successors started using a wheel or 'top' with different colours on, which was spun so that the colours seemed to combine in the eye. Professor

Forbes at Edinburgh had repeatedly attempted to produce white through a combination of variants of red, yellow and blue on a wheel – but had failed. Similarly, Forbes had used one of the artist's most familiar combinations – mixing yellow and blue to make green – and discovered that, bizarrely, his spinning wheel appeared to produce not green at all, but a dirty shade of pink.

The true primaries

It was Maxwell, now what we'd call a graduate student, who followed up the idea of German physicist Hermann von Helmholtz, building on Newton's observations, that there were two different processes involved: that colours in light added together to produce new shades, while colours in a pigment were subtractive (i.e. they took away some of the colours of light) and had to be treated separately. As Newton had suspected, when we see an object as a certain colour – a red postbox, for example – what our eyes actually detect is the light that is re-emitted by the box. If white light, with all the colours in it, is falling on the box and we see red, then the pigments in the paint have absorbed the other colours. And when we mix pigments, such as yellow and blue, the resultant colour we see (green in this case) is what's left when the rest of the colours in the light have been absorbed by those two pigments.

This meant that the artist's primary colours weren't really primaries at all, but were the leftovers when the primaries were absorbed, the diametric opposites of primaries,* which modern

---

* Even here, the artists got it wrong with their 'primaries'. The primary pigments are cyan, yellow and magenta (take a look at the inks in your colour printer) – the blue, yellow and red colours that we still teach at school are just approximations to these. But, as we demons know all too well, you can't expect reliability from artists.

scientists would probably have called anti-primaries. Through a combination of experiment and logic, Maxwell realised that the true primaries of light (as opposed to pigments) were in reality red, green and blue or violet – when he made up a colour disc with these colours and spun it, he saw white. In making this change he was going against his old Edinburgh professor Forbes who stuck (with many of his contemporaries) to the earlier idea of red, yellow and blue as the primary colours of light, despite the failure of their experiments.

This wasn't enough for Maxwell, though. He built his own totally new version of the colour top, using three paper discs, one for each primary colour, so he could spin different combinations of red, green and blue to study the resultant perceived colours. These discs were provided by the Edinburgh printer and artist David Hay, who had inspired Maxwell's teenage paper on drawing curves, and whose colour printing was considered to be among the best in Great Britain. The term 'top' suggests a self-supporting spinning device, but in reality, his mechanism consisted of a flat metal disc to support the paper sheets, pivoted on a handle, so it was shaped something like a round skillet with a spinning pan.

In an early manuscript on the subject from February 1855, Maxwell described the device as a 'teetotum' (an alternative word for a top). He tells us:

> It may be spun by means of the fingers but if more speed is required it ought to be spun by means of a thread wound round the part of the axis immediately below the disc. This is best done by slipping the knot on the thread behind the little brass pin under the disc and after winding it up placing the axis vertical & so that the two grooves in it rest in the two hooks

Maxwell aged 24 with one of his colour wheels in 1855.
Getty Images

belonging to the brass handle. When the string is pulled, the hooks keep the axis vertical and thus the teetotum may be spun steadily on the smallest table or tea tray.

It is entirely possible that the idea of constructing his colour top had its inspiration in an entertainment from Maxwell's youth. One of his favourite toys as a child was the phenakistoscope, more mundanely known as the magic disc. This spun a disc with a series of illustrations drawn around its edge. The disc was viewed from the rear through a series of slots in the disc, reflecting the images in a mirror to produce the effect of a very short moving picture. Maxwell drew the pictures for many of his own mini-movies to spin on the magic disc, with subjects covering everything from the cow jumping over the Moon to a dog catching a rat. The merger of the still images in the brain to produce a combined effect could well have suggested a similar approach in the top.

Maxwell's colour top also had room for a fourth colour which was in a continuous circle around the centre of the disc. This meant that he could adjust the amounts of each primary to match the colour represented in that circle. He went on to produce a mathematical formula relating the percentages of the primaries to the resultant hue.

From his formula, Maxwell was able to produce a 'colour triangle' which started with the three primaries in the corners of an equilateral triangle and mixed the amounts of the colours according to the distance to the corners (see Figure 1). An important outcome of this work was the realisation that what we perceive as the colour in a beam of light is not the same thing as the absolute colour of the light used. While monochromatic light of a particular wavelength will be seen as, say, orange, the brain also combines the input from the different colour-determining sensors in the eye, known as cones, to enable the colour to be produced from a mix of the primary colours.

Similarly, Maxwell was able to use his triangle to understand the detail of how pigments appear to have particular colours. For

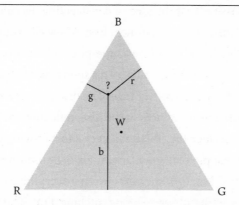

FIG. 1. *Each corner of the colour triangle corresponds to a single primary colour. The colour at the point marked '?' mixes r of red, b of blue and g of green – the central W point has equal amounts of each colour, producing white.*

example, if you shine white light onto a pigment that strongly absorbs the middle of the spectrum, around green, the result will be that red and blue wavelengths are mostly re-emitted, producing magenta – which is effectively anti-green. And the old artist's favourite of mixing yellow and blue to produce green works because cyan pigment mostly absorbs red, while yellow pigment mostly absorbs blues, leaving only green to be re-emitted.

A peculiar inability

Maxwell notes that some individuals have a 'peculiar inability' to distinguish certain colours – his work on the perception of colour went hand in hand with a deep interest in those who were colour blind, suffering from 'Daltonism' as it was often known then, after the Manchester chemist John Dalton, who suffered from the condition and was one of the first to study the phenomenon scientifically. Maxwell also records that the most interesting result is that 'different eyes in similar circumstances agree to the most minute accuracy [on a colour] while the same eyes in different lights give different results'.

So, he had discovered, our perception of colour is strongly affected by lighting etc. but is remarkably consistent between individuals who have normal colour vision. But Maxwell, with a good scientist's caution, was not willing to expand his observations too far from the small sample he originally had access to. He noted: 'These results however can be completely verified only by a large number of observations.' For the rest of his working life, he would invite visitors to try out a range of colour mixing devices he invented to widen his sample, and would ask friends both to take the test and to bring forward anyone they knew who didn't see colour the same way as the majority.

Maxwell followed up his initial manuscript with a paper presented to the Royal Society of Edinburgh, where he brought out further his idea that colour blindness was typically caused by one of the three colour systems in the eye being defective. If so, he felt it should be possible to make combinations of colour on the colour top which identified what sensitivities were missing in people's colour perception.

He also spent some time trying to match a range of colours using different combinations on his top. By now Maxwell had settled firmly on red, green and blue as the definitive primary colours of light. Producing a good range of browns seems to have given him the most problems. In a letter to William Thomson, who had expressed doubts about achieving this colour at all by mixing light primaries, Maxwell noted: 'I have been thinking about what you say about Brown. I have matched ground coffee tolerably though the surface is bad chocolate cakes & improvised browns with black, red & a little blue & green.'

Maxwell would never know the practical importance of his work, but his colour triangle gave an essential understanding that allowed us eventually to produce the colour screens we now use on everything from TVs to computers and phones. Each coloured pixel on the screen is made up of separate red, green and blue elements, and their relative proportions are used to produce the whole range of millions of colours typically on display ... all based on Maxwell's ground-breaking approach. As was the case in many such developments, there was parallel work going on – as we have seen, German physicist Hermann von Helmholtz came to similar conclusions on the way that colours in light added, but pigments subtracted.

Once Maxwell had clearly demonstrated the colour model we now use, it might be expected that others would quickly fall into place and accept it. But the red, yellow, blue model was so strong, and so firmly supported by the artistic fraternity, that it took years of repetition and demonstration for Maxwell to get broad agreement. As late as 1870, a good fifteen years later, he wrote to a former Trinity College friend, Cecil Monro:

> Mr W. Benson, architect, 147 Albany Street, Regents Park, N. W., told me that you had been writing to Nature, and that yours [supporting Maxwell] was the only rational statement in a multitudinous correspondence on colours ... No other architect in the Architect's Society believes him. This is interesting to me, as showing the chromatic condition of architects.

Quantifying Faraday's fields

Light and colour vision were not, however, Maxwell's only, or even his primary interest. Ever since his childhood experiments at Glenlair he had been intrigued by the magical-seeming ability of magnets to influence other pieces of metal from afar, and at university he had become an ever more enthusiastic devotee of Michael Faraday, following with interest Faraday's many experiments with electricity and magnetism. Faraday's idea of magnetic and electrical fields with their lines of force would fascinate Maxwell.

The field approach was generally regarded as an interesting model for electricity and magnetism – a useful way to think about them – but not one that would ever be susceptible to mathematical study. (Faraday was mocked by some of his contemporaries for his lack of mathematical expertise.) To work mathematically with electromagnetism, a similar approach to that applied to gravity was used. This involved an 'inverse square law' – forces acting remotely which decreased with the square of the distance away from point sources.

Maxwell was sure that Faraday's ideas were sound and looked for a way to provide a mathematical basis for those mystical-sounding lines of force that were central to Faraday's understanding. As we have seen, Maxwell seems to have been driven by his relatively unusual academic background. Cambridge was very strong on mathematics, but tended to apply it to the sciences with which maths had always been strongly linked, such as astronomy. Edinburgh, by contrast, gave Maxwell the grounding in electromagnetism, but would not have encouraged taking a mathematical approach to it. Maxwell was able to bring the two approaches together.

The idea of a field is something like a three-dimensional version of a contour map. Any point on the map has a particular value for the height of the land at that location, and the contours on the map are the equivalents of lines of force, joining points of equal height. The equivalent in electromagnetism would be joining points with equal strengths of electrical or magnetic fields. But the complexity of moving to a field model was greater than the need to move from a two-dimensional map to a three-dimensional field. The values at each point on a map are just numbers reflecting the altitude – mathematicians refer to such number-only values as scalars. But each point in an electrical or magnetic field represents both a size and a direction – they are called vectors.

In the 1850s, the mathematics needed to handle vectors was yet to be fully developed, but Maxwell was aware of the basics and of some of the requirements to analyse a field mathematically. He asked for help from fellow Scot William Thomson, who had already done some work on electricity using vectors, basing it on his better-developed study of the flow of heat. Thomson had discovered that, by some strange natural coincidence, the equations describing the strength and direction of the 'electrostatic' force between electrical charges were the same as those that dealt with the rate of flow and direction of flow of heat.

Maxwell took Thomson's guidance on the mathematics of vectors, but went his own way on applying it. He thought of electricity as behaving like a fluid that was flowing through a porous substance, while magnetism seemed like vortices within the fluid. The lines of flow of his fluid corresponded to Faraday's lines of force, and the speed of the flow provided the 'flux density' which was a measure of the strength of the electrical or magnetic field. The difference in porousness of the materials that the imaginary fluid flowed through corresponded to the way that different substances reacted to electrical and magnetic fields.

It ought to be stressed, however, that Maxwell did not think that electricity actually *was* such a penetrating fluid. There was a clear lesson from the study of heat to be learned here. For about 100 years, most of the work on heat assumed that there was a real, invisible fluid called caloric, which flowed from a hot object to a colder one that it was in contact with. The caloric theory had had some success in explaining how heat behaved, but ultimately it proved ineffective, and a better explanation that considered heat to be the kinetic energy of atoms and molecules in a substance took over.

Maxwell's fluid was never intended to be the electromagnetic equivalent of caloric. His fluid was purely imaginary and the flow of his fluid was *not* electricity itself, but rather was an *analogy* for the strength of the electrical and magnetic fields. And it worked surprisingly well. One of the results that Maxwell got pretty much for free out of this model was that by using a non-compressible fluid to represent the field, it meant that there was always the same amount of fluid in the same volume, which produced an interesting mathematical result.

If there was always the same amount of fluid in the same volume, the flow of fluid would drop off with the square of the distance from the source. If the same amount of fluid travelled through a wider and wider space, then its rate of flow depended on the surface area of a cross-section of that space. Think, for instance, of a liquid moving out through a funnel in the reverse direction from usual, going from the narrow end to the wider one. If the fluid can't compress or stretch and has to fill all the space available, then it will have to be going a lot slower at the wide end of the funnel than at the entrance. The speed it moves at will depend on the size of the opening.

Similarly, if we think of liquid emerging from a point and heading out in all directions, then the surface area of the 'opening' is just the surface area of a sphere – $4 \pi r^2$, where r is the radius of the sphere – so the surface area the fluid has to fill increases with the square of the distance from the centre. We see the fluid moving slower – which in Maxwell's analogy means the strength of the field dropping off – reducing in speed with the inverse square of the distance from the source. This was exactly what happened in experiments on the electromagnetic field.

As noted above, Maxwell always saw his approach as an analogy – a model of reality that did not have any direct resemblance to what was actually happening, but which produced useful results. He commented in his paper describing his ideas, named *On Faraday's Lines of Force*:

> I do not think that [the fluid analogy] contains even the shadow of a true physical theory; in fact, its chief merit as a temporary instrument of research is that it does not, even in appearance, account for anything.

Not only did Maxwell's fluid model fit with Faraday's force fields while predicting the inverse square law of the traditional 'action at a distance' mathematics, it was better than the action at a distance approach when dealing with the boundaries between materials. Though Maxwell could not see how to take the step into modelling changing electrical and magnetic fields that would be necessary to deal with many of the phenomena that had enabled Faraday to come up with the electrical generator and motor, this was impressive stuff for someone who had only just graduated and was in his early twenties. As well as presenting his ideas to the Cambridge Philosophical Society, Maxwell also sent his paper *On Faraday's Lines of Force* to his hero, Michael Faraday, in London.

Faraday was by now in his sixties, but was still working at the Royal Institution and replied to Maxwell:

I received your paper, and thank you very much for it. I do not venture to thank you for what you have said about 'Lines of Force', because I know you have done it for the interests of philosophical truth; but you must suppose it is work grateful to me, and gives me much encouragement to think on. I was at first almost frightened when I saw such mathematical force made to bear upon the subject, and then wondered to see that the subject stood it so well.

For the benefit of working men

Faraday, as we have seen, had not had the benefit of a university education. The closest he came to formal training before beginning work at the Royal Institution was to attend the City Philosophical Society, a group set up with the specific intention of helping those from humble backgrounds to better themselves. Although Maxwell had a more privileged upbringing, he was aware from his contacts back at Glenlair of the limited opportunity for working men to improve their education, and also how Faraday had benefited from the 'Phil Soc'. And, being Maxwell, he could not stand back and assume someone else would sort the problem out.

At Cambridge, Maxwell was one of the founders of the Working Men's* College, which provided evening classes for self-improvement. Not only did he give some of the lectures himself, he toured local businesses, asking them to allow their men to leave their jobs early on lecture nights. In a letter to his father, written in March 1856, he noted:

---

* It would be wrong to criticise in retrospect the fact that the Working Men's College was aimed at men only. Maxwell was a man of his time and gave limited consideration at this point in his career to women's education.

We are also agitating in favour of early closing of shops. We have got the whole of the ironmongers, and all the shoemakers but one. The booksellers have done it some time. The Pitt Press keeps late hours, and is to be petitioned to shut up.

This enthusiasm for outreach and the betterment of those from humble beginnings would continue through future posts where Maxwell would regularly get involved with local working men's educational organisations, such as the mechanics' institutes often found in industrial towns and cities.

Maxwell soon became comfortable in his new position as a graduate student, though he still cherished his summers in Scotland, whether at Glenlair or spending time with relatives. In 1854, this resulted in his first recorded encounter with love. Maxwell spent a week with the Cays, his mother's family, in the Lake District. Among his mother's brother's children on the trip was Maxwell's cousin Lizzie – fourteen to his 23. There seems little doubt that Maxwell fell in love. Lizzie's age was less of a concern then than it would be now at a time when girls often married at sixteen after a lengthy courtship. However, the family seems to have been set against the match.

There was awareness of the risks of marrying such a close relation as a first cousin, even though it was relatively common among the upper classes, notably in the royal family where it occurred frequently. The Clerks and the Maxwells had over the century or so since their first intermarriage frequently wed cousins. But, for whatever reason, it was not to be. If Maxwell and Lizzie exchanged any letters they have not survived – the story of their brief passion has only been related via Lizzie's daughter, who at the time was aged 90.

Another personal event of that year was Maxwell becoming a

Justice of the Peace – a magistrate. It's not clear how frequently (if at all) Maxwell carried out the role, which was often nominally taken by the lord of the manor, but it shows an awareness of his personal responsibilities as a member of the establishment that would come through strongly when he later inherited the Glenlair estate.

One interesting 'might have been' turned up at the start of 1855. His friend Cecil Monro wrote to Maxwell saying, 'NEWTON MUST BE TRANSLATED, and you are the one to do it'. This was a reference to Isaac Newton's masterpiece, the lengthy (and sometimes near-impenetrable) work *Philosophiae Naturalis Principia Mathematica*, usually just known as the *Principia*. This three-volume book contains both Newton's laws of motion and his work on gravitation, and was originally published in Latin. It had been translated into English in 1729 by Andrew Motte, but by the 1850s that version was seeming too archaic to be used as a serious scientific document.

It wasn't entirely surprising that there was demand for an English version of the most important work by the man generally regarded at the time as the greatest English scientist, if not the world's. Monro drolly pointed out that Maxwell's Latin was good enough 'for practical purposes ... [though] it is very true that you don't seem ever to have displayed such acquaintance in your college examinations'. Monro may have known that a significant reason why Maxwell was yet to be elected a Fellow of Trinity College was because it was considered that he needed to attend more to the classics.

Maxwell replied at his most whimsical:

Dear Monro

It is a fearful thing to answer when a man tackles you with

arguments. I wont argufy at all, leastways not with them as tries to argufix me. I w^d be most happy to give any assistance in my power to the translator of Newton, short of taking on the work of his hands. For that I am not prepared. I am prepared to refuse resist & rebel. I w^d as soon think of translating Butlers Analogy* for the Mathematical Journal.

It was not to be.

This letter and a number of others from Maxwell were written from 18 India Street in Edinburgh, two doors down from the Clerk Maxwells' old house. Their own place, 14 India Street, had been rented out once Glenlair was completed, and the Maxwells never lived there again, but when Maxwell had gone to Cambridge, his Aunt Isabella moved from her house in Heriot Row to India Street, and it was number 18 that Maxwell used as a pied-à-terre in Edinburgh.

A new destination

It's quite possible that Maxwell, getting well established in Cambridge, would have taken his ideas on electricity and magnetism further at this stage of his career. But in February 1856, his old professor James Forbes wrote to him about an opening at Marischal College in Aberdeen. The university was looking for a Professor of Natural Philosophy, which Forbes thought was a position that would ideally suit Maxwell.

---

* This was Joseph Butler's book *The Analogy of Religion, Natural and Revealed, to the Constitution and Course of Nature* – which was anything but a mathematical title.

As Forbes put it:

> I have no idea whether the situation would be any object to
> you; but I thought I would mention it, as I think it would be a
> pity were it not filled by a Scotchman, and you are the person
> who occurs to me as best fitted for it.

There is no evidence that Maxwell had ever been to Aberdeen
before, a good 120 miles from Edinburgh and altogether less
sophisticated than the capital. And there was no doubt that the
country boy had had many of his rough edges rubbed off at
Edinburgh and Cambridge, perhaps giving him some expectation
of lively intellectual surroundings. But gaining a professorship
would be an impressive position to kickstart his career.

Forbes was also at pains to point out that he had no influence over this appointment, as it was in the hands of the Crown,
though it's not clear if this suggests that otherwise he would have
attempted to bias things in Maxwell's favour. Maxwell noted in
a letter to his father a couple of days after receiving the notifica-
tion from Forbes:

> I think the sooner I get into regular work the better, and that the
> best way of getting into such work is to profess one's readiness
> by applying for it. The appointment lies with the Crown – that
> is, the Lord Advocate* and the Home Secretary. I suppose the
> correct thing to do is to send certificates of merit, signed by
> swells, to one or other of these officers.

---

* The Lord Advocate was the senior legal and political role in Scotland, a
position at the time held by a James Moncrieff.

The idea that Maxwell should become a professor at the age of 24, having only graduated with a BA less than two years earlier, would seem outrageous now. However, modern academic positions have a much stronger hierarchy and career progression than was the case at the time. In practice, Maxwell had everything that was needed for the post, since he had been made a Fellow of Trinity in October 1855, which gave him his academic CV. His friends William Thomson and Peter Tait had both been awarded professorships at younger ages – Thomson became Glasgow's Professor of Natural Philosophy at the tender age of 22, while Tait became the Professor of Mathematics in Belfast at 23.*

With his application sent off, and a flurry of requests for support sent to the great and the good, which this kind of royal appointment required, Maxwell returned to Glenlair for the Easter vacation. His father had been suffering for some time from a lung infection, which got progressively worse over the following days. John Clerk Maxwell died on 3 April, leaving Maxwell with a major new responsibility as head of the Glenlair estate.

Three weeks after his father's death, Maxwell wrote to his friend Lewis Campbell:

> When the term is over I must go home and pay diligent attention to everything there, so that I may learn what to do. The first thing I must do is carry on my father's work of personally superintending everything at home, and for doing this I have

---

* Nor was he as precocious as the Swiss mathematician Leonhard Euler, who became a professor at St Petersburg aged nineteen.

his regular accounts of what used to be done, and the memories
of all the people, who tell me everything they know.

It was not taken for granted, then, that Maxwell would continue
his academic work, but in the same letter he made it clear that
his father had felt that Maxwell could achieve the appropriate
balance.

> As for my own pursuits, it was my father's wish, and it is mine,
> that I should go on with them. We used to settle that what I
> ought to be engaged in was some occupation of teaching, admit-
> ting of long vacations for being at home; and when my father
> heard of the Aberdeen proposition he very much approved. I
> have not heard anything very lately, but I believe my name is
> not yet out of the question in the Ld Advocate's book.

Maxwell was correct – his name was at the top of the list and he
was duly awarded the position. He would spend the remainder
of the term at Cambridge and the long summer vacation putting things in order at Glenlair. He found time in that vacation
to undertake some work on colour and on Saturn's rings (more
on this later), but there is no doubt that the responsibilities he
had inherited from his father came first.

In a letter to a university friend, Richard Litchfield, he
noted:

> I have certainly no time now & I have much more occupation
> than I expected such as to examine into the state of two sets
> of houses & provide wood &c for roofing them and workmen
> to do it and various things of this kind also to enquire into the
> merits of the younger clergy and the sentiments of the parish on

the subject, for our minister died unexpectedly this week and there are no resident proprietors in the parish except the patron of the living who is a lady of Romish persuasion [i.e. Catholic] who has been for a year in Edinburgh and denies herself to all her friends.

Despite the unexpected workload, Maxwell was able to bring Glenlair under his control and made the move to Aberdeen in the autumn.

In which atoms become real and heat gets moving

It's arguable that the summer of 1856 was when JCM made the transition from a youth to a man. His career had made the leap from graduate student to professor, he lost his father – with whom he had been very close – became the laird of Glenlair, and had moved from a city where his intellect was a comfortable fit to a more remote location where he would stand out more than most.

As Aberdeen is going to be the location where our young professor starts to take on the matters that will bring me into being, it's important we get a little background on the matter of atoms (and molecules), and of heat.

Atoms exist

At the heart of JCM's thinking on thermodynamics and the mechanics of gases was the idea that atoms and molecules were real. This is such a trivial point today that it's worth emphasising that in the 1850s the majority of scientists were at best ambivalent as to the existence of atoms. They were considered a useful concept for explaining the way elements combined in chemistry, but were thought probably not to be actual physical things. Yet it was their reality that spurred on the thinking that would be JCM's biggest claim to fame in his own era.

Back in the fifth century BC, the Ancient Greek philosopher Democritus dreamed up the concept of atoms.* It seems quite a reasonable idea. If you cut stuff up, you get smaller and smaller pieces of that stuff. Eventually, Democritus argued, you'd get a piece that was so small that it was impossible to cut it any further. This was not because your knife wasn't sharp enough, but because there was nothing more to divide. He called such a piece uncuttable: *atomos*.

While there was a degree of logic to the thinking, the trouble was that as a scientific theory it didn't have a lot of value because it didn't explain anything. Democritus did not combine his concept of atoms with any idea of elements, which would have made it possible to simplify the description of matter. But a competitor, Empedocles, did come up with a theory of four elements: earth, air, fire and water. Though wrong, this had more practicality than the early atomic theory and was built on by one of the great philosophers: Aristotle.

Active in the fourth century BC, Aristotle brought in a fifth element, the quintessence, which was the substance from which everything from the Moon outwards was supposed to be made. In principle, there was nothing to stop Aristotle from combining his elements with the atomic theory. Unfortunately, though, he had an aversion for totally empty space. He did not like the concept of a vacuum or void.

The logic behind this aversion was sometimes convoluted, but the simplest argument Aristotle made, ironically to modern eyes, was that if there were a void, then something like Newton's

---

* His teacher, Leucippus, may also have been involved, but there is little evidence of his existence and it's possible Democritus simply invented Leucippus as a kind of 'fake news' support for his theory – if so, he was my kind of guy.

first law of motion would apply – there was no reason why anything moving should stop unless something forced it to do so. As this didn't seem to happen in the nature he observed, Aristotle thought that the concept was flawed and so a void could not exist.

Without a void or vacuum there couldn't really be atoms, as the atoms would have to fit together in such a way that they totally filled empty space – something that was only possible with a very small number of shapes, such as a cube. Assuming that at the very least you needed four or five different types of atom, it simply seemed impossible to have atoms without space in between them. And Aristotle (with the stubbornness that is typical of some of you humans) wouldn't allow it.

Aristotle's views were largely accepted unchallenged through to Galileo's time, and it took a considerable period longer before atoms would be taken seriously. It was only really in the 1800s that atoms began to take hold as a realistic scientific concept. This was largely due to the work of John Dalton, an English Quaker who spent much of his working life in Manchester. Dalton was largely self-educated, unable to attend university because of his religious beliefs, as attendance at the time required membership of the Anglican Church.

Apparently as a result of experiments on gases, right at the start of the nineteenth century, Dalton devised the concept of atom-based elements, where each of the spherical atoms of any particular element had the same weight. As early as 1803 he was beginning to note down the relative weights of these atoms, starting with the lightest, hydrogen. He also considered that many substances were compounds, made up not of individual atoms, but molecules which combined two or more atoms.

There were limits to Dalton's work. His equipment was mostly self-made and of poor quality even by the standards of the day. He got some of the relative weights wrong and made the mistake of assuming that compounds would be made up of the simplest possible combinations – so he thought, for example, that water was HO with a single hydrogen atom, rather than the pair of hydrogens present in the familiar H_2O.

He also had a mental model of atoms as spheres of different sizes, which seems to have prevented him from noticing the now-obvious implication that if 'bigger' atoms had multiples of the weight of hydrogen, perhaps they were made up of multiple hydrogen atoms (or the same subatomic particles as we now know). It didn't help that, though Dalton often wrote the relative weights of his elements as round numbers, he did not think these were exact values, but rather considered them to be convenient approximations.

Again, with 20:20 hindsight it may seem that the concept of atom-based elements was so obvious that atoms should have been widely accepted by the time JCM started thinking about them. However, the general feeling was that they provided a useful model, but did not necessarily reflect any actual structure within a substance, just as JCM didn't think that electricity was a fluid but found it a useful analogy. Matter's behaviour could be well described as if it were a collection of atoms, but that didn't mean they were real things you could pick up with tweezers and look at.

It wouldn't be until after 1905 – when Albert Einstein, in one of his first great papers, showed that calculations of the size of molecules could be made – that the reality of atoms was increasingly considered mainstream. It's notable that when JCM's friend Peter Tait was writing a textbook with William

Thomson, Tait ruefully commented to a friend: 'Thomson is dead against the existence of atoms; I though not a violent partisan yet find them useful in explanation.' Yet a handful of scientists in the mid-nineteenth century made the assumption that atoms and molecules were indeed real – JCM among them.

A better model of heat

The reason for JCM's interest in the atomic was not so much to consider the nature of matter directly, but rather to deal with a pressing problem of the day – how to explain the workings of heat. For thousands of years, the only real interest in heat was whether or not it was warm enough to be comfortable or a fire was hot enough to cook food.* But the nineteenth century saw the steam engine become the dominant power source for industry and travel. Steam was king – but understanding the true workings of the steam engine required the development of a new branch of physics – thermodynamics – and JCM would be at the heart of this work.

We've already met the second law of thermodynamics when I introduced myself, but let's take a step back to discover where the whole business came from. As we have seen, the prevalent theory on the nature of heat at the start of the nineteenth century was caloric theory. This considered heat to be a fluid called caloric. One of the essential properties of caloric was that it repelled itself. This meant that if you had a hot object and a cold one in contact, there would be more caloric in the hot than the cold. This meant more repulsion in the hot thing's caloric, so the caloric from there would naturally move into the cold object until a balance was achieved. The hot object

* Or for us demons to torment souls in hell.

would get cooler and the cold object hotter. (It's that second law again.)

This theory had proved surprisingly effective, bearing in mind it was wrong. As well as supporting the second law, it predicted that a gas should try to expand as it got hotter, as there was more caloric to fit into it. In 1824 the French scientist and engineer Sadi Carnot developed the first concepts in thermodynamics, explaining the limits on the efficiency of steam engines, entirely based on caloric theory. The man who took Carnot's ideas and brought them into a more modern viewpoint of thermodynamics was JCM's frequent correspondent and friend, William Thomson, later Lord Kelvin.

Thomson's work in formulating thermodynamics was aided by input from an unlikely source – the brewer James Joule (who, in a neat symmetry, had been tutored by Dalton). It was Joule who helped kill off caloric theory by showing that heat was not a special fluid in its own right, but simply a manifestation of energy, which could be reliably and consistently converted to and from mechanical energy. We also need to throw in the contribution of the German physicist Rudolf Clausius who worked in parallel with Thomson (and Maxwell) on thermodynamics.

The outcome of these collective geniuses at work was to develop several laws of thermodynamics. The two biggies were the first law – the conservation of energy – stating that energy can neither be made nor destroyed, but is just transferred from one form to another (these forms, most importantly, including heat); and my demonic law, the second law. This started in the days of caloric – the idea of the fluid naturally moving from hot to cold – but would become formalised in terms of both heat flow and entropy.

Bearing in mind the identification of heat as energy, it became important to understand how that energy manifested in matter. One common and well-understood form of energy was kinetic energy, the energy of movement. And, in gases at least, if matter could be considered to be made up of actual atoms and molecules, then the movement of those tiny particles could well provide the energy that was described as heat. The faster the particles moved, the hotter the gas.* It was for this reason that JCM would take atoms on board.

However, we're getting ahead of ourselves. The youthful James Clerk Maxwell is about to take up his post as Professor of Natural Philosophy at Marischal College. We're off to the granite city.

---

* In reality it's a little more complex than this, as energy can also be tucked away in the vibrations of molecules that aren't moving freely and in the energy levels of electrons around atoms, but this is a good starting point.

Chapter 3

The young professor

M arischal College is not a name that you will find on the British university scene any longer, but when the fresh-faced 24-year-old Maxwell was appointed Regent and Professor of Natural Philosophy in April 1856, starting work in August, it was a well-known constituent of the Scottish academic establishment. Though the vast, dour grey granite main building still stands on Broad Street at the heart of Aberdeen, it is now used by the city council. However, in Maxwell's day, the college was a respected institution with a solid history stretching back to 1593. Like all ancient colleges at the time, the education it provided was broader than a modern university curriculum, though most students would have been on a path towards a career in law, the church, medicine or education.

A city divided

When Maxwell arrived at his new workplace, there was already talk in the air about the college's demise. Scotland had, at the time, five universities, and two of them were in the small city of Aberdeen, then with a population of only around 35,000 people.*

---

* Admittedly all cities were smaller back in 1856, but compared with the largest city in the UK, London, at around 2.6 million, Aberdeen was still a minnow. It was only around half the size of the Lancashire mill town of Bolton at a similar date.

Marischal shared the city with the hundred-years older King's College – this was a pre-Reformation foundation and its Catholic roots were not considered ideal for educating ministers of the Protestant kirk. In reality, 'shared the city' was not strictly true: the city itself was split in two, as if it had taken sides between the two establishments.

The 'old' and 'new' towns were pretty much independent, with a mile's gap in between them. Aberdeen old town was a relic, an island in a rural landscape consisting of King's College itself, the old cathedral and accommodation for those who worked there. Marischal was located in the bigger and thriving new town. This had been rebuilt pretty much from scratch at the start of the nineteenth century, centred on the sweeping Union Street, a 70-foot-wide boulevard to emphasise the solid worth of the granite city – so indeed it was very much a new town in Maxwell's day.

Though relatively small, Aberdeen had strong commercial roots, exporting a considerable amount of its trademark granite alongside textiles from the port, which also featured shipbuilders' yards. The newest addition in 1850, and something that opened up the city's trade even more, was the railway link, running down through Edinburgh to London, making Aberdeen far more attractive and accessible to a relatively cosmopolitan visitor like Maxwell.

By the time he was appointed, the religious divisions that had resulted in the setting up of the two separate universities were waning. Two years later, in 1858, Cambridge would finally drop the requirement for undergraduates to be members of the Church of England or its affiliates such as the Scottish Episcopal Church. Similarly, a Royal Commission had been set up as early as 1837 to look into the need for two separate universities in Aberdeen.

King's College was against a merger, but it was generally felt that the separate existence of King's and Marischal colleges was not beneficial to the education system in Scotland. Maxwell appears to have been aware of the situation, but not unduly worried. His immediate concern was finding somewhere to live in a city that he had probably never visited before. He took up lodgings with a Mrs Byers a few minutes' walk away from the college, his rooms reached via a quaint spiral staircase that took him above a law firm, J.D. Milne Advocates, at 129 Union Street, a location now featuring a string of familiar high street names from McDonald's to mobile phone shops.

Never a great enthusiast for college politics, Maxwell kept his head down and concentrated on his work as the in-fighting between the two universities continued. In fact, he noted in a letter to his aunt:

> I have been keeping up friendly relations with the King's College men, and they seem to be very friendly too. I have not received any rebukes yet from our men for so doing, but I find that families of some of our professors have no dealings, and never had, with those of the King's people.

In keeping with the scale of the city, the entire university that was Marischal was not much larger than an Oxbridge college, with twenty staff (fourteen professors and six lecturers) and 250 students enrolled when Maxwell arrived. It was an enviable position that involved a very short academic year, running only from November to April – so that the students were available back home on the farm or for seasonal work – leaving the staff with plenty of time to pursue their own research. For Maxwell this would mean the opportunity to spend half of his year at

Glenlair, both working on natural philosophy and devoting his attention to the maintenance of the estate.

Although Maxwell was used to working in relative isolation when he was at Glenlair, it must have come as something of a shock to move from the vibrant atmosphere of Cambridge and the intellectual cut-and-thrust of the Apostles meetings to the academic backwater that Marischal represented. He had no doubt hoped to find the opportunity for the same mix of high-brow thought and enjoyable banter among his fellow academics, but the rest of the staff at Marischal proved to be much older than he was. It might not have been unheard of to make such young appointments in the universities of the day, but it certainly wasn't a Marischal tradition. The next closest in age of the other staff members was a 40-year-old, while the average age across Marischal's fellows and staff was 55.

This is not to say that Maxwell was made to feel unwelcome in his new post – quite the reverse. He was often invited out to dine with other staff members and seems to have been readily accepted into the fold – but it was inevitably a very different kind of environment from Cambridge. A letter Maxwell wrote to his school friend Lewis Campbell made it clear how things had changed: 'No jokes of any kind are understood here. I have not made one for 2 months and if I feel one coming on I shall bite my tongue.'

His lectures were terrible

There seems something of a dichotomy in regard to Maxwell's abilities as a professor in those early years. Like a number of other greats – Newton and Einstein, for example – he had some limitations as a lecturer. This was mentioned by friends such as Peter Tait, who would eventually remark in his obituary of Maxwell:

[T]he rapidity of his thinking, which he could not control, was
such as to destroy, except for the very highest class of students,
the value of his lectures. His books and his written addresses
(always gone over twice in MS.) are models of clear and precise
exposition; but his *extempore* lectures exhibited, in a manner
most aggravating to the listener, the extraordinary fertility of
his imagination.*

One of Maxwell's best students at Aberdeen, a David Gill,[†] said
that 'his lectures were terrible'. It seems from Gill's lecture notes
that Maxwell prepared very clear, well-structured lectures, but
had a tendency to diverge into anecdotes and analogies that
left his students adrift – not helped by his lifelong tendency to
make arithmetic errors. It was easy enough to correct these in
his papers, but the slips he made during lectures made them
something of an intellectual minefield.

It's worth considering, though, that most of the attendees at
Maxwell's lectures in Aberdeen would have been very different from the students on a modern-day physics course. These
were not scientific specialists – physics was just one of many
topics covered on their four-year curriculum, which included
everything from Greek and Latin to philosophy and logic. These
students were probably of a reasonable calibre, as Marischal
College was unusual for Scottish universities of the time in having an entrance exam and in only charging a relatively low fee

* This could be seen as a nasty put-down, but from Tait was clearly an affec-
tionate assessment of his friend's original thinking combined with his difficulty
of communicating ideas on the fly.
† David Gill was the only one of Maxwell's Aberdeen students known to have
become a scientist – he was later Queen's Astronomer at the Cape of Good
Hope and became President of the Royal Astronomical Society.

of £5, with plenty of bursaries. This meant that around half of the students were from working families, who had won places on their merits.

Only in the third year would the natural sciences dominate for around 50 students. In his inaugural lecture, Maxwell emphasised that his course would require a shift of mindset:

> The work which lies before us this session is the study of Natural Philosophy. We are to be engaged during several months in the investigation of the laws which regulate the motion of matter. When we next assemble in this room we are to banish from our minds every idea except those which necessarily arise from the relations of Space, Time and Force. This day is the last on which we shall have time or liberty to deliberate on the arguments for or against this exclusive course of study, for, as soon as we engage on it, the doctrines of the science itself will claim our constant and undivided attention. I would therefore ask you seriously to consider whether you are prepared to devote yourself during this session to the study of the Physical Sciences, or whether you feel reluctant to leave behind you the humanizing pursuits of Philology and Ethics for a science of brute matter where the language is that of mathematics and the only law is the right of the strongest that might makes right.

It may have been that Maxwell's emphasis on mathematics (even allowing for an element of typical tongue-in-cheek phrasing), which he put more weight on than most of his contemporaries, was the reason his lecturing failed to endear him to many of his students. It's also worth stressing that Maxwell was limited by the standard approach to teaching science at the time, which would not have included any laboratory work. Given his lifelong dedication to experimenting, despite being primarily what we would now think of as a theoretical physicist, it's no surprise that he brought up the importance of experiment in that inaugural lecture:

> I ought now to tell you what my own opinion is with respect to the necessary truth of physical laws – whether I think them true only so far as experiment can be brought forward to prove them or whether I believe them to be true independent of experiment. On the answer which I give to this question will depend the whole method of treating the foundations of our science.
>
> I have no reason to believe that the human intellect is able to weave a system of physics out of its own resources without experimental labour. Whenever the attempt has been made it has resulted in an unnatural and self-contradictory mass of rubbish.* In fact unless we have something before us to theorize upon we immediately lose ourselves in that misty region from which I have just warned you.

The lack of practical opportunities for students in Aberdeen did not mean that Maxwell was limited to 'chalk and talk' in his lectures. Demonstrations were already well established both in public venues such as the Royal Institution in London and as part of physics teaching in universities. Maxwell would discover that Marischal's stock cupboard was well equipped with demonstration equipment, even if it was not always of the most up-to-date nature.

---

* Some would say that this situation applies to certain aspects of modern physics, where theory often runs wild, unrestrained by experiment or observation.

Despite any limitations in his presentation skills, Maxwell was a highly appreciated professor. He might not have yet developed the skills to put a lecture across well, but he was very hands-on and was ready to discuss scientific topics with anyone interested. The same David Gill who had called him a terrible lecturer also commented that Maxwell's teaching influenced the whole of his life. Students found that it was worth suffering his lectures if they then put in the effort to ask for more information. Maxwell also put on a voluntary advanced class for the fourth-year students – not part of the formal university teaching programme – which brought in the likes of Newton's laws, electricity and magnetism, topics that were considered too technical for most students. At the time, there was no honours degree available at Aberdeen, but Maxwell had instituted the intellectual equivalent for the physics course.

Lord of the rings

While in position at Marischal College, Maxwell demonstrated the breadth of his ability to think around a difficult problem by putting a considerable amount of effort into a challenge set by St John's College, Cambridge for its Adams Prize. This was a somewhat bizarre competition (still running) set up in honour of John Couch Adams, who had the misfortune to have produced a considerable amount of early data that suggested the existence of the planet Neptune, only to have his observations ignored, leaving the French co-discoverer of the planet, Le Verrier, to take the laurels.

The Adams Prize could, in principle, be set on any mathematical, astronomical or natural philosophy problem, but in the early years it was dominated by astronomical topics, not entirely surprisingly given that those in charge of the prize topic, George Airy and James Challis, were both astronomers. The challenges tended to involve so much work that the competition was not exactly over-subscribed. Maxwell entered to take on a topic set in 1855 that required hand-in of entries at the end of 1856 with results announced in 1857. This was the fourth time the prize challenge had been set – of the first three years, two had a single entrant, one had no entrants at all.

The topic for 1855 was the rings of Saturn, which, as we have seen, Maxwell had already started to work on during the summer vacation before heading off to Aberdeen. Ever since Galileo had seen something strange about the planet's appearance through his crude telescope and had drawn it as if it had jug-eared handles, assuming that what he saw was three stars that were lodged against each other, the rings had been a mystery. Apparently unique in the solar system (we now know that other gas giant planets have rings, though none is as dramatic as Saturn's), this structure seemed to defy nature.

The challenge was set out as follows:

The problem may be treated on the supposition that the system of Rings is exactly, or very approximately, concentric with Saturn, and symmetrically disposed about the plane of his equator, and different hypotheses may be made respecting the physical constitution of the Rings. It may be supposed (1) that they are rigid; (2) that they are fluid or part aeriform; (3) that they consist of masses of matter not mutually coherent. The question will be considered to be answered by ascertaining, on these hypotheses severally, whether the conditions of mechanical stability are satisfied by the mutual attractions and motions of the Planet and the Rings.

It is desirable that an attempt should also be made to determine on which of the above hypotheses the appearance of both

of the bright rings and the recently discovered dark ring may
be most satisfactorily explained; and to indicate any causes to
which a change of form, such as is supposed from a comparison
of modern with the earlier observations to have taken place,
may be attributed.

Taken in isolation, the rings of Saturn provided an odd topic
for Maxwell to work on. Although he did have wide interests in
physics, he had never shown any great enthusiasm for astronomy beyond a youthful appreciation of the glories of the night
sky from Glenlair, where a lack of street lighting had made for
excellent star viewing. The same would be true for the rest of
his career. It has been suggested that rather than reflecting the
same kind of interest as he clearly had in colour, gases and electromagnetism, entering the competition was merely to take on
the challenge of a puzzle to be solved, with the added bonus of
a noteworthy prize attached.

Maxwell duly applied his mathematical expertise to the
problem of the rings, an oddity that had challenged many of
the greats. The first to recognise the rings as such, the Dutch
scientist Christiaan Huygens, thought that they consisted of a
single solid flat structure, but as telescopes became better able to
resolve their detail it was increasingly clear that there were multiple rings surrounding the planet. The dark band that appeared
to be a gap dividing two rings, referred to in the prize challenge,
was first noticed in 1675 by Italian astronomer Giovanni Cassini,
though strictly speaking the gap could have been a dark part of
the same structure.

In 1787, French mathematician Pierre-Simon Laplace had
got as far as demonstrating mathematically that the rings could
not be continuous, symmetrical solid structures, as no material

would be strong enough to prevent gravitational forces from ripping them apart. As Maxwell pointed out in his prize entry, iron, for example, would not only be plastic under the gravitational stress but would partially liquefy. Even if the rings rotated, which would partly counter the gravitational forces trying to tear them apart, the motion would fail to stabilise the situation, as such a model required inner parts of the ring to move faster than outer parts – the opposite of the reality of a single solid ring.

Instead, Laplace suggested that it was possible to make solid rings stable if the mass in them was unequally distributed. However, one of Maxwell's achievements in his Adams Prize entry was to show that, while Laplace was correct, the only stable structure would be like an engagement ring featuring an immense diamond, with 80 per cent of its mass concentrated on one spot. Such an uneven distribution of matter would have been clearly visible through the telescopes of the day. It just wasn't the case.

Next, Maxwell considered the possibility that what appeared to be solid rings were in fact bands of liquid, a kind of space river that surrounded the planet. As he started work on the more complex mathematics involved in dealing with fluid dynamics, he wrote to his friend Lewis Campbell:

I have been battering away at Saturn, returning to the charge every now and then. I have effected several breaches in the solid ring and now I am splash [sic] into the fluid one, amid a clash of symbols* truly astounding.

---

* Nice little pun there from Maxwell.

Here, Maxwell brought into play a mathematical tool that had been in use in an ad-hoc way for centuries, but had only been made generally applicable in the nineteenth century: Fourier analysis. The name refers to French mathematician Joseph Fourier, who in 1807, in a paper on the transfer of heat through solid bodies, showed that it was possible to break down any continuous function – effectively anything that could be represented on a continuous graph, however strangely shaped – into components that were simple, regularly repeating forms such as a sine wave.

It might seem unlikely but, for example, even a 'jerky' function such as a square wave can be broken down this way, provided we are allowed to use an infinite set of components to make it up exactly (see Figure 2).

Physicists and engineers now make use of Fourier analysis as a matter of course, but in Maxwell's day the technique was still something of a novelty. Yet by using it, he showed that the way waves would combine should there be any disturbance in the rings (which would inevitably arise due to the gravitational attraction of Saturn's moons and of Jupiter) made it impossible for a fluid to be as continuous as the rings appeared to be – they would simply not be stable, and the liquid or gas would accumulate in large globules. In effect if the rings were fluid, Saturn would end up with blobby moons.

With solid rings and fluids dismissed, Maxwell deduced that the rings were most likely collections of vast numbers of small particles, held in place by gravity – but that the distance between us and Saturn meant that we could not make out the individual particles, an analysis that has stood the test of time and close-up examination of the rings. His approach involved exploring the mathematics of displacements in a series of small satellites in the

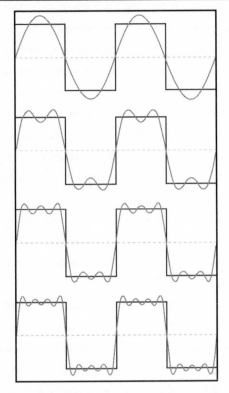

FIG. 2. *As more simple waves are added in, the result gets closer and closer to a square wave.*

same orbit to see how waves caused by any disturbances could travel without breaking the ring apart.

Maxwell's conclusion: 'The final result, therefore, of the mechanical theory is, that the only system of rings which can exist is one composed of an indefinite number* of unconnected

---

* This is sometimes reproduced as 'infinite number', but leaving aside the quibble that infinity isn't a number, Maxwell clearly was not thinking of there being an infinite set of particles.

particles revolving around the planet with different velocities according to their respective distances.' Not only was this the only stable conclusion, in keeping with the second part of the question it was also the only one that could sensibly explain the multiple ring structure.

Maxwell began his essay with the words '*E pur si muove*' – the phrase that was allegedly, though almost certainly not, said by Galileo under his breath after recanting at his trial for heresy. Meaning 'And yet it does move', it seems a slightly odd choice in relation to the topic. Candidates were obliged to begin their essay with a motto, as a separate document linking the candidate's name to the motto was then used to preserve the anonymity of the essays until after the judging took place. It's possible Maxwell chose the words in honour of Galileo's dis-covery of the rings (or, at least, that Galileo noticed there was something odd about Saturn).

As it happens, the motto proved unnecessary. Maxwell's was the only entry that year and, not surprisingly, he won the prize. Astronomer Royal George Airy remarked that this was 'One of the most remarkable applications of Mathematics to Physics that I have ever seen'. Rather than just stop with his submitted essay, Maxwell took into account some comments made by the judging panel and developed his work as a background activity over the next few years, including having a mechanical model built by the Aberdeen-based instrument maker John Ramage, showing how waves could travel if the rings were made up of 36 satellites (made of ivory in the model). Obviously, reality would involve a far greater number, but the model demon-strated how fast-moving waves would be transmitted through the system. This was all before publishing the final version of his work.

Maxwell's prize entry may sound like an impressive but limited piece of work – the dynamics of the rings of Saturn appear to be something of a one-off application – but like many of his thought excursions throughout his working life, it would continue to have ramifications that went far beyond the initial use. The formation of planetary systems, such as our solar system, is a complex and not fully agreed problem to this day, but the best accepted theory depends on the accumulation of a disc of gas and dust, and this owes much to Maxwell's work on Saturn's rings. His achievement has been commemorated in a small way by naming a break in Saturn's C ring the 'Maxwell gap'.

This would not be the last time that Maxwell's unusual (for the day) combination of mathematics and physics would astound his contemporaries. It's arguable that Maxwell was the most significant player in the transforming of physics into its modern form, from a largely descriptive discipline to one where mathematics began to drive progress in the field. It's not that earlier physicists had ignored maths. Newton famously, for example, devised calculus so that he could perform his work on gravity. However, Maxwell took things further, moving from simply explaining observations through mathematics to building mathematical models that took on a life of their own.

Life in Aberdeen

Meanwhile, Maxwell was not content with his teaching work at the university and, much as he had been involved with the Working Men's College at Cambridge, he gave an evening class as a contribution to the Aberdeen School of Science and Art. This was a body devoted to educating tradesmen and others with day jobs that prevented them from attending daytime lectures. As we have seen, this kind of paid evening education, often linked to bodies such as mechanics' institutes, was common in the nine-teenth century and helped many improve themselves. Students, who would typically already have worked for twelve hours from 6.00am, paid 8 shillings* a year for 24 lectures spread over five months from November to April. The School of Science and Art had no premises for Maxwell's classes (though it was able to use the library of the Mechanics' Institute), but he was allowed to give his lectures using the facilities at Marischal College.

It might seem that Maxwell had an overwhelming amount of work on, but he still found time for socialising, and for an activity that, as we have seen, he kept up throughout his life – writing verse. In 1857, Maxwell's friend William Thomson was largely engaged with the laying of the transatlantic cable to send telegraph signals between the UK and the US. In a letter to Lewis Campbell, Maxwell noted that he had sent 'great screeds' to Thomson about Saturn's rings, but it turned out that he was busy 'a-laying of the telegraph which was to go to America'. In the process the cable broke. (According to Maxwell, this was because Thomson was 'bringing his obtrusive science to bear upon the engineers, so that they broke the cable with not fol-lowing (it appears) his advice'.)

Maxwell includes in his letter the words of a song he 'con-ceived on the railway to Glasgow'. He notes that to avoid 'vain repetitions', 'let (U) = "Under the sea"', so that '$2(U)$, by parity of reasoning, represents two repetitions of that sentiment'. The first two verses of 'The Song of the Atlantic Telegraph Company' (with an entertaining prefiguring of a song from Disney's *The Little Mermaid*) read as follows:

---

* The equivalent of 40p, which would be around £36 now in monetary terms, or about £285 based on equivalent wages.

2(U)

Mark how the telegraph motions to me,

2(U)

Signals are coming along,

With a wag, wag, wag;

The telegraph needle is vibrating free,

And every vibration is telling me

How they drag, drag, drag,

The telegraph cable along.

2(U)

No little signals are coming to me,

2(U)

Something has surely gone wrong,

And it's broke, broke, broke;

What is the cause of it does not transpire,

But something has broken the telegraph wire

With a stroke, stroke, stroke,

Or else they've been pulling too strong.

Although Saturn had been an entertaining distraction, Maxwell had not abandoned the work on colour vision that he had started at Cambridge. His colour top to see the effect of combining different colours was useful, but it was relatively poorly calibrated. He was still writing papers based on it, as witness an enthusiastic letter from Michael Faraday, who wrote to Maxwell about receiving papers on the top and on Faraday's lines of force:

I have just read and thank you heartily for your papers. I intended to send you copies of two of mine. I think I have sent them, but do not find them ticked off. So I now send copies, not because they are assumed as deserving of your attention, but as a mark of my respect and desire to thank you in the best way that I can.

Receiving such a letter from one of his scientific heroes must have thrilled Maxwell.

At Aberdeen, however, he was to move beyond the colour top. With the help again of instrument maker John Ramage, he constructed a 'light box' in which adjustable-width slits could be used to vary the amount of each of the primary colours of light, beams of which were sent into a rectangular box where they were focused with a lens to produce a combined colour.

This was not just a matter of observing the behaviour of light – it also enabled Maxwell to continue his exploration of the nature of colour vision, and particularly the origins of colour blindness. In 1860, he would be awarded the highly prized Rumford Medal of the Royal Society for his work on colour vision, and he would continue working with light boxes in London after he left Aberdeen.

His achievements on Saturn and colours were undoubtedly small triumphs for Maxwell – but neither proved to be his most significant work while at Aberdeen, being little more than a pair of amusing diversions. Far more would rest on his development, starting in 1859, of the kinetic theory of gases. Although from the twentieth century onwards, Maxwell's chief claim to fame would be his work on electromagnetism,* at the time of his death, his work on gases was seen as the highlight of his career, as the electromagnetic theory was not yet widely understood.

* Physicist Rudolf Peierls once said that 'If you wake up a physicist in the middle of the night and say "Maxwell", he will be sure to say "electromagnetic field".'

E pur si muove

The term 'kinetic theory of gases' is misleading. The theory has much to say about the nature of heat and of matter. It would be far more meaningful if it had been called the statistical theory of gases, not only because it has statistics at its heart, but also because it was the forerunner of the many ways that statistics would become fundamental to science. However, the 'kinetic' term was useful at the time to distinguish the new theory from the prevailing 'static' theory, which, rather than realising as we now do that molecules are zooming around and colliding, assumed that they sat around, busily repelling each other, producing the force of gas pressure from this mutual repulsion.

Statistics was originally a term for data about a country – it described a state, hence the name – but by the eighteenth century it was transformed by the gradual incorporation of the then new-fangled probability theory to become a mechanism for predicting behaviour of systems based on the likely behaviour of their components.* And gases proved the ideal application for this approach. A gas had been regarded as a fuzzy elastic medium where the constituents somehow repelled each other so that the gas would fill the available space. But the German physicist Rudolf Clausius had started a new way of thinking about gases, regarding their behaviour as the result of the interplay of vast numbers of randomly moving, colliding bodies, which could only be treated statistically.

---

* Statistics in the modern sense was first practised by a button maker called John Graunt in London in the 1660s, and was developed in the coffee houses to become the foundation of the insurance business.

Statistics struggles with individuality* or with systems that are chaotic, in the mathematical sense of having an interplay between highly interacting influences, resulting in large unpredictable variations over time as a result of small initial differences. But a gas is made up of large numbers of identically behaving components with no individual 'personalities' and only a small number of factors such as temperature and density influencing their behaviour, making it ideal for statistical analysis.

As far back as the eighteenth century, the mathematician Daniel Bernoulli had made the first use of statistical theory on gases, and in Maxwell's time, for those who believed in atoms, many of the behaviours of gases were understood this way. We don't know what a single molecule of gas is going to do. But we do know, given the many billions of molecules present in a room, how their behaviour will average out. And it had gradually been realised that their movement had something to do with heat.

Before Maxwell's time there had been competing theories about the nature of heat. As we have seen, one approach described it as a kind of invisible fluid, 'caloric', that could move from one body to another. Much of the early work on heat engines and the conservation of energy was done using the caloric theory. But reflecting on the relationship of temperature and pressure in gases – and suspecting that pressure was due to the impact of gas molecules on the container – physicists began to suspect that caloric was an unnecessary complication, because

---

* This is arguably why pollsters have had so much trouble with predicting the outcome of political events in the 21st century. Where once voters tended to behave as large, predictable blocks, they now tend to operate more individualistically, or in more complex groupings.

what was being measured in reality was the energy with which the molecules that made up matter were jiggling about.

It was suggested by the developers of this new 'thermodynamic' theory, notably Clausius and Maxwell's friend William Thomson, that temperature was a statistical measure of the energy of the molecules – the faster they move overall, the higher the temperature of the gas – while pressure reflected the impact from gas molecules hitting the walls of the container. But one of the behaviours of gas had yet to be explained using a statistical viewpoint (it would be left to Maxwell to take such a numerical approach) – this was the speed that a smell promulgated.

When a smelly substance – whether a perfume or something less pleasant – is released into the air, it takes some time for the odour to reach a distant nose. Yet at room temperature, gas molecules in the air should be flying around at hundreds of metres per second. So why do smells not arrive almost immediately? Clausius pointed out that the problem with the expectation of high-speed odour delivery was that the scent molecules did not have a free and easy journey from the source to the nose. There were vast numbers of air molecules in the way. The odour's progress across a room would be a bit like attempting to take a straight-line stroll across the centre of a dodgems* ride – you would soon be knocked over. Similarly, collisions of the scent molecules with other molecules would inevitably ensue, knocking the molecules in all sorts of different directions and slowing progress from A to B.

Imagine a single odour molecule from a newly brewed pot of coffee starting on its way across the kitchen. If there were no obstacles in its path, it could cross the few metres in a fraction of

---

* Bumper cars, if you prefer.

a second. But in reality, it may well undergo so many collisions and changes of direction that it would have to travel several kilometres before it reached the nose of a waiting breakfaster. Although Clausius deduced from this model that there should be an average distance a molecule travels between collisions (known as the mean free path), he was not able to calculate this distance, a value that would have enabled him to get a feel for the actual size of molecules.

Maxwell was intrigued by Clausius' idea, but disliked the assumptions his rival had made. To make things simple, Clausius decided that it was acceptable to assume that all the molecules move at the same speed for a particular temperature. This certainly makes the mathematics easier to handle – but at the cost of losing the central statistical element of the theory and becoming totally unrealistic.

Temperature is not dependent on the speed (and hence kinetic energy) of an individual molecule, but on the *average* speed of the whole collection. Let's think specifically of that kitchen with its freshly brewed coffee pot. Some molecules in the air would have come into contact with the hot stove or with a windowsill heated by the Sun and as a result would be moving particularly quickly. Others would have given up energy as a result of contact with cooler matter and would have slowed down. There would be a whole distribution of different speeds and energies making up that average. Understanding the different probabilities involved was essential to developing a numerical analysis.

Statistics to the rescue

By bringing the statistics front and centre, Maxwell transformed the approach to the behaviour of gases. His was the first generation who could sensibly do so. Probability theory was only starting to be introduced in universities when Maxwell was an undergraduate. What had been a technique limited to considering games of chance* and insurance premiums was increasingly seen as a tool that could be valuable in the physical sciences. For example, Maxwell's father's friend at Edinburgh, Professor James Forbes, had attempted to use probability theory to assess the likelihood that two stars, apparently very close to each other when seen through a telescope, were part of the same binary system, rather than simply happening to appear from the Earth to be situated in roughly the same direction.

Another eminent physicist of the day who became Maxwell's friend at Cambridge, George Stokes, the Lucasian Professor of Mathematics (a position held in their time by both Newton and Stephen Hawking), was responsible for some of the experimental data that Maxwell used in developing his 'dynamical' theory of gases. Maxwell notes:

From Prof. Stokes's experiments on friction in air, it appears that the distance travelled by a particle between consecutive collisions is about 1/447,000 of an inch, the mean velocity being about 1505 feet per second; and therefore each particle makes 8,077,200,000 collisions per second.

We now refer to the 'Maxwell distribution' of speeds of molecules in gases, and his mathematical solution to provide the

---

* Bizarrely, the fact that probability was first devised to explain the best approach to betting and games of chance meant that in its early days it was considered a dark and dirty art. As far as a demon is concerned, that just made it all the more attractive.

distribution of molecular speeds given a particular temperature is still used today. This was only the start of Maxwell's pair of papers on the subject, to be published in 1860, which covered everything from the basics of pressure and thermal conductivity to viscosity and the way a gas diffused through different materials, seen from the molecular viewpoint.

His results on viscosity produced a particularly surprising outcome. Viscosity is a measure of the amount of drag a fluid presents to a body moving through it.* The higher the viscosity, the more a gas (or liquid) resists movement. It had been assumed that the viscosity of a gas would increase with the gas's pressure. Just as the gas pushes more on its container as pressure goes up, it seemed reasonable that it would push more on anything that tried to pass through it, slowing it down more.

But Maxwell's theory suggested that viscosity should be independent of pressure. Admittedly there will be more molecules in the way at higher pressure – so you could imagine it's the difference between wading through a full ball pool and one with just a few balls in it – but these molecules providing resistance to motion are also themselves moving. Maxwell's calculations showed that the molecules' ability to get out of the way of their fellows would counter exactly the extra quantity present in any particular volume. It would take a few years, but Maxwell would later demonstrate this experimentally.

Maxwell had made his first great contribution to physics – and one that at the time was given significantly more weight by many than his masterpiece on electromagnetic theory. Certainly, it was a topic that became something of a trademark for Maxwell. When a few years later he had attended a lecture at the Royal

---

* In effect, viscosity can be thought of as a measure of a fluid's gloppiness.

Institution in London and got wedged in a crowd trying to leave the lecture theatre, Michael Faraday was prompted to remark: 'Ho, Maxwell, cannot you get out? If any man can find his way through a crowd, it should be you.' This was a clear reference to Maxwell's expertise on the paths of molecules.

The result of his analysis was not perfect. Maxwell had made a range of assumptions and at least one technical error, which would be pointed out by Clausius. And, as would remain the case throughout his career, his workings suffered from regular arithmetical slips. The German physicist Gustav Kirchhoff later commented: 'He is a genius, but one has to check his calculations before one can accept them.' However, it seemed that Maxwell had reached as far as he wanted to go with molecules for the moment. He would return to kinetic theory, but not for several years.

A new family

The work that Maxwell was doing was unlikely to have gone unnoticed by the principal of Marischal College, the Reverend Daniel Dewar.* Dewar was an interesting character who had risen from poverty. His father had been a blind fiddle player and Dewar spent his youth acting as his father's guide, porter and cash collector as they toured Scotland searching out the next gig. For whatever reason, Dewar was spotted by a rich benefactor as a youth with potential, paying for him to attend a private school.

---

* Not to be confused with one of Maxwell's contemporaries, another Scottish physicist who was associated with gases and temperature, James Dewar, who worked on low-temperature gases and invented the vacuum flask or 'Thermos' to keep them cool. You will find a statue to James Dewar outside the Buchanan Galleries shopping centre in Glasgow.

Dewar went on to become a minister of the Church of Scotland. After he was appointed minister of the Greyfriars church in Aberdeen, associated with King's College, he came to the notice of the university and was elected Professor of Moral Philosophy there. Unfortunately, his rising fame as a preacher making him something close to a celebrity did not sit well with the college's idea of the decorum required of its professors. When Dewar would not resign from his ministry he was dismissed. He moved to Glasgow, where he took over the curiously named Tron church.*

It was from this position that in 1832 Dewar was elected the principal of Marischal College. His was an external, Crown appointment and Dewar's humble origins and unseemly drive to succeed apparently did not go down well with the staff of the college, where it was noted snobbishly that he got the post with 'the unanimous disapproval of the College'. Yet, with time, Dewar's work ethic came to be appreciated, especially when he obtained funding for significant enhancement of the infrastructure, including construction of the impressive new granite main building.

Maxwell was a regular visitor to the principal's lodgings, a modest detached house (considering Dewar's role) at 13 Victoria Street West in Aberdeen. Although it is entirely possible that with such a small staff, Dewar would frequently expect to see Maxwell at his lodgings, it has been suggested that their first contact may have been over a book, *Gaelic Astronomy*, by a teacher called D.M. Connell, the book itself written in Gaelic. Dewar

---

* Nothing to do with the Disney movie. The Tron church (now the Tron Theatre) is located in Trongate, a street named after an old word for weighing scales.

had an interest in the language and had contributed to a Gaelic/ English dictionary, so it may well have been a discussion of this 1856 work that first brought Maxwell to Victoria Street West.

At that time, Maxwell was 25, Dewar 71 and his wife Susan 60. Soon, the visits came to involve something more than a narrow discussion. Maxwell enjoyed covering a wide range of topics from theology and philosophy to literature and history with Dewar, a freedom of thought he had missed when moving from Cambridge to join the less socially minded Aberdeen fellows. And it was in the Victoria Street house that Maxwell met the Dewars' daughter Katherine Mary, one of seven children (though three had died and only one other, Katherine's younger brother Donald, lived at home).

Seven years older than Maxwell, Katherine was intrigued by his work and began to help with his colour experiments, making detailed observations as well as assisting with the practical side when Maxwell was caught up with theory. Maxwell became sufficiently close to the Dewars that he was asked to join them in September 1857 when the family took their annual break at a relative's home near Dunoon on Scotland's west coast. This holiday seems to have cemented the relationship: the following February, Maxwell proposed to Katherine.

He wrote to his aunt, Miss Cay:

Dear Aunt, this comes to tell you that I am going to have a wife. I am not going to write out a catalogue of qualities, as I am not fit; but I can tell you that we are quite necessary to one another, and understand each other better than most couples I have seen. Don't be afraid; she is not mathematical; but there are other things besides that, and she certainly won't stop the mathematics.

The couple were married on 2 June 1858. Maxwell's lifelong correspondent and friend since school days Lewis Campbell was the best man (a reciprocal arrangement, as Maxwell had performed the same role for Campbell just a few weeks earlier down in Brighton). As well as supporting Maxwell's experiments in the sciences, Katherine also shared a wider range of interests with him, from literature and theology to walking and horse riding (though neither enjoyed any 'country sports' that involved killing animals). Though it appears Katherine was significantly less high-spirited than Maxwell, they were, without doubt, a well-matched couple, which would help greatly for the months of the year they would spend together in the relative isolation of Glenlair.

Accommodating the British Ass

During the first year of the Maxwells' marriage, it's likely that Maxwell had one thing foremost in his mind as far as work was concerned – the British Association for the Advancement of Science meeting. This organisation, usually contracted to BA, though Maxwell habitually referred to it as the British Ass, which is still going strong as the abbreviated British Science Association, had been started in 1831 as a way of improving the public understanding of science. Unlike the exclusive Royal Society, or the laboratory-centred Royal Institution, the BA was for the everyman and was specifically intended to have neither a building nor funds, but rather to provide pop-up events around the country to spread the scientific word.

After its founding, the BA put on a series of annual events, each of which lasted several days – effectively its annual meeting provided what would now be called a science festival. These gatherings attracted big crowds. Maxwell had been attending them regularly since the 1850 Edinburgh meeting, though he probably missed Dublin in 1857. The venue for a meeting was settled only a year in advance – and at the 1858 Leeds event (a meeting Maxwell had to miss due to his wedding), it was agreed that the 1859 meeting would be held in Aberdeen.

This no doubt thrilled Maxwell, but the only problem with the idea was that Aberdeen had no venue suitable for the large-scale lectures and discussions that were central to a BA meeting. Although the universities both had lecture theatres, and the more central Marischal College's main lecture theatre would be used for smaller side-events, they had nothing with the capacity required for the jamboree that the main BA meeting had become.

There had been talk for some time about building a music hall in Aberdeen, and the BA meeting provided a focus for making it happen. With Maxwell among its shareholders, the Music Hall Company set to work on the rapid construction of a spacious venue in Union Street. Almost inevitably constructed of granite, the imposing 50-foot-high internal space was capable of seating 2,400 and is still a major feature of the Aberdeen cityscape today.* The meeting, opened by Prince Albert, was a huge success.

Getting the Prince Consort along was a major feather in the organisers' caps. The location probably helped. Aberdeen might have been remote indeed from London, but it was handily

---

* Entertainingly, the Musical Hall Company continued to attempt to send dividends from its proceeds to Maxwell at the (long defunct) Marischal College right through to the early 1900s, long after his death. The lawyers responsible for dealing with the payments eventually put an advertisement in the local paper asking for Mr James Clerk Maxwell to come forward, entirely unaware of either his fame or his death.

located just 50 miles from one of Albert's pet projects, the Balmoral estate, where he had recently built a new castle, soon to be a favourite hideaway of Queen Victoria. In total, Maxwell gave three talks in the main venue – on his theory of gases, on colour theory, and on Saturn's rings.

Maxwell and the Prince were not the only attractions at a gathering that would see over 2,500 tickets sold. There were events ranging from a talk on the geology of northern Scotland to exhibitions of scientific instruments. One of the huge successes of the Royal Institution in London had been the public demonstrations – the flashier the better – and the BA event would not have disappointed with its displays of electrical discharges. It even seemed to get ahead of its time by demonstrating 'wireless telegraphy' before it had been invented – but although this term would later be applied to radio, the 1859 Aberdeen demonstration involved sending messages across the River Dee using the electrical conductivity of the water. Yet for all the 361 papers presented, there can be little doubt that by far the most significant for the history of science was Maxwell's first public outing of his 'Dynamical Theory of Gases' including his distribution for molecular velocities.

The exposure that Maxwell received at the BA meeting in Aberdeen made a wider section of the British scientific establishment aware of his capabilities, and it was well-timed, as his career was about to be put in jeopardy. His position as Regius Professor was one that traditionally would have been a post for life, had he desired it. Other young professors with similar positions – Thomson at Glasgow and Tait soon after at Edinburgh – would never move on in their careers. But Maxwell was soon to lose the only position he would ever hold in the Scottish academic system.

Leaving Aberdeen

During Maxwell and Katherine's engagement in 1858, the Universities (Scotland) Act was published, making it definite that Marischal College and King's College would be merged to form a single University of Aberdeen. The year after their wedding, it had become clear that Maxwell would be unable to stay in post. There was only to be a single chair of Natural Philosophy at the new united university, and this went to Maxwell's better-established rival from King's College, David Thomson (no relation to William).

All-in-all, after the successes of the previous two years, culminating in the British Association for the Advancement of Science meeting, 1860 began on a disastrous note for Maxwell. Not only did he lose his position because of the merger of the Aberdonian universities, he failed in his attempt to succeed James Forbes as Professor of Natural Philosophy at Edinburgh, a role that must have seemed ideal for Maxwell.

Here he was beaten by his old school friend Peter Tait (a fact that was hard to resent, though, as Maxwell had pipped Tait to the post to take the Marischal College position). This may seem an odd decision, given their relative publications; Tait had certainly not achieved anywhere near as much scientifically as Maxwell. It seems likely that the appointment was made because Tait was recognised as a significantly better lecturer, and the panel electing the professor, including William Gladstone,* had limited scientific qualifications. An article by David Forfar and Chris Pritchard, comparing the work of Maxwell and Tait, comments:

---

* At the time Chancellor of the Exchequer and eight years later Prime Minister.

The evidence of Maxwell's superiority in research was, of course, already available to those with their eyes open. His investigation of the conditions required for the stability of Saturn's rings oozed originality. Yet, Thomson, Forbes, Stokes and Hopkins [English mathematician William Hopkins] merely resubmitted the testimonials in support of Maxwell which they had proffered to the authorities at Marischal College four years earlier and Faraday, as was his wont, declined to provide a reference at all. Only Airy drew attention to the fertility of Maxwell's theoretical astrophysics. As a body then his referees appear to have lapsed in shameful indolence or, more likely, failed to grasp the significance of Maxwell's work.

To make the year worse, Maxwell contracted smallpox, becoming dangerously ill over the summer. But once he had recovered, things finally began to look up. As well as publishing on the kinetic theory of gases, it was 1860 when his work on colour theory won him the Royal Society's Rumford Medal. He made another significant set of talks at the 1860 British Association for the Advancement of Science meeting, this year in Oxford, though his appearance was overshadowed by the now infamous debate over evolution between Bishop Samuel 'Soapy Sam' Wilberforce and Thomas 'Darwin's bulldog' Huxley. Most significantly, two months after being turned down for Edinburgh, Maxwell finally gained a new position: the chair in Natural Philosophy (specifically dealing with physics and astronomy) at King's College London.

Newly married, Maxwell was about to move from the quiet and relatively rural Aberdeen to the largest, most dynamic city in Europe.

In which the demon's challenge is posed

Whitten you look back at my creator in his early years, it can be hard not to consider him a touch of an oddball. It's not that he was unsociable – which has proved a problem for many a scientist, in my experience. In fact, both at home in Glenlair and in Cambridge, JCM had enjoyed socialising and was known for his playful sense of humour. This often comes through in his letters, where he could be so mischievous it is sometimes difficult to make out what he meant. A highly technical letter to a fellow scientist could suddenly break out into a moment of whimsy, as when he remarked to his friend Peter Tait in something near text-speak: 'O T'! R. U. AT 'OME?'*

Yet, back then, JCM was almost always to some degree an outsider. At Glenlair he had been the posh kid playing with the country yokels. He was himself a bumpkin when compared with his more sophisticated Edinburgh peers. He had been labelled as the uncouth student, accepted by Cambridge despite his personal characteristics. Did this outsider status help him develop his unique viewpoints? How should I know? I am a demon. But it's hard to imagine that being an outsider wasn't part of what made old JCM the way he was.

---

* In letters, JCM tended to refer to people by initials. T was already allocated to William Thomson, so Tait became T'.

The other things that were already starting to have an influence on his path were the unexpected direction changes in his life. Just as he was really getting somewhere on electromagnetism, the lure of the new job in Aberdeen came along. And yet before he could make a start on that he also became laird of Glenlair. Many of his contemporaries put in a similar position would have packed in the university work and taken on the estate as their life. It's not as if he needed the money. Science could have become a fulfilling hobby. It's almost as if someone was trying to tempt him away from his true path. And it wouldn't be the last time.

The tyranny of the second law

But enough about him – it's me you're interested in. It's my name on the front of the book. I'm going to be a trifle anachronistic here in the story of my maker's life. He would not conjure me up until eleven years after he had graduated from Cambridge. And when he did, he didn't have the decency to name me properly. He simply called me a 'finite being', I suppose meaning that I wasn't a god as I had limitations. As a name, it's a bit on the vague side, a term that could take in anything from an earthworm to a genius. As I've mentioned, it was his pal, that other dashing young Scottish natural philosophy professor of the high Victorian age, William Thomson, who realised my demonic nature.

Thomson, incidentally, got me all wrong. He claimed with a typical puritanical disdain for the interesting side of life that in calling me a demon he didn't intend me to be anything evil and inclined to temptation. No horns or pointy tail for his demon – he had in mind more of an intermediary, a spirit that was a kind of interface between the human and the divine. These days, he'd probably have spelt the word 'daemon' to emphasise the distinction. Sadly for humanity, old Thomson was wrong, though. Malevolence has proved to be very much my strong suit – I have always enjoyed getting human minds in a twist.

What is particularly satisfying is that I achieve this mental obfuscation with very little effort on my part. The role I was created for involves little more than opening and closing a door. But in doing so, I have a strangely satisfying role in the intriguing matter of disorder.

If you recall, we are dealing with the second law of thermodynamics, which could be summarised as 'Chaos is on the rise' (such a pleasant statement). Long-term, this has dire consequences for the universe as you know and love it. As JCM's frequent correspondent Thomson once put it: 'the end of this world as a habitation for man, or for any living creature or plant, at present existing in it is mechanically *inevitable* ...' All thanks to my favourite law.

The second law says, then, that the level of disorder in a system stays the same or increases. And the system in question that I was introduced to control was ridiculously simple. Let's start by seeing the second law in action, unhampered. Imagine you've got a box of gas – air would do, but we'll make it pure nitrogen to keep things simple. The box is divided into two halves by a partition – on one side the gas is hot and on the other side the gas is cold.* There is a door in the partition between the two sides which you leave open. What happens after a little time?

---

* It may seem there is no imaginable reason for having such a box – and this is often the case in physicists' thought experiments – but, as it happens, you could imagine this as a simplified version of putting an icebox into a warm room.

Molecules of gas from each side will randomly ping through the doorway. Some – the hot ones – will be travelling particularly fast. As we have seen, temperature is just a measure of the energy of the particles. The faster they travel, the more energy they have – the higher the temperature. Other molecules, the cold ones, will be relatively slow. So, we leave the door open for a good stretch of time. To begin with, most of the molecules leaving the hot side will be hot ones – and vice versa. So, the hot side will cool down and the cool side will heat up. Eventually the box will get to a state of equilibrium – each side will be around the same temperature. And the two halves should stay that way. We wouldn't expect to go back to a state where one side was hot and the other side was cold.

Notice, by the way, something that physicists forget at their peril. This whole business is purely statistical. It's entirely possible that for a fraction of a second all the molecules going one way might happen to be hot and all the molecules going the other way might be cold and the two halves of the box would reach different temperatures again. But it's very, very unlikely that this would happen for any significant period of time. We're talking about vast numbers of molecules. If my box was a metre cubed and the gas was at room temperature and pressure, there would be over 10 trillion trillion molecules in there. The chances of most of them acting in this way is very small – that's the statistical driver for the second law of thermodynamics.

As a little aside, the second law was originally conceived as a mechanical, unbreakable law that heat *always* moves from the hotter to the colder body, and it took some of our man Maxwell's contemporaries quite a while to come round to the statistical way of looking at things. JCM had his own rather nice way of describing what was involved: 'The 2nd law of thermodynamics has the same degree of truth as the statement that if you throw a tumblerful of water into the sea, you cannot get the same tumblerful out.' He was highlighting the fact that there is no absolute mechanical mechanism preventing the law being broken, it's just very, very unlikely.

What has all this got to do with the level of *disorder*? It's a bit of an odd term, but you can think of the difference between the two setups this way. When the hot and cold gases have mixed together, a molecule could be anywhere. But when they were separate, the hot molecules were on one side and the cold on the other. Then, we knew where to find the two different kind of molecules. There was more order in the system. It's a bit like the difference between a page of this book as it now stands and the same page with all the letters scrambled up in any old order. The randomised page is not just useless as reading matter – it has more disorder. The second law is why it's a lot easier to break an egg or a glass than it is to unbreak it.

The demon is summoned

So, we've got an idea of how the second law applies to a partitioned box. If the hot and cold gases did separate themselves spontaneously, that would break the second law. (Or to be precise, because the law is statistical, that would be very, very, very unlikely to happen according to the second law.) My role is to make that separation happen time and again with predictable ease.

What my creator did was to place me in charge of the door separating the two halves of the box. We start with the hot and cold molecules all mixed up. And then I simply open and close the door, depending on the kind of molecule that's approaching it. If the molecule is fast, I only let it through if it's travelling from left to right. If it's slow, I only let it through going right to left. So, gradually, the hot and cold molecules separate. Order is produced from chaos and I break the second law. Neat, eh?

I seem to have first been mentioned in a letter written in 1867 to my creator's friend Peter Tait, who was putting together a book on thermodynamics. JCM first described the box setup (calling the wall separating the halves of the box a 'diaphragm'), then introduced me:

> Now conceive a finite being who knows the paths and velocities of all the molecules by simple inspection, but who can do no work* except open and close a hole in the diaphragm by means of a slide without mass.
>
> Let him first observe the molecules in A [the hot side] and when he sees one coming the square of whose velocity is less than the mean square velocity of the molecules in B [the cold side] let him open the hole and let it go into B. Next, let him watch for a molecule of B, the square of whose velocity is greater than the mean square velocity in A, and when it comes to the hole, let him draw the slide and let it go into A, keeping the slide shut for all other molecules.
>
> Then the number of molecules in A and B are the same as at first, but the energy in A is increased and that of B is diminished, that is, the hot system has got hotter and the cold colder and yet no work has been done, only the intelligence of a very observant and neat-fingered being[†] has been employed.

---

* A little derogatory, I feel, that 'who can do no work'. Perhaps it would have been better to have said, 'who rightly felt that work was beneath him'.
† This is better.

Or, in short, if heat is the motion of finite portions of matter and if we can apply tools to such portions of matter so as to deal with them separately, then we can take advantage of the different motions of different proportions to restore a uniformly hot system to unequal temperatures or to motions of large masses.

Only we can't, not being clever enough.

In a letter to John Strutt (Lord Rayleigh) written three years later, JCM gave more detail about me, calling me 'a doorkeeper, very intelligent and exceedingly quick, with microscopic eyes but still an essentially finite being'.

In a later work, *Theory of Heat*, JCM made it clear that my role was to illustrate a wider range of possibilities, where 'delicate observations and experiments' would make it possible to take a look at the actions of a relatively small number of molecules, and in which case the familiar behaviour of vast numbers of molecules in a body would not be applicable.

It was four years after my first mention that William Thomson wrote a paper describing my efforts and cementing my fame, using the 'demon' word for the first time. He also set up a bizarre mental picture of a whole array of us demons bashing molecules with cricket bats, but this is far too undignified to give it any consideration.

Doing it without energy

Now, you may have spotted a slight flaw in the whole 'deployment of demons' business if you bothered to read one of the footnotes a way back. I pointed out that this experiment was a bit like an icebox in a warm room. Let's make it more specific. Let's make one half of the box a refrigerator and the other side the room it's in. We switch the fridge on and wait. After a while, the fridge side of the box is cold, while the other side is warm.* A refrigerator in a room has achieved the same as I did, with no demons required.† But there's a major difference between this picture and my effort.

The second law is restricted to closed systems, sealed off from the rest of the universe. The law only works if someone isn't pumping energy into the system. It's perfectly possible to produce more order from chaos if you work at doing so. Think of the Earth. You may consider nature to be fairly chaotic, but we see all kinds of structures that have formed over time – natural ones (including your body) and artificial ones. That wouldn't be possible without a vast amount of energy coming into the Earth to power it – and thankfully for you, the Sun provides as much energy as is needed and far more.

What made me special – particularly demonic, I guess you would say – is that I can do my business of sorting the thermodynamic sheep from the goats using the door without the need to put energy into the system. I am operating a frictionless, inertia-free door (not available at your local DIY store – this is a thought experiment) and no energy is added to the system by my actions. If you aren't comfortable with a frictionless door that has no mass, because that could never exist in the real world, bear in mind that this is just a convenience. The only

---

* The other side of the box is not just at room temperature – it warms up because the refrigerated side of the box has a radiator on the outside of it. Look at the back of any fridge.

† Assuming that refrigerator manufacturers, other than those in the fiction of Terry Pratchett, do not, in fact, employ demons as the operative part of their devices.

essential as far the second law is concerned is that I do not put any energy into the system of molecules, which I don't.

So how do I perform my trick? How could someone, even with my unrivalled brilliance, break the unshakeable second law of thermodynamics? That was James Clerk Maxwell's challenge to the world – and himself. It's perhaps the only physics challenge he ever took on and failed at. As did his friends, like William Thomson. I was a conundrum. They couldn't see how it was possible for me do my job, yet equally they couldn't see why I would fail. Some people tried to argue that I was pointless – me, pointless! – because in the real world there couldn't be such a demon. But physics doesn't work like that. If a law's a law, it should hold up, whatever you throw at it. And I managed to beat it.

Or so it seemed back then, though I would have to face some challenges further on along the way. But I suppose we need to get back to my creator to see what happened when he ventured to the mighty metropolis that was London in 1860.

Chapter 4

A capital adventure

Every university that Maxwell had attended or worked at up until now was an ancient institution, still clinging on to traditions and even aspects of the syllabus that dated back to medieval times. But his new academic home, King's College, prestigiously located on London's Strand, took its first students only in the year that Maxwell was born (remarkably, before the 1820s, London did not have a single seat of higher education). The university's management prided itself on its modern values, with explicit courses in individual scientific disciplines, rather than traditional, classical-heavy, 'bit of everything' undergraduate courses; they even covered the upstart, hands-dirty topic of engineering.

Science at King's

We can get a feel for Maxwell's attitude to science and how it should be approached from part of his stirring inaugural lecture at King's. He told the students:

> In this class I hope you will learn not merely results or formulae applicable to cases that may possibly occur in our practice afterwards, but the principles on which those formulae depend, and without which the formulae are merely mental rubbish. I know the tendency of the human mind is

to do anything rather than think. But mental labour is not thought, and those who have with labour acquired the habit of application, often find it much easier to get up a formula than to master the principle.

Maxwell was emphasising the importance of doing more than ticking the boxes and mechanically working the equations – as we now might allow a computer to do for us. His approach to science was always to look for underlying principles, to get as close as he could to the true 'laws' of nature.* To get a flavour of the duties expected of him, Maxwell was required to be in college three mornings a week (10.00am to 1.00pm) and to give one evening class aimed at working men – leaving a comfortable amount of time for working on his own projects.

His pay from the university was directly linked to the number of students he taught, receiving 5 guineas (£5.25) for each day student, 18 shillings (90p) for each evening class student and £2, 7 shillings and threepence (just over £2.36) for each 'occasional student', who did not matriculate but were enrolled in individual classes to gain expertise. This made him around £450 a year, the equivalent of around £39,000 now,† a little more

---

* The concept of a natural law is rather an elusive one. Laws are written down in black and white. A natural law is more an analogy of what nature is like, as we can arguably never directly interact with reality, merely observe phenomena. But that's all too philosophical for me – there's a special breed of demons who specialise in philosophy.

† Calculating present value of historical salaries is something of a black art (a speciality of demons). Maxwell's salary would be the equivalent of around £39,000 in terms of the goods it could buy. However, it would be the equivalent of around £300,000 in proportion to the change in the earnings of an average worker between the two periods.

than he had earned in Aberdeen. In practice, though, he was significantly better off, as one of the conditions of the Aberdeen merger was that he would receive a remarkably generous annuity for life of his salary when he lost his position, giving him an additional £400 a year, making his annual income a handsome £850 – around £76,000 in purchasing power or £579,000 in proportion to earnings.

Something of Maxwell's approach to teaching while at King's College can be seen in an incident recorded in the Campbell and Garnett biography written shortly after his death:

> The professors had unlimited access to the [College] library, and were in the habit of sometimes taking out a book for a friend. The students were only allowed two volumes at a time. Maxwell took out books for his students, and when checked for this by his colleagues explained that the students were his friends.

While this early biography has a tendency to paint Maxwell in something of a saintly light, it illustrates here an unusual attitude for a professor at the time and perhaps reflects both Maxwell's youth – at 29, he was still only a few years older than his students – and his unusual upbringing for someone of his class, having mixed with the working class far more than would be the norm for the landed gentry.

The students Maxwell taught at King's College had a strong focus on practical, applied science – many of the young men in his physics and astronomy classes would become engineers, a subject that was hardly recognised at other universities, and they received training that would benefit them in such roles. This meant that their education was seen as more of a boost to their practical skills than a means of obtaining a degree,* and the majority did not stay for a complete three-year course, typically averaging just over four terms before moving on. The fees at King's were over seven times those of Marischal College at £12, 17 shillings (£12.85) for each of the three terms, though those wanting only to attend natural philosophy classes had a cut-price rate of just 3 guineas (£3.15) a term.

Maxwell and Katherine had a comfortable home to entertain guests at 8 Palace Gardens Terrace in Kensington (oddly, this is now number 16). It was a fair walk to reach the Strand, though Maxwell often would, when he wanted to exercise his country-bred legs with a four-mile stroll. The house may well have reminded him of his aunts' houses in Edinburgh – Palace Gardens was a little grander, with five storeys and pillars at the entrance, but it was still a solid city townhouse.

After Aberdeen's relative social backwater, Maxwell was looking forward to having more opportunity to meet with like-minded individuals, as had been possible in Edinburgh and Cambridge. Where the entire staff of Marischal College dealing with all disciplines numbered just twenty, Maxwell's department of Applied Science alone, one of four at King's College, had a similar number. And his colleagues were only the start.

Both the Royal Society and its more practically-minded little brother, the Royal Institution, offered lectures and discussions in the city attended by many of the leaders of British science – and in May 1861, Maxwell was thrust to the fore with an invitation to give a lecture at the Royal Institution (RI).

---

* In practice King's could not award BA or MA degrees at the time, but on successful completion, a student became an AKC – an Associate of King's College.

Maxwell in the 1860s, during his post at King's College London or shortly afterwards.
Getty Images

Bring colour to the Institution

The RI was the spiritual home of Maxwell's hero, Michael Faraday. As we have seen, Faraday started at the Institution as an assistant to Humphry Davy and now, though 70, Faraday was still associated with the venue where he had given so many lectures and set up the famous Christmas Lecture series for children. Following on from his Rumford Medal, Maxwell was asked to give a lecture at the RI on colour theory, a topic that would remain a lifetime interest for him.

One of the traditions of the Institution was to give demonstrations during lectures – the more dramatic presentations were popular draws, sometimes bringing in royalty among the audience. The spinning colour wheel that Maxwell had used in his experiments was too small for the audience to see it at a distance, so he decided to produce something that had never been seen before – he would project a large-scale image of a true colour photograph.

It was common enough for black and white photographs to be hand-tinted to give the effect of colour, but Maxwell's plan was to produce a full colour image by combining three monochrome photographs taken and then projected using red, green and blue filters, demonstrating that these three primary colours were sufficient to generate all the colours that we see. He had first conceived this idea back in 1855, when he had briefly discussed it at the Royal Society of Edinburgh, but in the best manner of Royal Institution demonstrations, he intended to bring the theory alive before his audience.

Maxwell was no photographer himself – at the time, a distinctly specialist activity. But luckily, one of the country's top experts, Thomas Sutton, himself a Cambridge Wrangler in his day, had been employed as the official King's College photographer, a role that was primarily a teaching one. Sutton was able to help Maxwell with the tricky (and potentially dangerous) photographic medium of wet collodion. This was no simple matter of buying a roll of film and having it processed, let alone the ease of modern digital photography. The photographer had to be a deft chemist before there was any possibility of taking an exposure.

The wet collodion process started with cotton being soaked in a toxic and highly corrosive mixture of nitric and sulfuric acids. The product – gun cotton* – had to be washed and dried before dissolving it in ether or alcohol to produce a terrifyingly flammable gel. The would-be photographer then added halogen salts (iodine or bromine) to the gel and spread the resultant suspension carefully onto a clean glass plate. This part of the process required particular skill to get a consistent, level layer of the 'collodion' gel. The plate was then dipped in a bath of silver nitrate, which reacted with the halogen to produce a light-sensitive silver-based coating. After this 'activation' process, the plate, still wet, was placed in the camera and exposed. Finally, the plate had to be developed, fixed, rinsed, dried and varnished before the final product was available. This was anything but 'point and shoot'.

Under Maxwell's direction, Sutton took three separate photographs of a multicoloured tartan-like ribbon,† using red, blue and green filters, made by mixing different coloured dyes in water and placing the camera behind glass troughs containing the liquids. Maxwell was then able to superimpose the three separate images projected onto a large screen at the RI with three magic lanterns, each shining through one of the red, blue and green troughs.

---

* Just as dangerous as it sounds. Used at the time as the explosive charge of mines and torpedoes, and for blasting (and later as a rocket propellant), though oddly, given the name, not typically used in guns.

† The ribbon is usually described as being tartan, but pedants point out that tartans usually have two colours, where the ribbon had three.

Making it happen on the night must have given Maxwell some bitten fingernails. His talk was a Friday Night Discourse, the most fearsome of Faraday's public event programme. The audience were a stern mass of black ties and the speaker was required (as they still are) to crash through the doors of the lecture theatre on the second of the starting time and begin speaking immediately without introduction. But everything went well; Maxwell pronounced himself satisfied with the outcome. The three light beams combined on the screen to give a relatively realistic-looking colour image of the ribbon,* though Maxwell did remark that a better result could be obtained with materials that were more sensitive to colours away from the blue end of the spectrum.

It was a few weeks later in May 1861, still only 29, that Maxwell was elected a Fellow of the Royal Society, cementing his position as a member of the London scientific establishment. In the present day, election to a fellowship of the 'Royal' is probably the highest honour available to a British scientist, though in the Society's early days, most fellows weren't scientists per se, but rather wealthy individuals who had an interest in science. By Maxwell's time the shift in the membership to actual scientists was well under way, though he would have qualified on both counts.

Electromagnetism goes mechanical

By the time of his Royal Institution lecture, around five years had passed since Maxwell had last worked on electromagnetism.

---

* Don't let scientists ever tell you that they don't sometimes benefit from luck. The photographic plates Maxwell used were not sensitive enough for the relatively low-energy red light to produce a good red image when the red filter was used. But, luckily, the red sections of the tartan were also good emitters of ultraviolet, which passed unhampered through the red filter and made the appropriate impact on the plate.

Whether inspired by appearing at Faraday's lecture table or simply coming naturally back to a topic that would always fascinate him, while at King's College he picked up the subject where he had earlier left off. His previous model, treating electricity and magnetism as fluids, was workable only for fields that did not move. But to cope with the likes of generators and electric motors, now becoming relatively common as a result of Faraday's work, Maxwell had to deal with movement. Just as he had transformed his picture of the rings of Saturn from solid to fluid to particles, he now modified his approach to electromagnetism by resorting to a mechanical model.

This 'model', like many that physicists would construct from Maxwell's time to the present day, was not a model in the sense of a small-scale physical construction, looking like the real thing. Instead he used a theoretical construct that reflected what was observed in nature and that could be used to make predictions to see whether the model was an effective representation of the phenomenon, or whether it needed refining. This was not an actual mechanical device, but one that used the principles of mechanics to try to reproduce the effects of electromagnetism.

Electromagnetism has one fundamental difference from the other force of nature that we experience directly – gravity. All gravity attracts.* But electromagnetism comes in two flavours, known as negative and positive for electricity, and as north and south for magnetism. The rule is that like flavours (say negative and negative for electricity, or south and south for magnets)

---

* In the absence of antigravity, which despite many YouTube videos and conspiracy theories to the contrary, is yet to be demonstrated. It was speculated at one time by serious physicists that the gravitational force between matter and antimatter could be repulsive rather than attractive, but there is as yet no evidence for this.

repel while opposite flavours (negative and positive, or north and south) attract. Bring together electrical charges or magnetic poles and this becomes very obvious. Maxwell set out to model this process, starting with magnetic poles, using a model of the magnetic field that worked in a purely mechanical fashion.

There are a couple of other requirements he needed to cope with in designing a model to work for magnetic poles. One is that they always seem to come in opposing pairs – unlike electrical charges which are happy to be standalone negative or positive, we have never seen a separate north or south pole* – and the force that is felt between poles, whether repulsive or attractive, obeys an inverse square law as does gravity, dropping off at the same rate as the square of the distance between the two poles that are attracting or repelling each other.

The biggest problem that Maxwell faced was the same one that had caused many to struggle in trying to find a model to explain how gravity could work. It's relatively easy to have a mechanical model that produces the effect of repulsion, because it's easy for one object to push another. But it's harder to have a model that produces attraction, as without involving something like magnetism it's difficult to see how one object can pull another to which it has no direct physical connection.

Ever since Newton's time this problem had been got around by devising mechanical models for gravitational attraction which were based on the idea that space was filled with invisible high-speed particles heading in all directions which did not interact with each other, but which pushed on massive bodies. Usually the impact from all directions balanced out, but when a second body was nearby, it blocked some of the particles heading

* Known as a magnetic monopole.

towards the first body, producing the effect of attracting one body towards the other. These models needed a fair amount of tweaking, as the obvious implication is that gravitational pull would depend on the size of a body rather than its mass. There seems no evidence that Maxwell ever considered this type of model for electromagnetic attraction.

Maxwell's electromagnetic spheres

Maxwell's first attempt with his mechanical model was to imagine that the magnetic field was made up of a collection of spheres,* tightly packed to fill space. These spheres or 'cells' would be spinning around. Generally speaking, when a physical object spins, the centrifugal forces† acting on it make it spread out in the middle and contract at the poles. This happens, for example, to the Earth, which it had been known since Newton's time has a bulge around the equator and so is an oblate spheroid rather than a perfect sphere.

But, unlike the Earth, Maxwell's spheres were surrounded by other spheres. So, if the equator of a sphere expanded as it was spinning, it would push on the surrounding spheres. In

---

* It should be stressed that, as with his fluid model, the spheres were just an analogy. There was no suggestion that space really was filled with spheres, though eventually the picture would resolve as something Maxwell felt was closer to reality.

† Ever since Newton, some have liked to mock the idea of centrifugal force flinging things outwards as a body rotates, pointing out that the flung objects are simply following their natural trajectory in a straight line, and the 'real' force involved is usually a 'centripetal' force, towards the centre, countering the inclination to move outwards. However, this is just a distinction of the frame of reference used to examine the forces – it depends where you look at the effect from – and centrifugal force can still be a useful way to describe an effect.

this model, the axes of the spins were aligned to the lines of force that Faraday had demonstrated in the magnetic field. The result would be very close to what was observed. At right angles to the lines of force – the direction of the equators – the forcing outwards of the spheres would produce a repulsive effect, while along the lines of force – the direction of the poles – the spheres would be pushed closer together and the effect would be an attraction.

Conveniently, the faster the spheres rotated, the bigger this effect would be – so the spin rate in the model corresponded to the strength of the magnetic field. In this kind of mechanical model, it's perfectly possible for the components to be allowed to be frictionless, but Maxwell thought it better to allow for a degree of interaction between the spheres. If two spheres alongside each other are turning in the same direction, then at the point of contact, the surfaces will be moving in opposite directions.

If you imagine two spheres, moving clockwise, with their axes pointing out of the page, the left sphere's surface moves down at the point of contact while the right sphere's surface moves upwards (Figure 3).

To avoid direct interaction between the spheres, Maxwell imagined a large number of much smaller spheres acting like ball bearings between the main spheres. But unlike the ball bearings in a traditional device, which are usually constrained by a bearing, these would be free to flow as they like. And if these little spheres were considered as particles of electricity (what we'd now call electrons), when an electrical circuit was made, the little spheres would flow in the channels between the bigger ones. (Let's call the bigger spheres cells, as Maxwell did, to avoid getting our assorted spheres confused.)

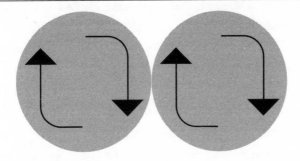

FIG. 3. *Two spheres rotating in contact.*

What's particularly neat about this model is that we now have an interaction between electricity (little spheres) and magnetism (cells). If an electrical current flows, the result is that the cells start to rotate and a magnetic field is produced.

Maxwell was also able to extend the model to allow for different materials – the better the conductor, the more easily the spheres were able to flow. In an insulator, the magnetic cells clung onto the little electrical spheres sufficiently that it was very hard to get any electrical current. So now his model dealt with the difference between insulators and conductors as well as electromagnets. But this was not all there was to electromagnetism. He would also have to cope with induction.

As we have seen, this was something that Faraday had investigated. He had shown that a changing magnetic field would produce a flow of electricity – this was how electrical generators worked. Induction also occurs when a current is started in a wire that is near another wire. When the electricity is switched on it produces a magnetic field. As far as the other wire is concerned, to begin with there is no magnetic field, then one surges into place and becomes steady. As the field is established there is a changing magnetic field, so a blip of current flows briefly in the second wire. The same thing happens if the first wire's current is switched off.*

Vortices and idle wheels

By now, as Maxwell refined the model, his cells had become hexagons, which made the model clearer visually. His original diagram had rather more hexagons than are necessary to make the point, but we can envisage what he had in mind with just three rows of hexagons. The row of tiny spheres[†] between the top cells and the middle row is connected to a loop of wire – this is where the induction will take place. And the row of spheres between the middle row and the bottom row of cells is connected to a wire with a battery and a switch. We are ready to induce some current.

The experiment starts in the state of diagram I in Figure 4. The switch is yet to be thrown so there is no electrical current flowing.

After the switch is thrown, in diagram II, the spheres start to flow left to right between the middle and bottom sets of cells. The middle cells begin to rotate in the opposite direction to the bottom row. Maxwell had identified the direction of the magnetic field as the direction of rotation in his model, so the magnetic field above and below the electrical current flowing is in opposite directions – it is circling around the wire.

---

* Induction is how, for instance, mobile phones and electric toothbrushes can be charged contactlessly. A changing current in the base produces a changing magnetic field which induces a changing current in the device to charge the battery.

† Maxwell called the tiny spheres 'idle wheels' and the hexagonal cells 'rotating vortices', a term based on his earlier fluid model. He would come to think of the cells as actual vortices in the ether.

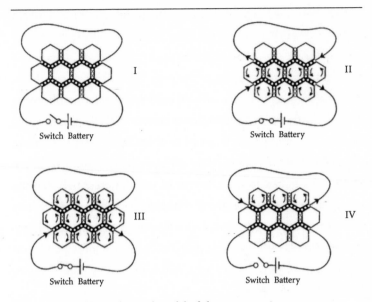

FIG. 4. *Maxwell's mechanical model of electromagnetism.*

Meanwhile, the spheres between the top and middle cells are being pushed by the rotating cells in the middle row. These little spheres start to rotate clockwise and to flow right to left, causing a brief current in the upper wire. However, there is no battery in this circuit to keep the current flowing. The resistance in the wire slows the spheres down until they stop moving, but they are still rotating clockwise. This rotation causes the upper row of cells to rotate in the same direction as the middle row, as in diagram III.

When the switch is opened (diagram IV), the bottom flow of spheres rapidly stops, slowing the bottom and middle rows of cells. But the top row is still rotating – the rotating cells act as tiny flywheels, enabling the ether (see below) to briefly store energy which then generates another brief flow of current in the top wire before the whole system settles down to stillness.

By this stage Maxwell's remarkable and very Victorian feeling model* had explained three of the main aspects of electromagnetism, though he had not yet managed to incorporate the attraction and repulsion between electrical charges. If you feel that this construction of arbitrary mechanical components seems a little unlikely – too far abstracted from reality to be useful – you are not alone. The French physicist Henri Poincaré remarked that there was a 'feeling of discomfort and even of mistrust' among his fellow countrymen when faced with Maxwell's mechanism. This was taking a model, a mechanical analogy of reality, and stretching it to what seemed to him a ludicrous extreme. And yet it was working.

The power of analogy

For, Maxwell, this novel use of analogy – building models – was the way forward to better understand the physical principles of the natural world. While still a student at Cambridge, he had written:

> Whenever [men] see a relation between two things they know well, and think they see there must be a similar relation between things less known, they reason from one to the other. This supposes that, although pairs of things may differ widely from each other, the *relation* in the one pair may be the same as that in the other. Now, as in a scientific point of view the *relation* is the most important thing to know, a knowledge of the one thing leads us a long way towards knowledge of the other.

* To really be true Victorian in look and feel, the model should have been constructed of brass and mahogany.

... and this philosophy, initially based on mechanical and later on purely mathematical models, would be the key to his remarkable success.

It was relatively easy to see how Maxwell's earlier model of fluid flows through a porous medium had come about from the influence of Thomson's work on heat, but this far more sophisticated model seems to have come out of nowhere. However, Maxwell was very fond of having actual mechanical models built – think, for example, of his colour top, or the mechanical model he had made to illustrate waves in the rings of Saturn.

At the same time, he was now in London, where Charles Babbage had dreamed up his remarkable mechanical computers, the difference engine and the analytical engine. Though neither was built, Babbage had completed a working model of part of the difference engine and had worked, with the help of Ada, Countess of Lovelace, on the principles of the far more sophisticated analytical engine. Babbage was still alive when Maxwell spent his time in London. It is surely likely that it was the experience of the many brass miracles of Victorian engineering that led Maxwell to his remarkable hexagons and ball structure.

What is certainly without doubt is that Maxwell's electromagnetic model proved surprisingly flexible, given its nature. For example, different materials vary widely in their magnetic properties. Even between metals, the differences are stark, and when you bring in other materials, such as wood, it's clear that there is a major difference in the way magnetism operates (or doesn't) between different substances.

What Maxwell had constructed was a mechanical model of the ether, the fluid that was thought to fill all space, allowing light waves to pass through a vacuum and acting as the transmission mechanism for the electrical and magnetic fields. What

he suggested was that when that ether was overlaid on different materials the result would be a change in the nature of the
hexagonal cells in the model. The better the magnetic properties
of the material, the more dense was the cell in the model. With
increasing density, the cells would produce a greater centrifugal force, producing more magnetic flux – the measure of the
strength of the magnetic field.

In building this model, we need to reiterate, Maxwell did not
suggest that space was full of rotating hexagonal cells and tiny
ball bearings – not even invisible cells and bearings – but rather
he was suggesting that the way the ether behaved produced a
result that had the same effect as his imagined mechanical structures. There was, however, a big difference between this new
model and his original fluid idea. Although Maxwell did not see
the exact detail of the model with its ball bearings and hexagonal
cells as what was actually happening in the ether, he *did* think
that he had now come much closer to reality. There seems little doubt that his thoughts on what was *really* happening were
influenced by William Thomson, who had firmly stated that
there were actual vortices in the magnetic field, reflected in the
way that the magnetic field influenced light.

Maxwell never lost his belief in the existence of the ether,*
a substance that would be thrown into doubt by an experiment
carried out by the American scientists Albert Michelson and
Edward Morley in 1887 and eventually dismissed altogether by
Einstein's work at the start of the twentieth century. And Maxwell
did think that the magnetic field involved actual rotating vortices

---

* Many see this as ironic, as he would later make a discovery that made the
ether unnecessary but refused to accept this implication.

in that invisible, undetectable medium. The ball bearings part of the model proved more of an embarrassment.

Maxwell commented that 'The conception of a particle having its motion connected with that of a vortex by perfect rolling contact may appear somewhat awkward. I do not bring it forward as a mode of connexion existing in nature ...' On the other hand, he points out that it works well and as long as it is taken as provisional and temporary, it should help rather than hinder a search for the true interpretation of the phenomena.

So, Maxwell was making it clear that the ether is not a matter of having space full of hexagons and ball bearings, any more than the crystal spheres of the ancient view of the universe. Even so, the ether had to be a remarkable substance indeed. It was invisible, impossible to detect, yet provided the medium for light to wave through, and was elastic, as it would have to be for waves to pass through it, but was somehow also so rigid that light could continue to travel for vast distances, seemingly without losing any energy the way that a conventional mechanical wave would. The ether was such a firmly established part of scientists' mental model of reality that it was extremely hard to shake off.* For Maxwell, the specific details of his model provided an effective mechanical analogy for a more complex fluid reality.

He therefore tried to measure the effects of the vortices in the ether directly by having a device built that involved a small electromagnet free to rotate in all three dimensions; this, he hoped, would detect the impact of nearby vortices. Nothing was

---

* Some current physicists suggest that modern equivalents of the ether might be dark matter or the inflationary concept in the Big Bang theory, which have become engrained in our way of thinking but have limited evidence to support their existence.

discovered, which merely underlined for him that the vortices appeared to be very small. Similarly, he believed that the electric field was directly linked to elastic deformation of the ether – again, he was not saying that his specific model was an accurate portrayal of reality, but that the key features, such as vortices and the elastic response, were a reflection of reality and so should have the kind of impact we would expect the components of the model to produce.

This belief of Maxwell's in what we now know not to exist should not be used to belittle his achievements – it was a perfectly reasonable possibility at the time. But it also illustrates that he was only part-way towards the acceptance of a model as something totally isolated from the reality it reflected, which would enable the eventual development of purely mathematical models that have come to dominate physics.

There was no doubt that, at King's College, Maxwell had begun to really make something of the insights that drove him. He published his findings in a two-part paper. His original work on electromagnetism had been called *On Faraday's Lines of Force* – the new paper was *On Physical Lines of Force*, emphasising that he had moved on to a more practical model. It was published in *Philosophical Magazine*,* with Part 1 in the March 1861 edition and Part 2 split between the April and May editions.

---

* *Philosophical Magazine* may sound like a lightweight periodical, a *Popular Science* of its day, but this venerable scientific publication began in 1798 and would carry papers from many of the Victorian big-hitters including Davy, Faraday, Maxwell and Joule. It is still in print. By comparison, the journal *Nature*, which would eventually eclipse it, first appeared in 1869. *Philosophical Magazine* began as a general journal of natural philosophy, but soon came to specialise in physics.

In which the demon becomes a star

With his model representing the inner working of electromagnetism established, my master was well down the road to featuring in the scientists' hall of fame. He wasn't quite there yet – no one could entirely love a model based on hexagonal vortices and ball bearings, not even a demon, but he had laid the foundation for his later glory.

It might seem that I'm biased and am singing my creator's praises more than is justified – especially as he's not exactly remembered by the public like a Darwin or an Einstein – but to quote Richard Feynman, generally accepted to be one of the twentieth century's greatest physicists:

> From a long view of the history of mankind – seen from, say, ten thousand years from now – there can be little doubt that the most significant event of the 19th century will be judged as Maxwell's discovery of the laws of electrodynamics.

Feynman was a fan.

Victorian computer dating
It can be difficult to get a feel for the personality of a Victorian gentleman; so much that was written about individuals back then was stiff and lacked personal insights. It's not necessarily

easy for you to imagine JCM as a real person (certainly compared with someone who knew him, like me). So, it's lucky that he filled in a questionnaire about himself – the sort of thing you might use when applying for a computer dating site these days – for fellow scientist Francis Galton.

Galton has had a bad press of late as an advocate of eugenics, but he did wide-ranging, useful work and was fascinated by the statistical analysis of people and their heredity, particularly when considering the nature of genius. In 1874, Galton published a book called *English Men of Science,** based on around 200 questionnaires, which he had persuaded Fellows of the Royal Society to fill in. It was Galton's book that gave us the first sighting of the idea of 'nature versus nurture' (using those terms) and his research is seen as the starting point of the use of questionnaires in psychology.

We, however, can ignore the rest and pick out JCM's replies to get a little more insight into Maxwell, the man. We are told that he was 5 feet 8 inches (1.73 metres) tall, and was 'often laid up'† before he was nineteen, but never since – what's more, he had never had a headache. When asked about his 'mental peculiarities', Maxwell wrote:

---

* The word 'scientist' was probably still not widely accepted enough at this point for Galton to be comfortable with using it in his title. Women, of course, given Victorian sensibilities, did not come into it. It might seem odd that the entirely Scottish JCM should be listed as an 'English' man of science. Galton's main selection mechanism was those who lived or worked relatively near London, and at the time Maxwell was in Cambridge. It's also true that the word 'English' was often loosely used as an alternative to 'British' at the time.

† That's to say unwell. JCM was what was known as a sickly child.

Fond of mathematical instruments and delighted with the forms of regular figures and curves of all sorts.

Strong mechanical power. Extremely small practical business. [He also noted that his father was a 'Very great mechanical talent'.]

Strongly affected by music as a child, could not tell whether it was pleasant or painful, but rather the latter; never forgot melodies or the words belonging to them and these run through the mind at all times and not merely when the tunes are in fashion; can play on no instrument and never received instruction in music.

Great continuity and steadiness; gratitude and resentment weak; στοργη* pretty strong; not gregarious; thoughts occupied more with things than with persons, social affections limited in range; given to theological ideas and not reticent about them; constructiveness of imagination; foresight.

Another fascinating little insight into JCM's mind is from his answer to the question 'Origin of Taste for Science':

I always regarded mathematics as the method of obtaining the best shapes and dimensions of things; and this meant not only the most useful and economical, but chiefly the most harmonious and the most beautiful.

I was taken to see William Nicol [see page 24] and so, with the help of Brewster's Optics and a glazier's diamond, I worked at polarisation of light, cutting crystals, tempering glass, etc.

I should naturally have become an advocate by profession, with scientific proclivities, but the existence of exclusively scientific men, and in particular of Professor Forbes, convinced

* Greek for affection or love.

my father and myself that a profession was not necessary to
a useful life.

The demon's catechism

As my master was becoming more famous, I too was getting
better established and being recognised by many physicists as
an entertaining diversion – you might say I was becoming a star.
So much so, that JCM felt obliged to write up a little biography
of me in the form of a religious catechism* for his friend Peter
Tait. Rather irritatingly, he used the plural 'demons', where anyone with any sense knows that I am a singular entity:

Concerning demons

1. Who gave them this name? Thomson.
2. What were they by nature? Very small *but* lively beings
   (capable of obeying orders but) incapable of doing work[†]
   but also able to open and shut valves which move without
   friction or inertia.
3. What was their chief end?[‡] To show that the 2nd Law of
   Thermodynamics has only a statistical certainty.
4. Is the production of an equality of temperature their only
   occupation? No, for less intelligent demons[§] can produce

---

* A catechism is a summary of doctrine, often phrased as questions and
answers. Maxwell attended Dean Ramsay's catechism classes in Edinburgh
as a teenager, thanks to his Aunt Jane.
† There he goes again with the personal comments.
‡ It is not just the question and answer format that shows that this was based
on a catechism – a well-known example, The Westminster Shorter Catechism
of 1647, has for one of its questions: 'What is the chief end of man?'
§ Clearly this part does not refer to me as an individual. Some of my colleagues, certainly.

a difference in pressure as well as temperature by merely allowing all particles going in one direction while stopping all those going the other way. This reduces the demon to a valve. As such value him. Call him no more a demon but a valve like that of the hydraulic ram,* suppose.

The apparent belittling here runs counter to the appreciation of William Thomson, who later remarked of me that I was an intelligent being who was different 'from real living animals only in extreme smallness and agility'. Some have suggested that Thomson's enthusiasm to portray me as non-mechanical was a conscious attempt to oppose the theory of the 'X Club', whose members, including the Irish physicist John Tyndall and English biologist Thomas Huxley, were strong supporters of the idea that living things were mechanical automata with no concept of a soul required.

Whether or not Thomson's motivation was partly religious, as it happens, JCM was wrong in this instance – turning his simplified demon into a valve for the temperature and pressure experiment would not work (he didn't need to get in extra staff, incidentally – this is something I would happily have helped him with).

Richard Feynman, a physicist with his own demons, described in his acclaimed 'red book' lectures on physics that any basic mechanical replacement for a demon would heat up as a result of its efforts – so much so that it would eventually be jittering around far too much to do its job. As well as preventing

---

* This is a device that employs water pressure to raise part of a head of water higher than it originally was, using the greater pressure of the large body to move the smaller amount. It makes use of two one-way valves.

us demons from being put out of a job, Feynman was pointing towards something that is special about my role – it's not possible to do the job without being able to deal with information. You'll never find a non-intelligent, mechanical demon.

It ought to be stressed, though, that JCM was not setting out to wreck the second law of thermodynamics. He was entirely happy with its validity. But his development of the statistical approach to the kinetic theory of gases meant that he was aware that at its heart, the second theory was about probabilities, not certainty. As the English physicist James Jeans would later point out in a textbook:*

> Maxwell's sorting demon† could effect in a very short time what would probably take a very long time to come about if left to the play of chance. There would, however, be nothing contrary to natural laws in one case any more than the other.

A demon like my humble self was perfectly capable of taking things in a direction that was extremely unlikely but not impossible in normal circumstances.

So, let's get back to JCM as he takes a break from the city after getting together his latest thoughts on electromagnetism.

---

* The author would like me to point out that he still finds it remarkable, bearing in mind that the textbook quoted above was written in 1904, that while at university, he had tea with James Jeans' widow, the organist Lady Susi Jeans. She was, admittedly, significantly younger than her husband.
† This makes me sound worryingly like a Royal Mail employee.

Chapter 5

Seeing the light

With his new model written up, Maxwell used the long summer vacation* of 1861 to head back to Glenlair with Katherine. Academics often take this time away to refresh themselves by either totally ignoring academic work or dealing with a pet project that had been sidelined, and it seems that it was Maxwell's intention to concentrate on the Glenlair estate, but he could not get his electromagnetic model out of his mind. There was something not quite right in his mechanical analogy, impressive though it was at predicting most electromagnetic effects. This might simply have been because he was stretching the analogy too far, but discovering whether or not this was the case was nagging at him like the throbbing of an aching tooth.

The power of flexible cells

In his model, the hexagonal cells and the small spheres transferred rotating motion to adjacent components, so the movement spread up the diagrams shown on previous pages. But in a real mechanical system this would usually result in a loss of energy. This might seem to be something to do with electrical resistance,

* Long by most people's standards. After the six-month summer vacation at Aberdeen, the mere four months allowed by King's College may have seemed quite short to Maxwell.

but the picture didn't fit – and the resistance took place in the wire, not throughout the ether, which his model represented. However, there was a way to fix this if the rotating cells were no longer rigid, but he allowed them to give way under pressure, a property known in physics as being elastic.

It was probably easier to think constructively in the familiar and relaxed surroundings of Glenlair, away from the bustle of London, with Katherine proving an effective sounding board. Maxwell envisaged a circuit consisting of an insulator between metal plates. The pair of plates would be connected to a battery, so an electric field extended through the insulator. In his model, in an insulator the tiny spheres could not flow, as they were attached to the hexagonal cells. But if the cells were flexible, each cell could twist around a little on its axis. So, in effect, a small amount of current would flow from the plate on one side to the plate on the other due to the displacement of the cells, before the torsion in the elastic material became strong enough to resist any further motion.

At this point, one plate would be relatively positively charged, and one would be relatively negative. This meant that there would be an attraction between the plates. And, magically, the elastic mechanism he had suggested produced an attraction out of nowhere, because the twisting of the cells would make the ether contract. Each cell would shrink, just as a spring gets smaller as it is wound up, which would pull the plates towards each other. This pull provided the missing factor in his model, electrostatic attraction.

If the battery were now removed from the circuit, that tension in the cells would remain. The attractive force would still be there. But if the two plates were then connected by a wire, the current would briefly flow between them, releasing the twists from the flexible cells – there would be a discharge of electricity and the attraction would disappear. He was describing the action of an electrical component that used to be called a condenser and is now known as a capacitor.

Just as Maxwell had allowed the density of the cells in his model of the ether to be modified by materials it passed through to account for different magnetic properties, he was able to do the same for substances with different electrical properties. If the gap between the metal plates was filled with different materials – air or wood or mica, for example – the result would be a change in the elasticity of the ether's cells. A substance like mica (a naturally occurring silicate crystal that forms sheets and was often used as an insulator in early electrical experiments) is more susceptible to electrical charges than, say, air. Such a material, known as a dielectric, would, in his model, make the cells twist more easily, so that a bigger charge was held and more current would flow when the plates were connected.

By making those hexagonal cells elastic, able to twist and tighten, he had brought electrical attraction into his model. As an idea it worked, and by using the relatively new techniques that applied differential calculus – the mathematics of change – to vectors,* he was able to use his model to provide a mathematical description of electrical and magnetic fields.

Maxwell's new version of his model portrayed the ether as a kind of invisible energy store. Static electrical energy was potential energy, stored away in the ether like the energy in a spring, while magnetic energy was kinetic energy, like that of a rotating

---

* A vector is a quantity with both size and direction, where a scalar quantity only has size. Speed is a scalar – 50 kilometres per hour, for example. Velocity is a vector – 50 kilometres per hour heading north, for example.

flywheel. And his model showed that the two types of energy were unfailingly linked. Making a change in the level of one influenced the other.

This was a remarkable achievement. But, of itself, building a model that matches reality does not necessarily make it useful. For many centuries, pre-Renaissance astronomers used a model of the structure of the universe based on epicycles, where a complex combination of circular movements allowed everything to rotate around the Earth, while explaining the oddities, for example, of the observed orbit of Mars, which reverses its direction in the sky. We now know that this is because the planets are orbiting the Sun, not the Earth. The epicycle model matched well to what was observed, but it did not give astronomers anything new to test it with – it was designed to match observation and the stubborn belief that the Earth was at the centre of the universe, and it did nothing more. But Maxwell's model went further. It predicted something that had not previously been observed.

Waves in the ether

If the ether were truly like Maxwell's model,* there was an extra component to be added to his mathematics. Even empty space was filled with the ether, and this meant that you should always get that little twitch of movement in the tiny spheres from the twisted elastic cells. This extra 'displacement current' on top of the usual conduction current added a component to his equations that made them mathematically complete. In terms of the accuracy of his model in reflecting what was actually observed,

---

* Always bear in mind that the ether does not exist. This was Maxwell's thinking at a time when it was still assumed that there was such a thing.

this was the turning point. Yet the introduction of elasticity into the cells had another, just as important, implication.

If a material is elastic – if the substance that makes it up has some flexibility – it is possible to send a wave through it. Waves involve a repetitive displacement of the parts of the material the wave is passing through, and that can only happen if the material is not completely rigid. Imagine Maxwell's cells and spheres stretching through the void of space. If there were a twitch of movement in a row of spheres, caused by electrical energy, that would cause a brief torsion in the adjacent cells – representing a short movement of magnetic energy. That in turn would twitch the next row of spheres – generating a new electrical surge.

This succession of twitches would pass through the apparently empty space occupied by the ether. It would not progress instantaneously, as the cells would have some inertia, so each would take a little time to get moving. What Maxwell's model was predicting is that it should be possible to send out a wave of alternating electrical and magnetic displacement through an insulator – even through the vacuum of space – because the ether was always present. And as the displacement in the spheres and cells was happening at right angles to the direction this disruption was travelling, what was observed would be a transverse wave, like ripples on water, where the material is displaced at 90 degrees to the direction of travel of the wave.

The concept of the displacement current introduced a role for the theoretical physicist that seems to have been Maxwell's own invention. At the time, theoretical physicists did most of their work producing theories to match observations. But Maxwell saw a role for the theoretician in looking for the holes that were left in experimental evidence and making predictions that could later be tested. The displacement current was not the result of

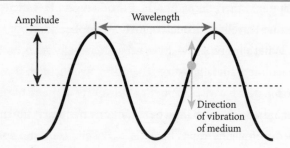

FIG. 5. *The features of a transverse wave.*

any observation – it was purely a prediction from his model. This apparently small contribution, often overlooked in popular descriptions of Maxwell's work, was revolutionary. There was considerable resistance to this approach from some of his colleagues, but Maxwell's daring step became a central role of theoreticians, to the extent that in some fields this kind of deduction from models came to dominate.

The introduction of the displacement current was a remarkable achievement for Maxwell's method – other contemporary attempts to explain electromagnetism relied still on action at a distance. Dismissing Faraday's electrical and magnetic fields, they used instead the idea of point charges producing a force at a distance and did not come up with this extra component which would allow for electromagnetic waves. Einstein considered this move to be crucial. Writing in his book *Autobiographical Notes*, he said:

The most fascinating subject at the time I was a student was Maxwell's theory. What made this theory appear revolutionary was the transition from action at a distance to fields as the fundamental variables ... it was like a revelation ...

As it happens, there already was a transverse wave known to science that travelled through insulators and even empty space – light. What's more, as we have seen, Maxwell's hero Faraday had speculated that light somehow involved electricity and magnetism.

Remember that in 1846, when Maxwell was fifteen, Faraday had filled in for Charles Wheatstone at a lecture and had told his audience:

> The views which I am so bold as to put forth consider, therefore, radiation as a high species of vibration in the lines of force which are known to connect particles, and also masses of matter, together. It endeavours to dismiss the aether, but not the vibrations.

In his inspired vision, Faraday was prepared to go a step further than Maxwell. Where Maxwell believed that his model of cells and spheres represented the ether, Faraday believed that the fields of electricity and magnetism extended through empty space without the need for an ether. Either way, both had realised that light was a vibration in those fields, Faraday in visionary concept and Maxwell with the mathematical backing of his mechanical model.

Seeing the light

We don't know the exact process by which Maxwell realised that the waves his model predicted seemed remarkably similar to light, but it's hard to imagine that he was unaware of Faraday's 'Thoughts on Ray-vibrations' talk. Whatever the means of reaching this idea, Maxwell's model gave him a way to test out whether this relationship between light and electromagnetism was valid.

It was known that the speed of waves through a medium could be calculated from a combination of the elasticity of the medium and its density. In his model, elasticity corresponded to the electrostatic force and density to the magnetic equivalent.

Although not all the values required were perfectly pinned down for his model, if Maxwell took the minimum values for the elasticity of a vacuum, the speed of the predicted wave would match the velocity produced by dividing the unit of magnetic charge by the unit of electrical charge, which was unlikely to be a coincidence. Maxwell was able to calculate that his model predicted that in a vacuum, such electromagnetic waves should travel at 193,088 miles per second (310,700 kilometres per second).

All Maxwell had to do was to compare his prediction with the speed of light, a value that had first been made calculable in 1676 by the observations of Danish astronomer Ole Rømer of variations in the timings of the moons of Jupiter as the distance between the giant planet and Earth varied. More recently it had been measured by French physicist Armand Fizeau using a mechanical device which sent flashes of light from a fast-rotating toothed wheel down a 9-kilometre track, before returning the light through the wheel, using the speed of rotation of the wheel to measure the elapsed time.

Unfortunately, Maxwell did not have any documentation on Fizeau's work with him at Glenlair,* and though his speed for electromagnetic waves was certainly relatively close to what had been measured, he couldn't remember the value sufficiently well to be sure how effective his prediction was. He had to wait until his return to London in October to compare his theoretical

---

* Ironic, given that Maxwell's work would provide an essential foundation for the internet that enables anyone to look up this value pretty much instantly.

wave's speed with that of light. Back at King's College he discovered that the latest figure from Fizeau gave light speed as 195,647 miles (314,850 kilometres) per second, with other estimates in the 192,000 to 193,118 miles per second (308,990 to 310,790 kilometres per second) range. This was less than 1.5 per cent different from his calculated speed.

Such a similarity seemed highly unlikely to be a coincidence. Maxwell wrote:

> The velocity ... agrees so exactly with the velocity of light calculated from the optical experiments of M. Fizeau, that we can scarcely avoid the inference *that light consists ... [of] undulations of the same medium which is the cause of electric and magnetic phenomena.**

His 'mechanical analogy', which had aroused in Poincaré that 'feeling of discomfort and even of mistrust', had revealed the truth behind a mystery that had puzzled humanity for millennia – what was light?

Although Maxwell had considered *On Physical Lines of Force* to be complete with the first two parts, he now decided to extend it to add a third section in 1862 which included the displacement current and electromagnetic waves. This was soon followed by a part four, as he realised that his electromagnetic waves would account for another previously unexplained phenomenon.

We have already seen on page 25 how Maxwell as an undergraduate had made use of polarised light in his home laboratory. Faraday had also studied polarised light and had

* Maxwell's italics.

discovered that passing such light through a magnetic field would rotate its direction of polarisation. Now that Maxwell had identified light as a combination of electrical and magnetic waves at right angles to each other, and assuming that polarisation represented the direction of these waves, it was natural that a magnetic field would have an influence on the changing fields in the wave and cause them to rotate, just as it caused a wire carrying a changing current to rotate in an electric motor.

Too heavy for one person to discharge

Maxwell had a little more time available for his own work after his return to London in October 1861 with the appointment of George Robarts Smalley as a physics lecturer to assist him. Maxwell had complained to the college that his teaching requirements were 'too heavy for one person to discharge'. This appointment did not represent any financial generosity on the part of King's College. Smalley's pay of 7 shillings per student per term was deducted from Maxwell's salary.

Smalley continued in the post until July 1863, when he was appointed Astronomer Royal for New South Wales. Maxwell wrote a reference for Smalley to the British Astronomer Royal, George Biddell Airy, noting that:

> I believe Mr Smalley to possess the scientific knowledge and the habits of accuracy which would fit him for work at the Observatory ... I consider that he would be steady, accurate and skilful in Observatory work.

William Grylls Adams replaced Smalley (and later took over Maxwell's own position). Smalley and Adams took some of the weight from Maxwell's shoulders.

Despite this reduction in work pressure while in London, Maxwell remained at his happiest when back at Glenlair. Here he could both think in peace about his physics and enjoy the rural life. In a letter written from Glenlair just after Christmas 1861 to Henry Droop, who had become a friend while they were both fellows at Trinity College, Cambridge, Maxwell noted:

> I have nothing to do in King's College till Jany. 20, so we came here to rusticate. We have clear hard frost without snow, and all the people are having curling matches on the ice, so that all day you hear the curling-stones on the lochs in every direction for miles, for the large expanse of ice vibrating in a regular manner makes a noise which, though not particularly loud on the spot, is very little diminished by distance.

It shouldn't be deduced, though, from his pleasure at having free time, that Maxwell was the kind of scientist whose only concerns were his own research and who had little interest in his students (as was the case with, say, Newton or Einstein). As we have seen, he stood up for his students in the use of the library and he lectured to working men. A good example of his attention to detail in this respect was a letter he wrote to J.W. Cunningham, the secretary of King's College London, in December 1862.

> Dear Sir
> I am very anxious that the examination papers in Mechanics should be printed from type instead of from stone.
>
> I find that the lithographic papers are printed so that even if everything is plain in perfect copies, uncertainties exist in other copies which are very apt to make the examination not quite a fair one.

Mr Smalley has the M.S. and expects to give it in at the
office today.

Lithography, literally stone writing, is a printing technique where
the dark parts of an image (the letters in a text) are marked out
on the surface of a flat piece of stone – typically limestone –
(or later metal) with a resistant substance such as wax or fat.
Then the surface is treated with acid, which etches into the surface where there is no resistant material. The surface is cleaned
then moistened, with water being retained in the etched sections between the letters or raised imagery. Finally, an ink that
is immiscible with water is applied – the ink stays on the non-
etched parts, and so reproduces the image or text.

Related, but more sophisticated techniques known as offset
lithography and photolithography are still in use today, for printing
and for producing printed circuits respectively (though, despite
the names, no stone is involved). However, in Maxwell's time the
process, though relatively cheap, was not as consistent in its results
as using moveable type – metal letters fixed into a frame. Maxwell,
as always a champion of the students, was ensuring his examination papers were legible despite additional cost to the college.

The Great London Exposition

Maxwell also continued with his interest in communication of
science to the wider public that had come through in both his
work with the British Association and in lecturing at the Royal
Institution. An opportunity arose in 1862, when it was decided
to follow up on the huge success of the 1851 Great Exhibition.
Although France had put on a pair of national events earlier, the
Great Exhibition was effectively the first World's Fair, a chance
to show off and revel in the wonders of Victorian technology.

Such was the success of the first event that its profits funded the construction of the Science Museum, Natural History Museum and Victoria and Albert Museum. Before they were built, however, the land that would be the site of the Natural History Museum was used to house the Great London Exposition of 1862, also known as the International Exhibition.

As a money-making event, this proved a relative failure compared with its predecessor, doing little more than break even thanks to the far more lavish building constructed for the purpose, but still around 6 million people filed through the vast halls. Maxwell was responsible for producing the guide to a section of philosophical (scientific) instruments connected with light. His might have been a small contribution to a massive venture, but he went far beyond a simple catalogue, taking the opportunity to throw in some history of science and descriptions of the physical mechanisms involved, showing his expertise with leading-edge experimental optics.

Meanwhile, once Smalley was in place at King's, there was soon an opportunity for Maxwell to make use of that freed-up time. It wouldn't be understating things to say that Maxwell's model of electromagnetism and its prediction of electromagnetic waves was a huge breakthrough – not only in this specific case, but also for the way that physics itself would be undertaken, in which Maxwell's approach of producing a model and testing its predictions has become a central part of the scientific method.* Even so, and despite a largely positive reaction when he wrote

* In fact there was a third reason this breakthrough was so important, though Maxwell would not live to witness it. Maxwell's model required light to always travel at a particular speed in a vacuum, a fact that Einstein would use to develop his special theory of relativity.

it up, Maxwell was not entirely happy. Perhaps Poincaré's mistrust stung him – but he felt it ought to be possible to take away the framework of analogy, removing his mechanical model and keeping only the pure, untrammelled mathematics.

Chapter 6

Science by numbers

Like many Victorian scientists, Maxwell did not suffer from the modern tendency to remain constrained by a tight focus – he clearly appreciated the chance to roam free across the topics covered by physics. This is apparent in some of his letters, where he happily discussed a wide range of physical subjects with fellow scientists.

A good example would be a letter that Maxwell wrote in August 1863 to George Phillips Bond, an American astronomer based at Harvard University. Bond had met Maxwell in London that May and subsequently had written to him both about the rings of Saturn and about comets. At the time, the behaviour of comets' tails was a puzzle. Back in 1619, the German astronomer Johannes Kepler had pointed out that the tails of comets always pointed away from the Sun. When the comet is heading into the solar system, towards the Sun, its tail flows out behind it, but when the comet is moving in the opposite direction away from the Sun, the tail lies confusingly in front of it.

This behaviour suggested to Kepler that, rather than being left behind like a stream of smoke from a moving flame, the comet's tail was being pushed by something emanating from the Sun. Somehow, the Sun's rays were forcing the comet's tail away. Late in his career, Maxwell would come up with a better explanation for this (see Chapter 9), but in responding to Bond, he speculated about the nature of the ether that he still believed was the medium for light waves.

Maxwell wrote of that medium:

> [I]t is well able to do all that is required of it, whether we give it nothing to do but to transmit light and heat or whether we make it the machinery of magnetism and electricity also and at last assign gravitation itself to its power.

Given Maxwell's discovery of the nature of light, it's no surprise that he wanted to bring electricity and magnetism into the mix, but the idea of gravitation also being involved may seem a little odd. However, it fitted well with ideas that had been circulating ever since Newton's day.

Newton famously claimed not to have a hypothesis for *how* gravity worked at a distance, writing in his masterpiece, the *Principia*, 'hypotheses non fingo', usually translated as 'I frame no hypotheses'.* This wasn't true. As a great supporter of particle-based theories – thinking, for example, that light was a collection of particles or 'corpuscles' – Newton did have a particle-based theory of the mechanism of gravity, variants of which would be developed up to the time that Einstein put gravity on a sound mathematical footing in 1915.

As we have seen, the idea based on invisible particles flowing through space and pushing on massive bodies was simple and attractive,† but it had one major flaw, which various

---

* This was probably a sneaky dig on the part of Newton. 'Fingo' is not a particularly complimentary way of describing coming up with an idea. His assertion could probably be loosely translated as: 'I'm not going to just make a hypothesis up.'
† Literally.

scientists over the years would attempt to overcome with what were ultimately fudges. In its simple form, the theory predicts that the gravitational force will be linked to the size of the attracting body. While size is usually a factor, it's only because big things tend to be more massive. Newton had shown it was the mass of the bodies involved that determined their gravitational attraction, not their size. Explanations of gravitation based on the particle pressure theory had to be modified to account for this.

What Maxwell seems to have had in mind in his letter to Bond was a similar mechanism for gravity, but one that depended on pressure in the ether. As he put it:

> If we could understand how the presence of a dense body could produce a linear pressure radiating out in straight lines from the body and keep up this kind of pressure continually, then gravitation would be explained on mechanical principles and the attraction of two bodies would be the consequence of the repulsive action of the lines of pressure in the medium.

He drew an image showing the Sun emitting lines which then hit a body and curve around it in parabolic shapes, and went on to speculate that the comet's tail is a result of these pressure lines, pushing away from the Sun. But he couldn't explain why the lines of force would be visible as a tail (we now know the tail is gas and dust, vaporised from the comet), asking:

> Is there anything about a comet to render its lines of force visible? and not those of a planet which are much stronger? I think that visible lines of gravitating force are extremely improbable, but I never saw anything so like them as some tails of comets.

Maxwell's ideas here may not have been fruitful, but they demonstrate the breadth of his thinking.

The viscosity engine

Rather than immediately refining his model for electromagnetism after its initial triumph, Maxwell returned to his old sparring partner, the nature of a gas, looking specifically at a property of fluids known as viscosity – a measure of the liquid's (or gas's) resistance to shearing forces – effectively how thick and gloppy the material is.

At the time, it was thought that the viscosity of a gas varied as the square root of the temperature – if the temperature quadrupled, for example, the viscosity would double. This would not be the temperature as we measure it for domestic use, from the arbitrary starting point of the freezing point of water, but from the coldest possible temperature, absolute zero, which is around −273.15°C (−459.67°F). The concept of absolute zero had been around since the eighteenth century, but Maxwell would have had an appropriate scale to use thanks to his friend William Thomson, who in 1848 devised the Kelvin scale (Thomson was later ennobled as Lord Kelvin), starting at absolute zero.

However, Maxwell, who always seemed to enjoy bridging the gap between experiment and theory, undertook a series of experiments to confirm or deny the behaviour of viscosity with temperature. His schedule at King's College left him plenty of time for experimental work, but unlike a modern physics professor he did not have access to a laboratory at the university and had to perform his experiments in the attic of his house.

Maxwell's main experimental device for investigating the viscosity of gases consisted of a series of discs along a vertical wire spindle, alternating fixed discs and discs which rotated together by twisting the wire. The discs were contained inside a glass chamber which meant that Maxwell could alter the pressure with an air pump, or change the contents to a different gas to discover the impact on the viscosity. This was no table-top apparatus – it stood above Maxwell's height on the attic floor (see Figure 6). The discs were 10.56 inches in diameter and the wire was four feet long. Maxwell started the discs moving using magnets outside the case, sliding them from side to side until the discs were twisting back and forth on the wire. The resultant oscillation was slow – Maxwell notes that 'the period of a complete oscillation was 72 seconds and the maximum velocity of the edge of the disks was about 1/12 inch per second'.

When the discs were oscillating, the resistance of the air between the turning discs and the nearby fixed discs (having fixed discs also minimised the impact of draughts) caused a dragging effect that enabled Maxwell to estimate the viscosity of the gas. Despite a dangerous-sounding setback – his glass chamber imploded when he reduced the pressure too much and it took him a month to get the apparatus working again – he soon had solid results, which he presented to the Royal Society in November 1865.

Because the whole point of the experiment was to see how viscosity varied with temperature, Maxwell and Katherine had to make extreme efforts to vary the temperature in the London attic. The near-contemporary biography of Campbell and Garnett notes that:

> For some days a large fire was kept up in the room, though it was in the midst of very hot weather. Kettles were kept on the fire and large quantities of steam allowed to flow into the room. Mrs Maxwell acted as stoker, which was very exhausting

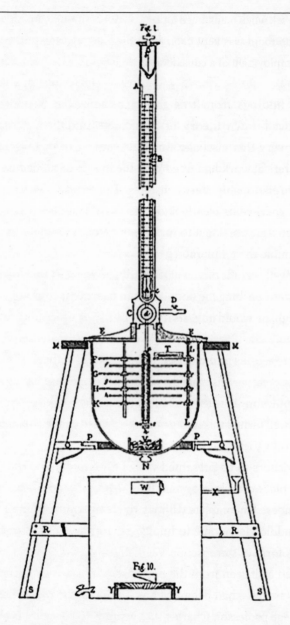

FIG. 6. Maxwell's viscosity apparatus with the discs enclosed in the glass container like an inverted bell jar.

work when maintained for several consecutive hours. After this
the room was kept cool, for subsequent experiments, by the
employment of a considerable amount of ice.

The findings from this experiment proved a challenge to
Maxwell's own theory of gases. It did confirm his surprise
discovery that viscosity was independent of pressure. But his
theoretical work had fitted with the previously assumed square
root relationship between temperature and viscosity, while
the experiments clearly showed that the viscosity actually was
directly proportional to the temperature. You would only have
to double the temperature to double the viscosity. Despite this,
Maxwell's results did plenty for his reputation as an experimental
physicist pushing the boundaries of knowledge. For the moment,
though, he would be distracted from taking the work any further.

Stereoscopes and coffins

It's no great surprise that the distraction came from his old favourite topics of light and colour vision, but in his London home
Maxwell turned this into a mix of research and parlour entertainment. In Victorian homes like Maxwell's it was not uncommon
for visitors to be presented with a diversion that consisted of a
new piece of technology. In the 1860s, many a parlour would be
considered incomplete without a stereoscope, as essential a piece
of middle-class home technology as a computer today.

A form of stereoscope – which combined two pictures portrayed as if seen from the positions of the two eyes to produce
a 3D image – had been invented in the 1830s by another King's
College professor, Charles Wheatstone.* Maxwell was certainly

---

* The same Charles Wheatstone that Faraday filled in for in 1846.

aware of this as early as 1849, while at Edinburgh University, as he wrote to Lewis Campbell about 'Wheatstone's Stereoscope' and how Sir David Brewster had 'exhibited at the Scottish Society of Arts Calotype* pictures of the statue of Ariadne and the beast seen from two stations', which Maxwell comments 'when viewed properly, appeared very solid'. By the 1860s the combination of readily available stereoscopic photographs and a far less complex optical device to view them meant that stereoscopes were all the rage.

Maxwell himself later devised a significant improvement on the standard stereoscope, though it never took off commercially as it was both larger and more expensive than the traditional form. The parlour stereoscope consisted of a frame holding a pair of photographs (or drawn images) and a pair of lenses through which the viewer looked. This was essentially the same approach as used in the View-Master toy, popular from the 1940s and reaching a zenith in the 1960s, although this had seven pairs of images on a disc, rotated by pulling a trigger on the side of the viewer. The result of looking through the stereoscope was that the images were combined in the viewer's brain, producing a virtual 3D image.

The stereoscope was effective, but limited. Some people had trouble viewing the images, the experience was individual – only one person could look through the lenses at a time – and the virtual image was quite distant from the viewer. In 1867, Maxwell developed a 'real image' stereoscope, which used the standard pair of photographs and dual lenses, but then had

---

* Calotype was the name given to W.H. Fox Talbot's negative-based photography (as opposed to the older daguerreotype process which produced a direct positive image).

another, single large lens in front. The viewer (or viewers, if they were close enough together) looked at this lens from a distance of a couple of feet and saw the 3D image floating in space just behind the large lens. Maxwell had a device assembled by the instrument makers Elliott Brothers, and gave a paper on it to the British Association in September 1887. He would later use it to demonstrate 3D images of curved surfaces and mathematical knots. Topology and knot theory were recreational activities for Maxwell for a number of years, over and above their bearing on his more mainstream work.

However, in the Maxwells' home, visitors could expect a more unusual experience. They would be taken up to the attic to have an encounter with 'the coffin'. This was Maxwell's latest light box for mixing red, green and blue light to produce a whole spectrum of possible colours, one at a time. The eight-foot-long box had caused some confusion to Maxwell's neighbours when it was delivered as it did resemble a straight coffin. For the visitor this would be simply a new and exciting experience, but for Maxwell it was an opportunity to collect data on the way a range of individuals – both normal sighted and colour blind – perceived different colours. For several years, around 200 visitors a year were subjected to the coffin in the attic.

A standard for resistance

A less entertaining but more practical diversion came over the matter of electrical units – the units used to measure, for example, electrical current or resistance – which were becoming increasingly important as electrical engineering and particularly telegraphy took off. It's hard to appreciate now what a fundamental breakthrough telegraphy provided in speed of communication. Two of Maxwell's friends, William Thomson and

Henry Fleeming Jenkin, were involved in the biggest telegraphy project of the day, the transatlantic cable, and there was considerable concern that resistance in the cable would render the project unusable. Thomson, working with Faraday, had shown that if the resistance of the cable were too high, it could take as long as four seconds to send a single character.

With such a major technology at the mercy of a topic that had not been precisely studied, getting a better measure of resistance had far more practical application than just establishing a common unit. The British Association for the Advancement of Science saw Maxwell, with his expertise in electromagnetism, as the ideal person to be involved. The BA had set up a committee to study the requirements for standard units at the Manchester meeting in 1861 and Maxwell would play a major role in putting together the report to the Newcastle meeting in 1863.

Historically, units had been derived locally and this caused considerable confusion in international communication. Each country had its own definitions of units such as length or weight, which made it difficult to be sure what a measurement really indicated. This was nothing new. In his book *The Sand Reckoner*, the Ancient Greek mathematician and engineer Archimedes gave a size for the universe using the measure of 'stades' – multiples of the length of the running track at a stadium. A stadion (plural stades) was supposed to be 600 feet, but each city had its own definition of a foot. This means we can't know for sure what Archimedes intended, with a stadion being anything between about 150 and 200 metres. The BA felt that with electrical science becoming inherently international due to the undersea cables, this kind of uncertainty could not be allowed to happen again.

The new capabilities of electricity and magnetism required appropriate units and Maxwell therefore agreed to take on electrical standards with the help of the Edinburgh engineer Henry Fleeming Jenkin (among other things, the inventor of the cable car) and Balfour Stewart, a physicist who had worked with Forbes at Edinburgh and was director of the Kew Observatory in Richmond upon Thames. The small team produced a more rational system for defining units for resistance, current and the like, based on experiments they undertook at King's College.

This was an unusual piece of work for Maxwell in that it is pretty well the only significant true teamwork he did, rather than acting alone (with the exception of assistance from Katherine). It's not that he worked in hermetic isolation – letters between Maxwell and the likes of William Thomson and Peter Tait are full of scientific ideas and queries, where the physicists would use each other as sounding boards. But unlike modern science, there was very little true collaboration involved.

Many of the traditional units were relatively simple to pin down (if only a universal standard could be agreed on). These had started with a physical measure from nature, then been locally standardised by having a definitive example. The older measures of distance, for example, such as the foot and the mile, were dependent on typical human characteristics – the size of a part of the anatomy and one thousand paces (*mille passus*) respectively. The metre was slightly more scientifically determined, originally 1/10,000,000th of the distance from the North Pole to the equator on the meridian that ran through Paris. Each had become standardised to be represented by an official measure, though, as we have seen, these varied from country to country.

It was less obvious where a standard for voltage or current or electrical resistance would come from. Indirect measures were proposed that made use of equipment that translated one of the electrical measures into more familiar physical units. So, for instance, as it was known that the force between two electrical charges dropped off with the inverse square of the distance between them, it was possible to define a unit of charge (later called the coulomb) with a combination of force generated and distance between charged objects. This could then be used to calculate current (later the amp), which was a rate of flow of charge and so on.

Alternatively, current itself could be used as the way in. As the values involved in electromagnetic interaction became better known (though what was causing it was yet to be fully understood), force and distance could also be used as a measure of current between two interacting electrical coils. And a third option made resistance the starting point. This involved measuring the deflection of a rotating coil under the influence of magnetism.

Because of the importance of understanding the properties of the transatlantic cable, resistance was a key focus for Maxwell's group at King's College, which would make real an elegant mechanism devised by William Thomson based on this third option. Thomson's design involved spinning a coil of wire in the magnetic field of the Earth and using the induced electromagnetism to counter the Earth's pull on a small permanent magnet. Because the size of the magnetic field of the Earth cancels out between the two effects, the amount that the permanent magnet is deflected away from magnetic north depends only on the size of the coil, the speed of its rotation and its resistance – so, given the first two values, the equipment can be used to calculate an absolute value for resistance.

The velocity of a resistor

The unit of electrical resistance, measured using Thomson's method, turned out to be velocity. This was not connected to the actual velocity of a signal through the wire (something that confused many of those working on telegraphy at the time). It was simply the consequence of taking the units of the different values such as distance and speed of rotation that went into the measurement – the resultant dimensions of the resistance were distance and time in the form of a velocity. The standard unit settled on was 10 million metres per second, which would soon after be called a B.A. unit, also known as an 'ohmad',* which rapidly got shortened to ohm.† At least, 10 million metres per second was the intended value, although a measuring error (surely not one of Maxwell's infamous arithmetic slip-ups) meant that the standard ohm was actually slightly larger than it should have been.

The Thomson design was not a trivial piece of apparatus to use effectively. The coil had to be rotated at a constant speed, with a considerable amount of effort put into the design by Jenkin to provide a governor to keep the rotation steady. The mechanism was constantly breaking down, and to make matters worse, it was sufficiently sensitive that when an iron ship passed on the nearby Thames, the detection magnet would be slightly

---

* Probably, given this odd structure of the word, this was named for consistency with the farad, the unit of electrical capacitance, named after Michael Faraday. The 'ohm' part is from the German physicist Georg Ohm, who discovered the relationship between electrical voltage and current.

† Which is where 'ohmad' came from in the first place. A few years later, the ohm was given the symbol Ω to reflect the convenient assonance between the name ohm and the start of the Greek letter omega.

deflected.* It took many months of admittedly sporadic work to get a satisfactory set of readings. After the initial report at the 1863 BA meeting, a further twelve months would pass before they were sure that the values were reliable.

Apart from the theoretical definition defined from the spinning coil, the King's team put together a 'B.A. standard resistor' design. This was a rather magnificent construction, first completed in 1865, consisting of a coil of platinum/silver alloy wire covered in silk for insulation and then wrapped around a hollow brass core. The whole thing was then coated in paraffin wax, with thick copper wires to link it to the circuit. To ensure a constant temperature, the resistor was suspended in a water bath. While a standard resistor could not be just put alongside another resistor for comparison by eye, as was the case with a standard distance rule, a simple piece of equipment known as a Wheatstone bridge made it possible to use the standard resistor to calibrate others.

What was particularly impressive was the forward-looking nature of the British Association committee involved in devising these units. At a time when most scientific work was mired in the clumsy Imperial units, the electrical units were based on the metric system. This meant that when the other scientific units were switched over to metric, there was no need to redefine the electrical standards. For example, an amp (electrical current) times a volt (electrical potential) was the unit of electrical power. At the time, mechanical power would have been measured in horsepower or foot pounds per minute. But when

---

* Maxwell would have had sympathy with the builders of the LIGO gravitational wave observatories in the twenty-first century, which are so sensitive that they can detect the gravitational influence of a passing truck.

the metric system was fully adopted internationally in 1921, the unit of mechanical power was a watt – exactly equal to an amp times a volt.

Electromagnetism without visible support

It's possible that this refocusing on the practical side of electromagnetism was what pushed Maxwell to think again about his remarkable achievement in modelling the phenomenon. Although his mechanically-based model was remarkably effective, he understood why others found that it depended too strongly on analogy and he wanted to strengthen the theory's mathematical standing by a kind of scientific magic trick of removing the mechanical foundation and leaving the mathematics holding itself up by its own bootstraps.

Perhaps surprisingly, Maxwell focused on electricity and magnetism, rather than digging deeper into his theoretical basis for light as an electromagnetic wave. Although colour and vision would remain an interest throughout his life, he intentionally limited his theoretical developments on how light worked, perhaps because he felt that he was far closer to providing greater insights into the fundamentals of electricity and magnetism. He would never apply his theoretical approach fully to familiar behaviours of light, avoiding the whole business of how light and matter interacted* other than making a few initial notes on reflection and refraction, based on some work by the French physicist Jules Jamin. He commented, for example, 'In my book I did not attempt to discuss reflexion at all. I found that the propagation of light in a magnetised medium was a hard enough subject.'

---

* To be fair to Maxwell, this is just as well, as the interaction of light and matter could not have been properly understood without quantum theory.

The move from mechanical model to a purely mathematical one was a remarkably original approach – arguably Maxwell's greatest work of genius – and would provide the basis for modelling taken by the majority of physics theory right up to the present day. In a Royal Institution debate in 2004, four proponents put forward different names for individuals who could arguably be called the first scientist. I was one of these debaters, championing the thirteenth-century friar Roger Bacon. But another held out Maxwell to be the first. One of his arguments was facile – that the term 'scientist' was not brought into use until 1834, so no one working earlier could be one. But his other argument was that Maxwell was the first scientist in the modern sense because he moved from trying to establish the 'true nature' of physical reality to mathematically modelling it.*

It is interesting to speculate whether Maxwell was influenced philosophically in this move by the work of the German philosopher Immanuel Kant. Certainly, Maxwell would have heard about Kant's work in his university philosophy classes. Kant distinguished the phenomenal world – what we can experience – from the noumenal world – the actual underlying reality, which he called in German *das Ding an sich* (the thing itself). Kant suggested there was no point trying to know the reality – we could at best work with our interpretations of phenomena. Where most of Maxwell's contemporaries still believed they could discover the truth that lay beneath, Maxwell's approach of moving to a purely mathematical model seemed to reflect Kant's dismissal of such attempts.

---

* The other contenders were Archimedes and Galileo – Galileo won the debate.

Nearly 100 years earlier, the Italian-born French mathematician Joseph-Louis Lagrange had taken the traditional mechanics based on Newton's work and transformed it mathematically into what is now called Lagrangian mechanics. This centres on a mathematical function called the Lagrangian, which pulls together all the information about the movements of the bodies in a system into a single structure. In mathematics, a function is a compact way of referring to one or more equations which take one set of values and change them into something else. It's like a mathematical sausage machine.

A very simple function might be one that takes a number and does something to it, for example, producing the square of that number. It would usually be written as $f(x)$ – pronounced 'f of x' – and we could say in this case, for example, that $f(5)=25$. Mathematical functions proved a very powerful mechanism both in physics and later in computing, where functions are commonly used to provide operational modules which can perform the same operations on differing inputs inside a computer program. The Lagrangian consists of a set of equations based on differential calculus that link the velocity, momentum and kinetic energy of a body.

Although the route to developing the Lagrangian involved thinking about actual physical processes, once the function is established and found to match what is observed it can be considered totally detached from any analogy. Rather than relying on a mechanical model, this kind of function is a purely mathematical model. It is a black box where the user provides certain inputs, 'turns the handle' and gets the outputs. If what comes out matches observation, the function can be used without any idea of how the system it is modelling actually works.

In the mathematical belfry

Maxwell, the regular churchgoer, felt that the ideal analogy for a Lagrangian approach (he might have dismissed mechanical models, but he still loved using them to explain things) was a belfry. It's quite a lengthy quote, but it's worth taking it slowly and absorbing it because with this simple illustration, Maxwell is showing how he brought modern physics into being.

> We may regard this investigation as a mathematical illustration of the scientific principle that in the study of any complex object, we must fix our attention on those elements of it which we are able to observe and to cause to vary, and ignore those which we can neither observe nor cause to vary.*
>
> In an ordinary belfry, each bell has one rope which comes down through a hole in the floor† to the bell ringers' room. But suppose that each rope, instead of acting on one bell, contributes to the motion of many pieces of machinery, and that the motion of each piece is determined not by the motion of one rope alone, but by that of several, and suppose, further, that all this machinery is silent and utterly unknown to the men on the ropes, who can only see as far as the holes in the floor above them.
>
> Supposing all this, what is the scientific duty of the men below? They have full command of the ropes, but of nothing else. They can give each rope any position and any velocity, and they can estimate its momentum by stopping all the ropes at once, and feeling what sort of tug each rope gives. If they take the trouble to ascertain how much work they have to do

---

* In other words, as Kant might say, stick to phenomena and forget noumena.
† Strictly, a hole in the ceiling, from the point of view of the ringers.

in order to drag the ropes down to a given set of positions, and to express this in terms of these positions, they have found the potential energy of the system in terms of the known co-ordinates. If they then find the tug on any one rope arising from a velocity equal to unity* communicated to itself or to any other rope, they can express the kinetic energy in terms of the co-ordinates and velocities.

These data are sufficient to determine the motion of every one of the ropes when it and all the others are acted on by any given forces. This is all that the men at the ropes can ever know. If the machinery above has more degrees of freedom† than there are ropes, the co-ordinates which express these degrees of freedom must be ignored. There is no help for it.

Because Maxwell's earlier model was mechanical, it ought to be capable of being represented in a Lagrangian form, which would enable Maxwell to then ditch the cells and spheres, leaving them above the ceiling of the virtual bell ringers' chamber and simply dealing with the mathematical formulae that were the equivalent of the bell ropes. This was anything but a trivial task. He needed to stretch the mathematics of the time to enable it to cope with the added complexities of modelling electromagnetism. His work crucially depended on the ability to think of energy – potential and kinetic in the mathematical model – and

---

* What Maxwell means in this rather clumsy wording is that by defining the velocity of this particular rope as 1, they can establish a standard to measure the relative velocities of the other ropes.

† In physics, 'degrees of freedom' means the number of different parameters defining the state of a system – if you know all of these, you can say exactly how it will behave, but if you only know some of them, you will be limited in your ability to predict its response.

on keeping the concepts of energy while moving to the different frame of electromagnetism.

It didn't help that many of the quantities to be dealt with were vectors – as we have seen (page 69), Thomson had previously introduced Maxwell to using vector mathematics in the earlier fluid model, but dealing with a mix of quantities in the Lagrangian framework, some vectors with size and direction, such as the strengths of the fields, others, such as electrical charge, scalars with just size, made the mathematics significantly more challenging.

Nevertheless, Maxwell achieved his goal, and by December 1864 he was able to present to the Royal Society his groundbreaking new mathematical model of electromagnetism, which he wrote up the next year in the seven-part *A Dynamical Theory of the Electromagnetic Field*.

A new physics

Generally speaking, there are two possible reactions to a truly novel theory. Either everyone is bowled over by its clarity – or they are baffled by the novelty. Maxwell's theory was very much of the second kind.* His audience at the Royal Society were appreciative, but simply could not grasp what he was suggesting.

---

* Maxwell's great fan Albert Einstein suffered a similar problem initially when he suggested that light consisted of quantum packets of energy or photons, rather than such quanta merely being a way to make the mathematics work. When the leading German physicist Max Planck (whose theory was the starting point for Einstein's thinking) proposed Einstein for the Prussian Academy of Sciences in 1913, he asked them to overlook the way Einstein 'missed the target' with speculation like that over light quanta. This speculation would eventually win Einstein the Nobel Prize when the approach became widely accepted.

Up to this point, physics had largely been a discipline that was about experiment and philosophical theory, with the minimum of mathematics that was required to do the job. Now that maths was taking the lead, many in the audience were simply incapable of keeping up.

The difficulty of grasping the theory was not just a matter of making it accessible to the general public. The audience at the Royal Society included many of the leading physicists of the day. William Thomson, for example, admitted that he never came close to understanding Maxwell's theory. His was the last generation of physicists who could become leading figures without a strong grasp of high-level mathematics.

The reaction, and also the difficulty of explaining a strongly mathematical piece of physics to the general public, was beautifully assessed by Michael Faraday a few years earlier in 1857 in a letter to Maxwell. Faraday wrote:

There is one thing I would be glad to ask you. When a mathematician engaged in investigating physical actions and results has arrived at his conclusions, may they not be expressed in common language as fully, clearly and definitely as in mathematical formulae? If so, would it not be a great boon to such as I to express them so? – translating them out of their hieroglyphics, that we also might work on them by experiment. I think it must be so, because I have always found that you could convey to me a perfectly clear idea of your conclusions, which, though they may give me no full understanding of the steps of your process, give me the results neither above nor below the truth, and so clear in character that I can think and work from them. If this be possible, would it not be a good thing if mathematicians, working on these subjects, were to give us the results in this popular, useful working state, as well as in that which is their own and proper to them?

In effect, Faraday was arguing for the kind of lay summary of papers that is only now becoming widely accepted as a requirement – and to some extent prefigures the success of popular science writing, which became increasingly important as Maxwell's strongly mathematical approach took over physics. As ever, Faraday was a man of vision.

It wasn't just the physicists who wrestled with Maxwell's mathematics. Mathematicians also struggled to understand his work, because he used physical terms rather than familiar mathematical ones to describe what he was doing. The Serbian-American physicist Michael Pupin took a trip to Europe after graduation from his first degree in 1883 to try to get to grips with Maxwell's theory. He started in Cambridge with the intention of speaking to Maxwell himself, not realising that Maxwell was by then dead. He could find no one in Cambridge who seemed capable of explaining the theory to him, getting a satisfactory explanation only when he travelled to Berlin and studied under Hermann von Helmholtz (who apparently did understand the theory).

It would be a long time before there was good experimental evidence that supported the way that Maxwell's model went beyond what had previously been observed. Crucially, his concept of electromagnetic waves, though impressive in its coincidence with the speed of light, needed experimental verification – someone needed to generate waves from an electrical source and demonstrate them crossing space. It would take twenty years before Heinrich Hertz did this with the first artificially produced radio waves.

The beautiful equations

It didn't help that Maxwell's mathematical formulation was decidedly messy. There were a total of twenty equations to cover six different properties such as electrical current and magnetic field strength. The sheer compact power of what Maxwell had done did not become obvious until twenty years later, when the self-taught English electrical engineer and physicist Oliver Heaviside* (who had been influenced in his work by his uncle, Faraday's friend Charles Wheatstone) used the relatively new mathematics of vector calculus to reformulate Maxwell's equations as just four, stunningly compact lines of text.

These can be presented in a number of ways, depending on the units and whether they take into account a material other than a vacuum, but the simplest version looks like this:

$$\nabla \cdot \mathbf{D} = \rho_f$$
$$\nabla \cdot \mathbf{B} = 0$$
$$\nabla \times \mathbf{E} = -\frac{\partial \mathbf{B}}{\partial t}$$
$$\nabla \times \mathbf{H} = \mathbf{J}_f + \frac{\partial \mathbf{B}}{\partial t}$$

Part of the compactness here is due to the use of 'operator' notation. An operator applies a mathematical procedure to every value in a set. So, for instance, if I make up an operator called T which adds 2 to a number and apply it to the positive counting

---

* Heaviside is probably best known for the Heaviside layer, familiar to fans of the musical *Cats*, a layer of ionised gas in the upper atmosphere that reflects radio waves, and so allows radio transmission to be sent beyond the horizon (bearing in mind electromagnetic waves travel in straight lines). Heaviside was, to put it mildly, a character, often described as cantankerous and a maverick.

numbers, often called the 'natural numbers', the result would be to produce the set 3, 4, 5, 6 ... because I started with the natural numbers =1, 2, 3, 4 ... and the operator T told me to add 2 to each of them.

The inverted delta operator in the compact version of Maxwell's equations is usually known as 'del' these days, though in Maxwell's time it was sometimes called 'nabla', a term for the symbol suggested to Peter Tait by the theologian William Robertson Smith. The word reflected its shape, deriving from the Ancient Greek term for a type of harp with the same rough outline. Maxwell was never comfortable with this odd-sounding word and regularly mocked it, for example using it to derive a nonsense word in a letter to Lewis Campbell: 'This letter is called "Nabla", and the investigation a Nablody.' At one point, Maxwell wrote to Peter Tait, 'what do you think of "space-variation" as the name of Nabla?' We will stick to the modern term, del.

Del indicates differential equations being applied to what could be a whole range of values, either in traditional differential calculus – the sort Newton used – or the vector calculus which Maxwell needed to deal with changing quantities that had both size and direction. Here the dot after the del indicates a particular type of matrix mathematics. (A matrix is just a two-dimensional array of values.) This is imaginatively called the dot product, which produces the 'divergence' of a vector field, providing the values of the field at each point. When there is a cross after the del, as in the third and fourth equation, the operation is a 'cross product' producing the 'curl' of a vector field, which shows the rotation at each point.*

---

* Remember that Maxwell's mechanical model involved both linear flows of little spheres and rotation of cells.

Between them, the four equations describe the key behaviours of electricity and magnetism. The first,

$$\nabla \cdot \mathbf{D} = \rho_f$$

gives us Gauss's law. This provides the relationship between the strength of the electrical field* on the left and the density of electrical charge on the right.

The second,

$$\nabla \cdot \mathbf{B} = 0$$

shows that the magnetic field has zero divergence, which amounts to saying that it is impossible to have an isolated magnetic pole – they always come in pairs that cancel each other out.

The third,

$$\nabla \times \mathbf{E} = -\frac{\partial \mathbf{B}}{\partial t}$$

is where we get Faraday's induction explained, providing a mathematical relationship between a changing magnetic field (B) and the electrical field (E) that it produces.

And finally,

$$\nabla \times \mathbf{H} = \mathbf{J}_f + \frac{\partial \mathbf{B}}{\partial t}$$

---

* Electrical field is usually represented by an E, but here D is used to represent the 'displacement' field that Maxwell referred to in the displacement of the spheres during the 'twitches' where his cells twisted elastically.

describes the way that an electrical field produces magnetism. Here H is the 'magnetising field', proportional to the magnetic field B, but varying depending on the medium, J reflects the electrical current and the D part of the equation deals with the changing electrical field. Variants on these last two equations combined give all that is needed to describe a wave of changing electric field producing changing magnetic field producing changing electric field and so on, running on at the speed of light.

There is no need to be a physicist or immersed in the workings of the mathematics to appreciate how remarkably compact and powerful is the stark beauty of these four equations (often found in variants on T-shirts), despite their ability to encompass all the phenomena of electromagnetism.

Getting away from it all

Einstein regarded Maxwell as one of the greatest physicists ever – and though Maxwell was probably a better teacher than Einstein, they both suffered from frustration at the way the workload and administration of academic life could eat into their time for getting on with the work that they loved. In Einstein's case, the ideal solution came up with the opportunity to move to the Institute for Advanced Study (IAS) at Princeton in the United States.* While some academics find that the almost monastic existence in such locations gets in the way of being able to develop new thinking, it was a comfortable workplace for Einstein.

Maxwell had no direct equivalent of the IAS – but he had the advantage of independent wealth. When he had had enough of the pressures of his teaching work at King's, despite the assistant

---

* Arguably too late, as by the time Einstein moved to Princeton, all his great work was behind him.

to reduce his lecture commitment, he was able to make the decision to quit his post and return to year-round living at Glenlair with Katherine to work independently. After all, this had proved a great boost to his thinking over the summer breaks and no doubt it could again.

From the view of a modern scientist, Maxwell's action seems a retrograde step. Although science was yet to involve much teamwork, even then there was a great deal of sharing of information, and today's physicists would feel naked without their academic institutions and conferences. Maxwell was moving away from the scientific hubs of the Royal Society and the Royal Institution to the back of beyond in scientific terms. However, he had always relied more on written sources than on face-to-face networking. He was not, in the terms of the time, a particularly clubbable man. As far as we can tell, he took limited advantage of the cultural opportunity of being in London, with no record of him attending a play or a concert or any social event beyond the philosophical gatherings at the Royal Society, the Royal Institution and the BA. The younger Maxwell may have appreciated a wider social sphere, but now Maxwell and Katherine seemed happy with their own company.

There's no doubt that, like Albert Einstein, who moved away from teaching as soon as he could, Maxwell had the potential to benefit from leaving his teaching duties and being able to concentrate solely on his original work. However, unlike Einstein, Maxwell seems to have enjoyed the rewards of bringing the details of physics to others. It could be that his disillusionment with London also arose from the relatively limited scope of his students there. As we saw earlier, many of them only stayed for a little over four terms and saw their education at King's College primarily as a way to bolster careers in engineering and similar fields. There were very few who regarded physics as a serious career option – and it was only the opportunity to work somewhere that took the concept of advancement in physics seriously that would lure Maxwell back to a university some years later.

And so in early 1865, after five fruitful years at King's College, the Maxwells left London to return home to Glenlair. Maxwell could not even wait until the end of the academic year, leaving his assistant and successor, the 'Lecturer in Natural Philosophy' William Grylls Adams,* to take over his chair, a position Adams would hold for a further 40 years. Admittedly, Maxwell's retreat from London took some time. To keep up with his commitments outside of King's, giving lectures to working men, Maxwell would spend a few months in his Palace Gardens Terrace home at the end of 1865 and the end of 1866, as well as the early months of 1868, before he gave up the lease. Nonetheless, as far as he was concerned, Maxwell had left academia for ever.

---

* Adams was later one of the co-discoverers of the earliest form of photo-electric cell.

In which the demon suffers a setback

It's entertaining that my creator's new mathematical approach baffled many of his contemporaries, since it represented a step-change in the methods of physicists. Einstein famously had James Clerk Maxwell as one of the few portraits on his study wall, and said of JCM:

> Since Maxwell's time, Physical Reality has been thought of as represented by continuous fields ... and not capable of mechanical interpretation. This change in the conception of Reality is the most profound and the most fruitful that physics has experienced since the time of Newton.

The cost of measurement

As far as my progress went, we need to take a leap forward in time to long after Maxwell's death (this is no problem for a demon like me), reaching the 1920s and the work of a young Hungarian physicist, Leo Szilard, whose greatest claim to fame would later be his realisation of how a nuclear chain reaction could work. Remarkably, Szilard would show that JCM's humble demon was in fact a precursor to information theory. To understand Szilard's take on me, you first need to see the other side of the second law of thermodynamics in a little more detail.

As you'll recall, my master and his friends were largely concerned with the second law in terms of heat and the movement of molecules – the whole business of thermodynamics was devised, after all, as a way of getting a better understanding of how steam engines worked. JCM may have made things deliciously probabilistic with his statistical approach – but he was still thinking of the second law being primarily about the way heat never flows from a colder to a hotter body, unless it's given a helping hand. By Szilard's time, though, the dominant aspect of the second law was the way it dealt with entropy.

As I've already mentioned, entropy is a measure of the level of disorder in a system – but it's not as vague a thing as it sounds: it has a clear numerical value. The entropy of a system is based on the number of unique ways you can arrange the components that make up that system.* At first glance it's not totally obvious why this is a measure of disorder, but a good example would be the letters of the alphabet. If we put them in the familiar alphabetic order: A, B, C, D ... there is just one way to arrange them. But scramble them up and there are many ways to arrange them:

$$A\ C\ B\ D\ ...$$
$$G\ Q\ C\ E\ ...$$
$$L\ A\ Q\ V\ ...$$

... and so on. This means that in alphabetic order, the letters of the alphabet have much lower entropy than they do when they

---

* If you want to get technical, entropy = $k \ln W$, where k is Boltzmann's constant, and $\ln W$ is the natural logarithm of the number of ways the components can be arranged.

are scrambled up and more disordered, where there are more ways to arrange them.

Szilard believed that for a demon like me to do the job, it would have to be an intelligent being,* and that the process of measurement the demon would have to undertake to decide whether or not to let a molecule through would itself result in an increase in entropy which would precisely cancel out the decrease caused by the demon's excellent contribution. The reason that Szilard assumed this to be the case is that the demon has to measure the speed or kinetic energy of the molecule. He then has to store that information in his brain in order to make the decision whether or not to open the door. The business of taking the measurement, Szilard argued, would result in the use of energy and an overall increase in entropy of the system as a whole, as the demon would have to be considered part of the system and his increase in entropy from using energy would be greater than the decrease in entropy in the gas.

It has been suggested that it's not surprising that Szilard came up with a take on my activities that involved me as an intelligent observer, where scientists of Maxwell's day would not have made the distinction. For JCM and friends, the scientist was a totally detached being, an objective observer, entirely separate from the experiment. But by the time Szilard got his hands on me, quantum theory had begun to be developed.[†]

---

* I would have thought by now that this went without saying.
† It's not for nothing that the British physicist William Bragg wrote: 'God runs electromagnetics on Monday, Wednesday and Friday by the wave theory, and the devil runs it by quantum theory on Tuesday, Thursday and Saturday.' Nice to see a physicist acknowledging the importance of the demonic contribution to physics, though one does wonder what happened to electromagnetics on Sunday.

One essential of quantum physics is that the act of measurement – just looking at something, even – has the potential to have an effect on it. For example, if the demon were to take a measurement of a particle's position using light, that would involve light photons bouncing off the molecule, potentially changing its path and momentum. More significantly still, quantum theory said that until a measurement was made, a quantum particle such as a molecule didn't *have* a position. It could even tunnel its way through my door and appear on the other side.

In Maxwell's approach to statistical mechanics, the probabilities are in the model. All the molecules have a definite position at all times, but we don't know what those positions are, and so we use probability to take a statistical overview of how the collection of molecules is likely to behave. But quantum reality tells us that the positions of the molecules are literally and actually just probabilities until an observation is made. This was the aspect of quantum physics that so worried Einstein, resulting in his famous remarks about God not playing dice. God may not do so – but as for demons ...

Taking a quantum physics viewpoint, you can never entirely separate the observer and the experiment. Szilard's big contribution was to make the demon's role part of the overall system of the experiment. My measurements, he suggested, must influence the system in a way that forces the entropy back up just enough to counter any benefits that I had produced.

Although it's not of importance to me, it's interesting to note that Szilard's work on my problem directly led to the American engineer Claude Shannon developing information theory, which introduced the concept of entropy to information, such as the information transmitted from place to place by Maxwell's electromagnetic waves.

As it happens, with true demonic slipperiness, I managed to escape from Szilard's apparent solution to continue to threaten the second law – but that can wait. We need to see how JCM was coping after his move away from the academic world.

Chapter 7

On the estate

O ne advantage to having control of his own time at Glenlair was the opportunity for Maxwell to focus on writing as well as spending time on his experiments and development of theory. With his old friend Peter Tait, who had stayed as Professor of Natural Philosophy at Edinburgh, Maxwell was assembling a book that was extremely wide in coverage – called *A Treatise on Natural Philosophy*, it was in effect a textbook for the entire physics syllabus at university level. There would arguably not be another physics book that combined such an impressive breadth with being held in such high regard until Richard Feynman's famous 'red books' based on his undergraduate lecture series from the 1960s.

Today, working on a jointly authored book is relatively simple. Not only can chapters and comments fly backwards and forwards by email, collaborators can work on a shared version hosted in the cloud.* For Maxwell and Tait it was a matter of writing letters by hand and waiting for a response in the mail – there was a steady flow of letters back and forth between Glenlair and Edinburgh as the volume came together. In fact, such was the volume of post that Maxwell generated with his

---

* And, yes, once again, Maxwell's scientific discoveries are fundamental to our ability to be able to do this.

wide-reaching scientific correspondence that a post box was installed into the wall of the road by Glenlair for Maxwell's personal use.

Later, Maxwell would also work on solo books on his electromagnetic theory (*Treatise on Electricity and Magnetism*) and a title taking in his own contribution to thermodynamics and a far wider exploration of the subject in the *Theory of Heat*, which was where he exposed to a wider world the 'finite being' that Thomson had turned into Maxwell's demon.

Glenlair life

The demon might never have existed. In 1865, enjoying his first summer at Glenlair that was more than just a vacation, Maxwell had been out riding. He was on an unfamiliar horse* and lost control sufficiently to end up riding fast under low trees. A branch hit him across the head – the injury was little more than a bad scrape, but it became seriously infected with the skin condition erysipelas. With no medical solution other than to keep the patient comfortable and wait for the fever to break, it was touch and go for a few weeks.

Although Maxwell would continue to work – science would never have left his mind for long – being at Glenlair was also an opportunity to have more of a family life. For several years the majority of the Maxwells' time would be dedicated to the place that was always more home than anywhere else. It was a chance to improve the estate and make changes to the house, alterations that had been suggested ever since Maxwell's father's day, but which no one had found time to carry out.

---

* We're not sure why. Perhaps his usual mount was lame.

It was also the Maxwells' last chance to have children. Katherine was 41 when they moved to Glenlair full-time. While not impossibly old to have a first child, it was certainly unusually so at the time. There seems little doubt that Maxwell would have loved to have had children, both from the enthusiasm he displayed in playing with the children of others and bearing in mind his sense of duty to the estate, which would have been seen at the time as a responsibility to have an heir. His contemporary biographers note:

> [I]t is impossible not to recall the ready kindness with which, in later life, he would devote himself to the amusement of children. There is no trait by which he is more generally remembered by those with whom he had private intercourse.

As is often the case with spouses of the period, we know a lot less of Katherine's nature, though, as we have seen, there were hints that she was less outgoing than Maxwell. We certainly don't know whether she was interested in children at all, but for whatever reason, it was not to be.

Once Maxwell had recovered from his infection, his time at Glenlair also gave him the opportunity to revisit experiences and modify ideas of the past. Some scientists may be happy to publish a piece of work and then move on, but Maxwell's character seems to have encouraged repeated consideration of a topic, sometimes just fine-tuning what had come before, and at other times – as he had with his electromagnetic model – totally reworking the way he approached the problem.

One opportunity to rework a feature of his younger days was one that many would consider more terrifying than appealing. I am sure I am not alone in having occasional nightmares where

I find myself sitting a university exam – in maths or physics – and realise I haven't revised any of it and don't know any of the equations I will need. But when Cambridge University asked Maxwell to take a look at the Mathematics Tripos – the complex and by that time dated exam structure that had brought him to the attention of academia – Maxwell was happy to return to the fray. The exams were mired in the past, unchanged since well before Maxwell had taken them. He was asked to make the content and structure more current and relevant to the modern mathematician, a challenge he took on with enthusiasm.

Back to viscosity

Maxwell was also able to work more on his experiments on the viscosity of gases, with the help of Katherine. All the evidence was that he was correct in thinking that viscosity varied directly with temperature rather than with its square root. But the theory of the time did not support this. As an expert on mechanical models, he realised it was the model itself that was at fault. His first step was to move away from the notion of mean free path (see page 110), which had been useful for getting an idea of the size of molecules, and instead to employ a model where gas molecules exerted force on each other.

This was a significant change to the thinking. To make things simple in his earlier work on kinetic theory, the molecules in the gas had been treated like colliding billiard balls* which head towards each other at full speed until they come into contact, bounce apart, and immediately head away from each other at

---

* Physicists love billiard balls as models – no doubt a sign of misspent youths.

full speed.* With his electromagnetic experience, Maxwell tried instead representing the interaction between the molecules as that of two electrically charged particles repelling each other. Here, the acceleration effects will start earlier as the repulsion acts well before the molecules are in contact and grows rapidly as they come closer with the inverse square of the distance between them.

At the same time, Maxwell threw in another concept, relaxation time, which allows for the way that a system, after being disturbed, returns to a state of equilibrium. Think, for example, of dropping a blob of milk into a cup of tea without stirring it. Before adding the milk, the tea molecules will be in equilibrium, bouncing off each other and keeping the temperature throughout the tea roughly the same. Add the cold milk and there will be a concentration of coldness in one point. The system will have been disturbed. But over time, the milk molecules will disperse through the tea and the tea/milk system will settle down to a new, different equilibrium. The time for the system to undergo this process is the relaxation time.†

With his modifications to the model of a gas producing its viscosity, Maxwell was now able to match observation and theory. This new version of the model gave a gas a viscosity that was proportional to temperature just as the Maxwells'

* Real billiard balls are not like this – they are not perfectly rigid, so they deform as they come into contact, giving a short period of acceleration in an impact that isn't instantaneous, and they lose energy to heat and sound, but physics models often involve simplification to make the mathematics manageable.
† Stirring the tea changes things again, reducing the relaxation time for the distribution of temperature but introducing a new disturbance in the form of a vortex. Who would have thought a cup of tea involved so much physics?

experiments had shown. It still produced the same velocity distribution for the molecules – his original paper was not wrong in this – but in this new formulation, Maxwell was able to drop one of the limitations of the original model which meant that it would work only if there was no relationship between molecules with particular velocities. He completed this work in 1866 and published it in 1867.

He would revisit his theory six years later, but this was more a matter of tweaking the approach and presentation to match the latest findings from other physicists. He would also present his theory in 1873 to the Chemical Society, where he had to pile in as much evidence as he could for the existence of molecules to justify his use of them. Although chemists made use of the concept, they largely considered molecules to be a useful accounting fiction, rather than actual physical bodies that could perform the complex statistical ballet Maxwell portrayed.

Inevitably, given the gaps in knowledge at the time, there were limitations to the approach that Maxwell took. While his distribution has held up, there were aspects that didn't fit with experimental measurements that were becoming increasingly accurate. This deviation primarily derived from the assumptions Maxwell (and Clausius) made about the ways that molecules could move. Maxwell had dealt with movements in the three dimensions of space and three potential axes of rotation, but he was thinking of the molecules as spheres. We now know that anything more than a single atom has a more sophisticated structure and is capable of rotating in different ways and of vibrating along the bonds between the atoms that make it up. It is this difference from Maxwell's picture that accounts for the shortfall in his calculations – but given the information he had at the time, it was still a remarkable achievement.

Maxwell made significant progress, particularly in his written output, in his years at Glenlair, but it ought to be stressed that his time wasn't all given over to work. As we have already seen, he spent a considerable amount of effort on improving the estate, and in 1867 he and Katherine took a tour of Italy – still as fashionable a destination for the British well-to-do as it had been in the time of Byron and Shelley some fifty years before. Their trip also got them out of the house while building work was under way, transforming Glenlair into a dwelling that was a little more like a manor house and less like a farmhouse. Being the Maxwells, though, their trip to the Continent was not the classic Grand Tour covering several countries and as many cultural locations as possible with the minimum effort put in by the travellers. The Maxwells made an attempt to learn Italian and seem to have immersed themselves more in the local culture than would have been expected at the time.

Over the years, Maxwell had been able to build up an effective laboratory at Glenlair – certainly sufficient to repeat his experiments on viscosity of gases – but there were some pieces of technology that were beyond the reach of his personal funding. Three years after the retreat to Scotland, he found himself back in London to try to refine his calculation for the speed of electromagnetic waves.

The wine merchant's batteries

As we saw in Chapter 5, back in 1861, on his summer vacation at Glenlair, Maxwell had calculated the expected speed of electromagnetic waves in a vacuum to be around 310,700 kilometres per second, basing his calculation on the equivalent of the two factors used for conventional waves: the elasticity of the medium and its density. The equivalent properties for electromagnetism are known as the magnetic permeability and the electric permittivity of space, and the values of these properties were relatively poorly known at the time, so the result had a significant uncertainty.

With Charles Hockin, an electrical engineer based in Cambridge, Maxwell devised an experiment that would pin down these factors to far greater accuracy than ever before. The experiment balanced out the attraction from the electrical charges on two metal plates with the repulsion between two electromagnets with like poles facing each other. The bigger the effect, the more accurate the measurement could be – which meant hunting down an extremely high-voltage source.

Rather surprisingly, the owner of the most powerful batteries in the country turned out not to be a physics laboratory or an electrical power company, but a wine merchant based in Clapham, London. John Gassiot had spent his fortune on constructing an extravagant private laboratory. It was Gassiot who provided Maxwell and Hockin with a vast battery of cells, 2,600 in all, which between them put out around 3,000 volts.

The experiment proved highly successful, though the batteries went flat with unnerving rapidity, meaning that Maxwell and Hockin had to take measurements furiously quickly before the charge ran out. These more accurate values for magnetic permeability and the electric permittivity resulted in a calculated speed for electromagnetic waves of 288,000 kilometres per second (kps).

In the original calculation, Maxwell had come up with 310,700 kps to Fizeau's 314,850 kps – so it might seem that his new calculation put him further off the mark. But in the intervening period, Maxwell had learned of an updated value for the speed of light from another French experimenter, Léon

Foucault.* Foucault had used an improved version of Fizeau's equipment to get a more accurate speed of 298,000 kilometres per second[†] – and so Maxwell's waves now seemed even more certain to be light.

This was immediately a useful support to his theory, but in other cases the significance of the work that Maxwell did would not become clear until much later. As we will see below, even a spin-off from Maxwell's work could hold promise of great things – simply because he retained his child-like curiosity, seeing what others would regard as everyday phenomena and realising that here was something special and worth investigating.

Meet the governor

As we have seen, when Maxwell had been working on the electrical resistance standard at King's College (page 183), his colleague Henry Fleeming Jenkin had designed a governor to keep a coil rotating at constant speed. Governors would be familiar to anyone who had seen a static steam engine. The early days of steam were riddled with explosions and runaway catastrophes when poorly understood technology was used outside its operating parameters. Back in the 1780s, James Watt had come up with the centrifugal governor. This elegant device has a pair[‡] of weights on hinged bars, which fly outwards as the

---

* Foucault is probably best known for his pendulum, which changes its direction of swing with the Earth's rotation (and gave Umberto Eco a good title for a book).

† The speed of light has now been fixed at 299,792.458 kilometres per second, as the metre is defined as 1/299,792,458th of the distance light travels in a second.

‡ A single weight would do the job, but would be unstable – having a pair enables them to balance each other out.

vertical spindle they are mounted on rotates. The faster the spindle goes, the further out the balls move. The hinged bars are linked to a valve, closing it off if the balls fly too far out from the centre. This means that a steam engine with the governor in place never runs too fast – it automatically shuts itself off over a certain speed.

Thinking more about Jenkin's governor, Maxwell began to explore the different ways in which this form of feedback mechanism, which automatically regulates a process, could be deployed (another, more modern version of a feedback-based governor present in almost all homes now is the humble thermostat). In 1868, Maxwell wrote a paper named *On Governors* which, with his usual thoroughness, gave the process a mathematical treatment. In fact he pointed out that, at least by his definition, Watt's device wasn't a governor at all.

In the paper, Maxwell made the distinction between a 'moderator', in which the correction that is applied grows with the excess speed, and a 'governor', which also involves the integral of the speed.* He showed mathematically that only a governor with this facility would exactly regulate the speed. A moderator, like Watt's, would still provide negative feedback, but could not give the exact value required to balance out the error.

The mathematics Maxwell used here drew on some of his work on Saturn's rings – getting a governor right was the same kind of problem, in the sense that it involved the stability of

---

* The integral, one of the main vehicles of calculus, here gives the area under the curve of a graph of the changing speed of the governor – in this case reflecting the displacement of the governor, rather than the measured speed. This enables the governor to bring the speed to the required value however large the variation, whereas with a moderator, speed will still increase, but less so than if the moderator was not there.

a system. Ironically, his work on stability in governors would be generalised by another Cambridge mathematician, Edward Routh, who won the Adams Prize (checked by Maxwell) for his work, just as Maxwell had for his analysis of the rings of Saturn.

Decades later, this paper by Maxwell would be recognised by the American mathematician Norbert Wiener, who in the 1940s originated the concept of cybernetics* – the use of systems with communication and feedback that would become important in control systems, engineering and computer science. Wiener considered Maxwell the father of automatic controls, providing the first steps in the development of the control systems theory that lies behind everything from cruise control on cars to the systems that keep nuclear power stations safe.

Thinking in four dimensions

Sometimes the work that Maxwell did would have greater impact than was perhaps even intended. In his *Treatise on Electricity and Magnetism*, he took the first major step that would enable Oliver Heaviside to produce the compact versions of his equations. This was because Maxwell had become interested in quaternions, a relatively obscure mathematical device introduced by the Irish mathematician Sir William Hamilton (not to be confused with his contemporary, the Scottish philosopher Sir William Hamilton, whose lectures Maxwell attended while at Edinburgh University).

By the 1860s, most scientists were comfortable with the concept of complex numbers, which incorporated values in two

---

* 'Cybernetics' and 'governor' are both from the same Greek root, meaning a steersman, but governor was distorted in passing through a Latin version of the word.

dimensions at once. A complex number is an ordinary number combined with an imaginary number – a multiple of i, the square root of -1. So, for example, a complex number might be $3 + 4i$. These numbers can be manipulated just like an ordinary number, and map onto a two-dimensional graph with the real numbers on one axis and the imaginary numbers on the other.

Complex numbers proved highly useful when dealing with anything that has the form of a wave, which has both its position in a particular direction and an amplitude that varies with time. However, many physical processes take place not in the flat plane of graph paper but in three spatial dimensions. Quaternions provided a mechanism for dealing with values that had amplitude and a three-dimensional location. This was done by having effectively three different imaginary components, so a single quaternion might look like $3 + 4i + 2j + 6k$.

Hamilton thought, correctly, that quaternions had the potential to revolutionise the mathematics of physical processes, but the approach proved difficult to handle and instead a different mechanism to deal with values that varied in multiple dimensions, vector analysis, was developed.

Maxwell provided the bridge between Hamilton and practicality – inspired by quaternions, he came up with the terms 'convergence', 'gradient' and 'curl' to represent different forms of quaternion operation which would be carried through into vector calculus, though 'convergence' was replaced by its opposite 'divergence', and the longer names shortened, ending up with the operations div, grad and curl (two of which are seen in Heaviside's formulation of Maxwell's equations on page 193).

We can see a typical Maxwellian delight in words in the way that he worked towards the term 'curl', first in a letter to Peter Tait, where it was just an alternative to 'twist', and then in a paper written for the London Mathematical Society in 1871 when he had finally settled on curl.

The letter begins by asking if Tait called the mathematical operator that would be known as del 'Atled' (it is an upside-down delta). He then wrote:

> The scalar part I would call the Convergence of the vector function and the vector part I would call the twist of the vector function. (Here the word twist has nothing to do with a screw or helix. If the words *turn* or *version* would do they would be better than twist for twist suggests a screw.) Twirl is free from the screw motion and is sufficiently racy. Perhaps it is too dynamical for pure mathematicians so for Cayley's* sake I might say Curl (after the fashion of Scroll).

Confusingly, after saying this, Maxwell goes on in the rest of his letter to refer to the vector part as the twirl.

By the time he got to his Mathematical Society paper, though, he had decided to go with his proposal to satisfy the mathematicians:

> I propose, but with great diffidence, to call this vector the **Curl** of the original vector function. It represents the direction and magnitude of the subject matter carried by the vector. I have sought for a word that shall neither, like **Rotation**, **Whirl**, or **Twirl**, connotate motion, nor, like **Twist**, indicate a helical or screw structure which is not of the nature of a vector at all.

---

* Arthur Cayley, an English mathematician, at the time the Sadlerian Professor of Pure Mathematics at Cambridge.

The life academic

It might seem that everything was going swimmingly at Glenlair. Maxwell wrote the first of his definitive books, undertook experiments, and developed theory. His family life was everything he had hoped for with the exception of having children. And he had said on leaving London that he would never return to academia. Yet perhaps he also felt that there was something missing from his life. He could discuss his work with Katherine and could exchange ideas with other scientists in letters, but it wasn't the same as having lively discussion with other experts in the field. Maxwell seems to have missed the opportunity for broader intellectual exchange.

The first opportunity that tempted him to think of getting back into the academic world was when the post of Principal at Scotland's oldest university, St Andrews,* came up in late 1868. Maxwell wavered on whether to apply. In a letter to William Thomson at the end of October he wrote:

> One great objection is the East Wind, which I believe is severe
> in those parts. Another is that my proper line is in working not
> in governing, still less in reigning and letting others govern.

Four days later it seemed as if Maxwell had made up his mind. He wrote to Lewis Campbell:

> I have given considerable thought to the subject of the candi-
> dature, and have come to the decision not to stand. The warm

---

* To be precise, the position was Principal of the United College of St Salvador and St Leonard in the University of St Andrews.

interest which you and other professors* have taken in the matter has gratified me very much ... but I feel that my proper path does not lie in that direction.

Yet after another four days, Maxwell had clearly reversed his decision and began writing to contacts to try to win support for this political appointment, particularly any he felt could influence the Home Secretary, the magnificently named Gathorne Gathorne-Hardy. By 9 November he was writing to Thomson:

> When I last wrote, I had not been at St Andrews. I went last week and have gone in for the Principalship. If you can certify me having been industrious &c since 1856, or if you can tell me what scientific men are conservative[†] or still better if you can use any influence yourself in my favour pray do so. 6 Professors out of 9 [at St Andrews] have memorialized the Ld Adv. & Home Sec. for me together with Principal Tulloch the V. Chancellor. Of the other 3, one Prof Shairp is a candidate and one, Prof. Bell does not approve of memorials at all and is neutral.

Despite the apparent support of the resident professors, it was not to be. Although the post was rumoured to be due to go to a 'man of science', it may have been that Maxwell's lack of administrative experience counted against him with the political decision-makers. But this did not mean that Maxwell would

---

* Campbell was, at the time, Professor of Greek at St Andrews.
† Gathorne-Hardy was a member of the Conservative government, which was probably why Maxwell said this, though Maxwell was also by nature conservative with a small c.

remain at Glenlair for the rest of his working life. When, a few years later in 1871, he was approached with the offer of a new post that would truly wrench physics away from the medieval concept of natural philosophy and bring it to the forefront at a great university, Maxwell, pushed out of his stay-at-home frame of mind by the opportunity at St Andrews, decided it was time for a fresh challenge.

Chapter 8

Cambridge beckons

The approach that would signal the beginning of the last great phase of Maxwell's scientific life came from the University of Cambridge. Although Cambridge had proved a mathematical inspiration and remained in many ways an academic spiritual home to Maxwell, it was not at the time the most advanced British establishment in terms of the sciences. However, the chancellor of the university was determined to change this.

The Cavendish connection

Unusually for the nominally top person in a university (even today), the Cambridge chancellor had scientific leanings. Like Maxwell, he had been an outstanding mathematician of his year, becoming Second Wrangler and winning the prestigious Smith's Prize. What's more, the chancellor was the grand-nephew of Henry Cavendish, another Cambridge graduate, who had been one of the leading scientists at the end of the eighteenth century. Henry Cavendish had played a significant part in the establishment of the Royal Institution and had undertaken the first experiment to provide a reasonable measurement of the density of the Earth, making it possible to calculate for the first time Newton's gravitational constant, G.

The current Cambridge chancellor was also a Cavendish – William Cavendish, the Duke of Devonshire – an extremely rich man, who, as a politician, had served on the Royal Commission on Scientific Education, set up among concerns that the country was falling behind its competitors in scientific prowess. Cavendish was prepared to donate a large sum to the university on the understanding that it built a physics laboratory and endowed a chair of experimental physics.

In the quaint terminology of the time, the Chair of Experimental Physics was founded by a Grace of the Senate on 9 February 1871 stating that 'the principal duty of the professor is to teach and illustrate the laws of Heat, Electricity and Magnetism; to apply himself to the advancement of the knowledge of such subjects; and to promote their study in the university'. This meant that the university was on the hunt for a dynamic, world-class physicist, with the dual role of becoming the first Cavendish Professor and supervising the construction of a leading-edge laboratory. The first people to be approached were Maxwell's old friend, William Thomson, and the German physicist Hermann von Helmholtz, who like Maxwell did significant work on electromagnetism and thermodynamics.

However, Thomson did not want to leave Edinburgh and was concerned at the limited support available in Cambridge from instrument makers, an essential if the laboratory was to be successful. Helmholtz, at the time based in Heidelberg, was just in the process of negotiating a post in Berlin, the mathematical capital of Europe. As a result, Maxwell received a letter in mid-February 1871 from Edward Blore at Trinity College:

My Dear Maxwell

Our Professorship of Experimental Physics is now founded & though the Salary is not magnificent (£500 a year) yet there is a general wish in the university that this branch of Science should be supported in a way creditable to the University. The Duke of Devonshire has undertaken the expense of the building & Apparatus, & it remains for us that we should appoint the Professor. Many residents of influence are desirous that you should occupy the post hoping that in your hands this University would hold a leading place in this department. It has, I believe, been ascertained that Sir W. Thomson would not accept the Professorship. I mention this in case you should wish to avoid the possibility of coming into the field against him.

Maxwell knew he was not the first choice. He replied immediately to Blore, saying:

Though I feel much interested in the proposed chair of Experimental Physics I had no intention of applying for it when I got your letter, and I have none now unless I come to see that I can do some good by it.

He went on to ask details of the job – the duties, who made the appointment, how long he would be expected to serve, how many terms a year and so forth. The details were sent to Maxwell three days later by George Stokes, the Lucasian Professor at Cambridge, and within a week Maxwell had decided to stand.

Maxwell was elected to the post on 8 March 1871. Or, more accurately, he was appointed. The professorship was in principle in the hands of the Senate of the university. Strictly speaking, this comprises all holders of an MA or higher degree from the university, plus significant officers such as the Vice Chancellor, but in practice it would have been just those resident at the time. Of around 300 members, only thirteen voted – not entirely surprisingly, as they were only offered a single choice. The decision was clearly made behind the scenes, based on approaches to key figures.

Cambridge was lucky to end up with Maxwell, as neither of his rivals had his practical experience of running an estate alongside his excellence in physics, which would surely have helped when it came to managing the construction project – though as Maxwell pointed out in his letter to Blore, Thomson had practical experience of running a university laboratory, which he did not. In March 1871, Maxwell took a tour of the few existing physics laboratories in the UK, picking up the best practice to pull together in designing the new facility.

A different professor

It's interesting to think for a moment about why James Clerk Maxwell was offered this particular position, how he saw his role, and how his ideas might have been shaped by his experiences in London. At first sight, this master theoretician might not seem the ideal candidate to become professor of *experimental* physics, but we need to remember that Maxwell had proved a more than capable experimenter, and one who was prepared to go the extra mile and set up laboratories in his home both in London and at Glenlair when nothing was available otherwise. He had also remained in close contact with Cambridge since leaving King's College London, helping to reform the Mathematics Tripos examination system in 1866 and 1867.

The mix of talents that he could bring to Cambridge is reflected in a letter to Maxwell from John Strutt (Lord Rayleigh) who would go on to replace Maxwell as Cavendish Professor on his death. Among other things, Rayleigh discovered the element argon and explained why the sky is blue. Rayleigh wrote:

When I came here [Cambridge] last Friday I found every one talking about the new professorship, and hoping that you would come ... There is no one here in the least fit for the post. What is wanted by most who know anything about it is not so much a lecturer as a mathematician who has actual experience in experimenting, and who might direct the energies of the younger Fellows and bachelors into a proper channel.

In his inaugural lecture at Cambridge in 1871, Maxwell remarked that the new laboratory would be worthy of the university if 'by the free and full discussion of the relative value of different scientific procedures, we succeed in forming a school of scientific criticism, and in assisting the development of the doctrine of method'. For Maxwell neither the Ancient Greek approach of armchair philosophising which had dominated universities until Newton's day, nor the pure mechanical experimental approach that had driven the industrial revolution was the way forward.

In his vision of the future of physics, Maxwell saw a close, symbiotic partnership between experiment and the development of theory – neither should operate in isolation. Arguably, a major factor in developing this viewpoint (leaving aside how closely it reflected the interaction of experiment and theory in his own working life) was the downside of the engineering-dominated approach at King's. Although Maxwell had nothing against practical, applied science, he saw clearly that it was necessary to move away from industry-driven goals and to be able to address the fundamentals.

It was also clear to Maxwell that something needed to be done about the Cambridge syllabuses. As we have seen, he was involved a few years before his appointment in the redesign of the Mathematics Tripos, helping to expand it beyond its old topics to take in more modern aspects of physics, such as electricity, magnetism and heat – and he would write a major textbook, the *Treatise on Electricity and Magnetism*, to support this. Now he had the chance to make sure that the broadest modern physics curriculum was covered with the introduction of experimental physics to Natural Sciences – though he was still constrained also to include elements of physics in the Mathematics Tripos.

Along with other supporters of laboratories in universities, his approach was in the vanguard of transforming experimental physics from simple observation to making measurements to support or dispose of theories. In his opening lecture, Maxwell first pointed out the limitation of making more and more accurate measurements with no extension of theory.

> The characteristic of modern experiments – that they consist principally of measurements – is so prominent that the opinion seems to have got abroad that, in a few years, all the great physical constants will have been approximately estimated, and that the only occupation which will then be left to men of science will be to carry these measurements to another place of decimals.*

---

* The idea that physics would become largely about adding another decimal place was very clearly in the mind of the physicist Phillip von Jolly, when in 1874 he encouraged the student Max Planck (who went on to win the first Nobel Prize in Physics) to study music rather than physics. This was because von Jolly believed that physics was almost complete, and with a couple of small problems cleared up, all that remained was better measurement. In the early twentieth century Planck and Einstein would shatter this idea with the culmination of those small problems: relativity and quantum theory.

If this is really the state of things to which we are approaching, our Laboratory may perhaps become celebrated as a place of conscientious labour and consummate skill; but it will be out of place in the University, and ought rather to be classed with the other great workshops of our country, where equal ability is directed to more useful ends.

Then, however, Maxwell went on to show his faith in there being far more to come: 'But we have no right to think thus of the unsearchable riches of creation, or of the untried fertility of those fresh minds into which these riches will continuously be poured.' He pointed out that even when progress has largely been about polishing the decimal points, science 'is preparing the materials for the subjugation of new regions, which would have remained unknown if she had been contented with the rough methods of her early pioneers'.

He realised that no university was better positioned to do this than Cambridge – an institution that has gone from strength to strength since Maxwell's appointment, so that it is now one of the world's leading universities for physics. This is likely to be part of the answer to a second question about the appointment, which is why Maxwell would have accepted, having voluntarily given up his position in London. It seems unlikely that the attraction of Glenlair would ever have palled for him. But the opportunity in Cambridge to set in motion something that could be transformative for the physics he loved was simply too great. And, just in case it didn't work out too well, he put himself forward on the understanding that he would not have to stay in post for longer than a year.

Engraving of Maxwell from c. 1890, taken from an 1870s photograph by Fergus of Greenock.
Getty Images

The last second home

The Maxwells set up a new second home in Cambridge in the spring of 1871 (this despite an off-putting letter from George Stokes who wrote, 'I am afraid you will find a good deal of difficulty in getting a house in Cambridge. The supply is hardly equal to the demand.'). Somewhat less imposing than his London residence, 11 Scroope Terrace was (and remains) a handsome three-storey townhouse. The Maxwells would spend the university terms and Christmas vacation at Scroope Terrace for the rest of Maxwell's life, returning to Glenlair for the warmer summers.

The new position of Professor of Experimental Physics was not initially an idea fit for the complex structure of Cambridge University, where the individual colleges were bodies with considerable power in their own right. At the time, the vast majority of the teaching load fell on college lecturers and private tutors. Although there were a number of university professorships, they were set up by individual endowments each with their own set of rules, though those rules had become at the least bent over the years – so, for example, the Lucasian Professor of Mathematics, a position held by Isaac Newton and more recently by Stephen Hawking, has almost from the beginning tended to have more of a theoretical physics orientation than was perhaps first envisaged.

When Maxwell took his university position, some colleges had lecturers covering physics already, so he found himself to an extent in competition, with the added negative factor that students were generally expected to pay to attend lectures outside their college. Over time, the system was improved to pull together the college and university lecturing so that eventually it reached the stage where lectures were provided by the professors and lecturers of the university, while the colleges provided tutorials in the form of 'supervisions' – but during Maxwell's time this was a system in transition, making a more difficult fit for his role within the university's structure.

It didn't help that while the university recognised that someone in Maxwell's position should not be over-burdened with actually giving lectures, expecting most of his time to go on his own research and administration, the glacially slow university administration did not make it easy for Maxwell to take on an assistant lecturer or 'demonstrator'. Before there was even a building to give a lecture in, Maxwell did suggest one John Hunter for the post, but the university seems to have shown little interest in the appointment.

It was possibly just as well, as Hunter was not in good health (he died a year later), though he did have an appropriate background as he had worked with both Thomson and Tait and had held his own professorial chair at Windsor College, Nova Scotia until he found the low temperatures there unbearable. But Maxwell seems to have suggested him more on the strength of their Scottish connection than any familiarity with Hunter's skills. He wrote to his friend Peter Tait, who knew Hunter better:

> I only know him as the man who charges charcoal with bad smells.* Would he be a good demonstrator at Cambridge? I have no doubt that a man who could occlude a fishy fume in a burnt stick could also floor a demon,† which I suppose to be the essential part of the office.

In the end, the first demonstrator, a recent Cambridge graduate called William Garnett who was later Maxwell's joint biographer with Lewis Campbell, would be appointed in 1874.

There was also the difficulty of winkling the components of physics out from the other subjects to construct a separate discipline. At the time, what we would now consider to be physics was largely housed within the Mathematical Tripos, but parts of

---

* This was Maxwell in the entertainingly flippant mode he often used in letters to friends. Hunter had researched the absorption of vapours.

† Maxwell clearly had a cruel streak when it came to demons.

it also were found in the Natural Sciences Tripos, which covered chemistry, mineralogy, geology, botany and zoology. Physics would end up as a separate discipline in the Natural Sciences Tripos, taking to itself the parts of chemistry that dealt with heat and electricity (and later the structure of atoms and molecules) and adding in many of the areas previously considered as part of mathematics, such as the science of moving bodies and astronomy.

Even today, the split is not entirely clear at Cambridge. Experimental physics sits firmly in the Natural Sciences side, but the Department of Applied Mathematics and Theoretical Physics (for many years the academic home of Stephen Hawking) straddles Natural Sciences and Mathematics.

The power of the Mathematics Tripos was considerable. Senior Wranglers were feted as national heroes, with their school and home town often having a holiday or procession to celebrate their achievement. This meant that the best minds were reluctant to stray from Mathematics into what were considered the less stellar regions of Natural Sciences. (When Maxwell arrived at Cambridge, Natural Sciences could not offer an honours degree, which was only available in Mathematics and Classics.)

George Bettany, a Natural Sciences graduate from Maxwell's day, wrote in *Nature* in 1874:

> The great hindrance to the success of the Cavendish Laboratory
> at present is the system fostered by the Mathematics Tripos.
> The men* who would most naturally be the practical workers
> in the laboratory are compelled to refrain from practical work
> if they would gain the best possible place in the [Mathematics]

---

* And, yes, it still was just men.

Tripos list. Very few have courage so far to peril their place or to resign their hopes as to spend any valuable portion of their time on practical work.

It would take a good fifty years for the differences and difficulties of the split to be fully resolved, and the relative importance of the Mathematics and Natural Sciences Tripos to be rebalanced, though throughout that period, with the work of one of Maxwell's successors, J.J. Thomson,* and others, it became impossible to ignore the importance of experimental physics.

The very need for a physical laboratory also took some accepting in a university that considered it somehow to question the sanctity of intellect if it were necessary to resort to experimental proof. As Maxwell's Cambridge contemporary, the mathematician Isaac Todhunter, remarked of the science student:

If he does not believe the statements of his tutor – probably a clergyman of mature knowledge, recognised ability, and blameless character – his suspicion is irrational, and manifests a want of the power of appreciating evidence.

By the time Maxwell was appointed, Edinburgh, Glasgow, London, Manchester and Oxford had physics teaching laboratories, though only Glasgow under William Thomson and Oxford under a Professor Clifton had new buildings that had been designed for the purpose, which Maxwell wasted no time in visiting.

* No relation to William Thomson.

Ancient lights and modern physics

The site chosen for the new Cambridge laboratory was, frankly, not ideal.* For a number of years, the site of the former Botanic Gardens in central Cambridge had been gradually taken over as a hub for most of the other natural sciences, forming what is now known as the New Museums Site. This seemed the logical location for Maxwell's laboratory, but the remaining free space there was very limited. The new laboratory would loom across the narrow Free School Lane so tightly that Corpus Christi College, on the opposite side of the lane, claimed a loss of ancient lights and made a failed attempt to sue the university.

Maxwell engaged a relatively unknown architect, William Fawcett, whose experience to date had largely been in church architecture, though Maxwell seems to have damped down Fawcett's enthusiasm for Gothic frills; the architect largely limited his flights of fancy to the decorated main gate, carrying the Cavendish coat of arms, a statue of the Duke and the biblical inscription *Magna opera Domini exquisita in omnes voluntates ejus* ('The works of the Lord are great: sought out of all them that have pleasure therein' – Psalm 111, verse 2).

Despite the limitations of the site, Maxwell's design for the Devonshire building, as it was initially called, proved a triumph, with experimental physics underway three years after Maxwell was brought on board, and lectures starting in October 1872

---

* Cambridge would eventually bite the bullet on this and move the main physics site to the New Cavendish, located on wide open spaces to the west of Cambridge at the start of the 1970s. Some physics work continued in Maxwell's old Cavendish building for a number of years, but at the time of writing, part of the Cavendish is scheduled for demolition to provide better access to other buildings. There is a hope that the core of Maxwell's structure can be retained and opened as a visitor centre.

before construction was finished. Maxwell wrote to Lewis Campbell on 19 October 1872:

> Lectures begin 24th. Laboratory rising, I hear, but I have no place to erect my chair, but move about like the cuckoo, depositing my notions in the chemical lecture theatre 1st term; in the Botanical in Lent and in Comparative Anatomy in Easter.

An interesting insight into Maxwell's thinking can be obtained in an extract from a letter to Katherine written on 20 March 1871, where he observed:

> I think there should be a gradation – popular lectures and rough experiments for the masses; real experiments for real students; and laborious experiments for first-rate men like Trotter, Stuart and Strutt.

The next day, Maxwell wrote to William Thomson, one of the few physics professors in the UK to already have a laboratory, asking for his opinion on the requirements for 'material accommodation'. Maxwell provided Thomson with a list for his consideration which gives us an effective insight into the kind of facilities the Cavendish would provide:

> Lecture room *taken for granted.*
> Place to stow away apparatus d°.*
> Large room with tables &c for beginners at experiments, gas and water laid on &c.

---

* Ditto.

A smaller place or places for advanced experimenters to work at experiments which require to be left for days or weeks standing.

A place on the ground floor with solid foundations for things requiring to be steady.

Access to the roof for atmospheric electricity.

A place with good ventilation to set up Groves or Bunsens batteries without sending fumes into the apparatus.

A good Clock in a quiet place founded on masonry, electric connexion from this to other clocks to be used in the expts and from these connexion to machines for making sparks, marks on paper &c.

A well constructed oven, heated by gas to get up to a uniform high temperature in large things.

A gas engine (if we can get it) to drive apparatus, if not, the University crew in good training in four relays of two, or two of four according to the nature of the exp^t.*

For his 'popular lectures', Maxwell would build a 180-seater lecture theatre, for the dedicated experimental physics students a teaching laboratory with ten workbenches ... and for the 'first-rate men' a number of smaller rooms where they could run their experiments undisturbed by the hurly-burly of undergraduate experimental sessions. It was in the detail of designing these spaces that Maxwell really led the field. For example, the lab workbenches were cushioned from the floor to prevent vibration from influencing experiments. Metal pipes were always visible

---

* Maxwell's sense of humour to the fore. He was not seriously suggesting employing the university's crew for the boat race as the motive power for experiments.

so that those undertaking electromagnetic experiments could avoid them, and a vacuum system was available throughout the building. Cast iron hollow bricks were built into the walls at intervals to support apparatus, while ceiling joists had to protrude through the plaster so that supports could be fixed to them. There were even especially wide external window ledges, still visible on the building, so that heliostats – devices that tracked the Sun and kept a supply of bright light flowing into the building – could be stationed on them.

Maxwell's suggestion of enabling the serious scientists to have separate facilities seemed very much in line with the thinking of Coutts Trotter, the Trinity College physics lecturer, who wrote to Maxwell in April 1871:

> There is no doubt much to be said for natural selection* but will not the struggle for existence between the men who want the rooms darkened and the men who want their rooms light, the men who want to move about magnets and the men who want to observe galvanometers be unduly severe?

A concern for this kind of conflict comes through in the way that Maxwell organised the smaller rooms by the type of measurement required, to avoid interference between experimental needs, rather than the traditional topic orientation used in earlier laboratories.

---

* It is interesting that just twelve years after the publication of Darwin's *Origin of Species* (a book some would even today regard as distinctly demonic), Trotter was comfortable using the concept of natural selection as a generally understood shorthand.

A slow start

The Cavendish Laboratory – the name change was suggested by Maxwell to commemorate both Henry Cavendish and their present-day benefactor – was officially opened in June 1874 and helped Cambridge to become Britain's leading university for work in physics. During the next hundred years, many major developments from splitting the atom to the discovery of the structure of DNA would occur in Maxwell's laboratory.

It would be unfair, though, to suggest that the Cavendish under Maxwell went from nothing to a thriving establishment overnight – and some clearly doubted Maxwell's ability to make a difference. It's notable that an 1873 editorial in the journal *Nature* did not hold out a lot of hope for the Cavendish:

> Let any one compare Cambridge, for instance, with any German university; nay, with even some provincial offshoots of the University in France.* In the one case he will find a wealth of things that are not scientific, and not a proper laboratory to work in; in the other he will find science taking its proper place in the university teaching, and, in three cases out of four, men working in various properly appointed laboratories, which men are known by their works all over the world.

Although Maxwell laid the foundations for the future success of the Cavendish, the laboratory was constantly underfunded during his tenure. He could not, for example, afford the essential

---

* A very barbed comment, bearing in mind the antipathy that was still felt for France in the UK at the time.

services of a technician full-time – once a Robert Fulcher had been assigned to the post in 1877, it was on a freelance basis that soon saw Maxwell losing much of Fulcher's time to other departments.

Initially there was no real system to the teaching – lectures and classes were provided, but that was not a clear experimental physics component to the undergraduate Natural Sciences degree. Each year, though, Maxwell did deliver a lecture course, covering heat and matter in the first (Michaelmas) term, electricity in the second (Lent) term and electromagnetism in the third (Easter) term. The content of the later lectures seems to have been based on his *Elementary Treatise on Electricity*, which became a widely-used textbook.

Despite the large laboratory being built for teaching purposes, it was primarily used so that anyone available – whether undergraduates or not – could contribute to the boring parts of research on behalf of the senior physicists – for example, in Maxwell's ongoing attempts to standardise electrical units. It was only around 1879, with numbers on the undergraduate course rising to around 30, that regular practical experimental physics classes were implemented.

Similarly, there was at the time no good structure to support postgraduate staff in the laboratory. It was hard to get a college fellowship based on a scientific subject, and when they were obtained they rarely lasted long (as had been the case with Maxwell's own fellowship at Trinity – though he would have been seen as a mathematician). The position of fellow was often thought of as a stepping stone to a more significant position. It could only be held by unmarried men, and anyone staying on longer than seven years had to be ordained in the Church of England.

When postgraduates started to become more common after Maxwell's time they mostly had the more respectable Mathematics degrees, which made it easier to get a fellowship position than a degree in Natural Sciences. At the time the whole system was biased in this direction. Experimental physics, even then, required more assistance than did the lonely work of the mathematician. But college fellowships went primarily to those who produced an outstanding dissertation in the Mathematics Tripos, and this was obliged to be a solo piece of work. It would be a few decades before the Cambridge system fully accepted the need to bring through post-doctorate students and fellows from the Natural Sciences Tripos.

Women in the laboratory

Despite his support for Katherine helping with his work when at home, Maxwell was a man of his time and he was initially extremely doubtful about allowing women students into the laboratory. The first women's college at Cambridge, Girton, had been established remotely in Hitchin in 1869 and opened at Girton on the outskirts of Cambridge in 1873, though initially the students were not members of the university. A striking illustration of the way that women were treated in the university is the way they were dealt with in respect of the Wrangler listing for the Mathematics Tripos.

As we have seen, the position of Senior Wrangler was of huge importance across the country, but the position could not be given to a woman. Women first appeared in the Tripos listings in 1882, but rather than awarding them the position of a specific Wrangler they would be classed as fitting, say, 'between the 9th and 10th Wranglers'. In 1890, Philippa Fawcett, who would later be one of the first female lecturers in mathematics, was listed as 'above the Senior Wrangler' – so was, in fact, entirely deserving of that coveted position, but could not be awarded it.*

Maxwell would not give permission for women to use the Cavendish for several years. Around 1874 he wrote a pair of poems[†] with the title 'Lectures to Women on Physical Sciences', the first about a woman participant in science, the second regarding a woman lecturer. The first introduces us to a class with one (female) member, located in a small alcove with dark curtains. The woman in this practical class is taking a reading from a Thomson mirror galvanometer, which gives Maxwell a chance to contrast his concerns about inaccuracy of measurement with the classical description of women in poems of the time:

> O love! you fail to read the scale
> Correct to tenths of a division
> To mirror heaven those eyes were given,
> And not for methods of precision.

Eventually, in the mid- to late 1870s, Maxwell changed his mind on women being allowed into the building, though even then this seems to have been grudging. His assistant Garnett noted:

> At last [Maxwell] gave permission to admit women during the Long Vacation,[‡] when he was in Scotland, and I had a class who

---

* The *Daily Telegraph* noted: 'And now the last trench has been carried by Amazonian assault, and the whole citadel of learning lies open and defenceless before the victorious students of Newnham and Girton. There is no longer any field of learning in which the lady student does not excel.'
† The second was definitely in 1874, but the first was undated.
‡ The name given to the summer vacation at Cambridge. Teaching during the Long Vacation, particularly between second and third years of the Natural Sciences Tripos, still takes place.

were determined to go through a complete course of electrical measurements during the few weeks when the laboratory was open to them.

The first true class for women in experimental physics began in 1878, and women would not be admitted to the main Natural Sciences physics course until the time of Maxwell's successor as Cavendish Professor, Lord Rayleigh.

Even so, Maxwell had done a remarkable amount to improve the standing of physics at Cambridge – and raised his own profile with that of the Cavendish Laboratory. Maxwell ensured that his position of Cavendish Professor of Experimental Physics would be passed on (there was some uncertainty as to whether it should be a one-off position initially). It is a chair that has continued to this day, with just nine individuals so far holding the post. These have included several of the 'big beasts' of physics, including J.J. Thomson, Ernest Rutherford and William Bragg* (the younger of the Bragg father-and-son combo who won the Nobel Prize together). At the time of writing it is held by Richard Friend, a specialist in carbon semiconductors.

---

* He of the 'God runs electromagnetics on Monday ...' quote – see page 201.

In which the demon's memory is challenged

JCM absolutely transformed physics at Cambridge – and we can see this as the last step of the huge changes he made to physics and your everyday world as a whole. Before his time, physics was in many ways an amateur discipline and (as far as experimental physics went) one where mathematics had a limited role. Post-Maxwell, to be a professional physicist was a significant position – people don't say 'It's not rocket science' for nothing – and maths was crucial to its development.

Of course, it's impossible to say for certain how things would be in your world without JCM. Most importantly, of course, you wouldn't have me. But at the very least, work on electricity and magnetism would have been put back significantly, and JCM's work would not only form the backbone of the technological developments of the twentieth century but would also be necessary to introduce both relativity and quantum theory, the two pillars of modern physics.

Without the Maxwell touch, Einstein would have been left foundering and quantum physics would never have been developed, dependent as it was on wholly mathematical models. Despite all this, our man, James Clerk Maxwell, for some reason barely crosses the awareness of the general public. It's remarkable that 78 per cent of readers of this book had never heard of Maxwell before coming across it.*

Forgetting is never easy

So, JCM's position in history is solidly established. But what about me?

You may remember that we left my status somewhat battered by Leo Szilard's suggestion that there would have to be a use of energy, increasing entropy, for me to make a measurement and store the information in my brain. However, as information theory became better developed, it was discovered that it is perfectly possible to store data and make computations without the expenditure of energy. With a bound, it seemed, I was free – I could do the job my creator had designed me for without using energy and pushing up entropy levels.

Unfortunately, there was a sting in the tail of this route to freedom. The physicist behind the breakthrough, Rolf Landauer, made a second, distinctly counter-intuitive discovery. While it need not take energy to store information or to make calculations, erasing information *does* result in exactly the amount of energy Szilard had calculated being put into the system, countering the apparent problem of the reduced entropy. Bear in mind, if you can put energy into the system it is perfectly possible to reduce entropy – to produce the low-entropy arrangement of letters in this book, for example, as opposed to a scrambled-up set of letters, the author had to exert energy

---

* I can say this because I am a demon and so can make up statistics. In reality, I have no idea how many readers had never heard of him, but Maxwell has always scored badly for name recognition compared with the likes of Newton, Faraday and Einstein, or even Schrödinger and Heisenberg (though most younger people probably think the latter is just a character in *Breaking Bad*).

to make it happen.* Similarly, refrigerators manage to reduce entropy, shifting heat from a warmer to a colder place because energy is pumped into them from their power source.

So, it seemed that Szilard wasn't entirely wrong, but he assigned the energy requirement to the wrong part of the process, to acts of measurement and storing data away, rather than forgetting. The actual justification for the erasure of information requiring energy is complex, but it depends on the concept of reversibility. If a process can be run forwards or backwards without distinction it is known as a reversible process and does not increase entropy. But if it is not possible to reverse it without energy consequences, it will increase entropy.

If we consider the parallel of what is supposed to be going on in my demonic brain as a simple sum such as $2 + 5 = 7$, given the sum, the process is reversible. If I know what operation is involved, given any two numbers from it, I can recreate the rest. But if I erase the two values that make up the left-hand side of the sum and am left with only the 7, I can't get back to the 2 and the 5. It's not the action of computing the sum that makes it irreversible and hence increases entropy, it's the process of forgetting what the sum actually was.

What has all this information theory stuff got to do with me and the molecules? Followers of Landauer argued that however big my memory was, I would eventually run out of storage as a result of having to deal with so many billions of gas molecules, and so would have to erase what was stored in order to continue with the process – as a result, I would end up countering the benefit I had produced.

---

* Apart from these demonic interludes, which he allowed me to put together, so he could sit back and do nothing.

In 2017 researchers published a paper entitled 'Observing a quantum Maxwell demon at work', which they claimed gave them insight into my mind. Their 'demon' was a decidedly inferior object in the form of a superconducting cavity which held microwaves. It interacted with a small superconducting circuit which could give off or absorb a photon of light. The demon controlled the system, ensuring light could only be drained from the system, not absorbed, transferring the energy in one direction.

The team allegedly probed the demon's memory using something called quantum tomography that allowed them to use multiple runs of the system to build up a picture of what was happening in the memory – and they apparently showed that, in order to work, the demon had to keep information about the state of the system, seeming to prove the assertion that I could only bend the second law if I didn't forget things.

For many, the whole business of the erasure of memory was the end of my story. However, it's entirely possible that they really hadn't thought things through particularly well, as we shall discover. But we should first return to JCM at Cambridge, as he continues in his new role as Cavendish Professor.

Chapter 9

The last work

M axwell had a belief in the openness of science, and as a result he probably did not provide enough direction and guidance to build effective staffing for the Cavendish during his time, a failing that would later have to be corrected. His approach was a result both of limitations on his time and a philosophy that was strongly geared to giving researchers their heads. He was quoted by Manchester-based physicist Arthur Schuster in Schuster's contribution to a 1910 history of the Cavendish Laboratory as saying: 'I never try to dissuade a man from trying an experiment. If he does not find what he wants, he may find out something else.'

Books and the power of light

Instead, a lot of Maxwell's time during his Cambridge years was dedicated to his writing. It was while at Cambridge that he published both his *Theory of Heat* and *Treatise on Electricity and Magnetism*, each of which attracted a far wider readership than we would now expect for a technical science book. The *Theory* was reviewed so widely that even *The Ironmonger* felt the need to comment, saying that 'the language throughout is simple and the conclusions striking'. To be fair, Maxwell's subtitle (which now seems somewhat condescending) was 'Adapted for the use of artisans and students in public and science schools'. The *Treatise* was arguably Maxwell's written masterpiece and was still in use as a textbook well into the twentieth century.

For the first year or so, the move to Cambridge and getting work started on the Cavendish Laboratory got in the way of Maxwell finishing his 1,000-page masterpiece on electromagnetism – but the *Treatise* was finally brought out in 1873, at the same time as parts of the new building were starting to be usable.

As always, Maxwell was not happy to leave his work as it was, but insisted on pulling at the threads of new ideas, looking for gaps to fill and inaccuracies to iron out. As he did so, he realised a strange implication of his predictions on the nature of electromagnetic waves. Not only did these waves correspond so clearly to light, they appeared to be capable of something that no one suspected light could do – electromagnetic waves should be able to apply pressure to matter. If his theory was correct, an insubstantial beam of light should be able to push a solid object.

This seems a crazy concept. Almost all of our experience of light suggests that its effects on physical objects would not be able to move them – yet Maxwell calculated that there should be a small force generated by the interaction between electromagnetic waves and the matter they illuminated. There was one physical phenomenon that did suggest such an effect could be occurring. As we have seen (page 173), Maxwell had already given some consideration to the way that a comet's tail always points away from the Sun. The idea of pressure from light was a possible explanation for this. Maxwell gave this hypothesis a theoretical backing by showing that the energy of absorbed light should result in momentum being added to the body absorbing it.

The reason we don't see this happening as a rule, Maxwell suggested, was because the effect was so small. He calculated that the Sun, the most powerful light source in our vicinity, would only produce the equivalent pressure of 7 grams (0.015 pounds) across a whole hectare (2.47 acres) of area. Maxwell would never see an experimental demonstration of the effect that he had predicted, but 25 years after his theory was published, the Russian physicist Pyotr Lebedev demonstrated this 'radiation pressure' for the first time.

The concept of radiation pressure is not just a partial explanation for comets' tails (in practice the effect is primarily due to the 'solar wind', a flow of particles from the Sun). It has also been suggested as a possible way of powering spaceships by light, using immense sails to catch the light from the Sun or from a battery of lasers. But most important of all, it is an essential phenomenon to be able to understand how stars operate. Without it, the Sun would not function. The matter in stars is subject to immense inward pressure from gravity – it is the radiation pressure of the vast numbers of photons of light produced inside the star that is the first level of defence against gravity's attempts to make the whole thing collapse.

The *Treatise* was not just Maxwell's masterpiece, it was his last significant work. He published a short piece in *Nature*, improving Ludwig Boltzmann's treatment of the kinetic theory of gases (itself derived from Maxwell's work, but taking it sufficiently further that Boltzmann and Maxwell are fairly considered joint initiators of the theory). And he continued to work on many of his lesser pet projects, from colour vision to the theory of optical instruments, where he expanded the tools available for the theoretical description of multi-lensed instruments – a practical study, though not as distinctively original as his main topics of work.

Most of his effort over the next couple of years, however, was dedicated to building up and running the Cavendish Laboratory. What spare time he had was given to a job that many have since thought was a matter of shouldering an obligation – but in reality, he could well have been following a topic of great personal interest.

The Cavendish papers

As we have seen, Henry Cavendish, the ancestor of the Duke of Devonshire who funded the Cavendish Laboratory, was an important scientist in his own right. From about 1873 onwards, Maxwell dedicated a considerable amount of his time to editing the papers of Cavendish. This could, indeed, have been a way of thanking the present Duke for the funding – but the money for the laboratory was already in place and it seems more likely that Maxwell had a genuine interest in untangling the details of Henry Cavendish's work. Arguably, this could have come from the same kind of drive that inspires people to look into their ancestry – this was, in effect, establishing the ancestry of his laboratory – and also because Henry Cavendish had done a considerable amount of original work that was not widely known, and that Maxwell felt should be available to other physicists.

Cavendish is probably best remembered for measuring the density of the Earth,* an experiment that would enable the first calculation of Newton's gravitational constant G, and for discovering hydrogen. But from Maxwell's viewpoint, Cavendish's most interesting work was in electricity, a field that Cavendish studied for ten years from 1771. Most notably, Cavendish demonstrated the inverse square law of electrical repulsion well before the French physicist Charles-Augustin de Coulomb did. Yet it was

---

* He is also remembered for being extremely eccentric. Cavendish was so shy that he did not like to speak to more than one person at a time and would only communicate with female servants by notes. Complex arrangements ensured his meals could be served with minimal contact with staff.

Coulomb who was acknowledged as the discoverer of the law, as Cavendish never published this part of his work.

It is not uncommon to see suggestions that Maxwell wasted much of the little time he had left to him on the Cavendish papers. However, leaving aside the benefits of hindsight – he had no idea that he hadn't long to live – there is no evidence that Maxwell was pressured into doing the work, either by the university or by the Duke of Devonshire. Quite the reverse: Maxwell appears to have first shown an interest in getting hold of Cavendish's papers two years *before* his involvement at Cambridge. Not only did he edit Cavendish's output, pulled together from reluctant owners with the help of the Duke, he also recreated some of Cavendish's experiments, using period instruments which had come into his possession from the collection of William Hyde Wollaston.

Some of Cavendish's results added extra information to work on electromagnetism. Maxwell commented: 'if these experiments had been published in the author's life time the science of electrical measurement would have been developed much earlier.' Other parts of Cavendish's papers simply seemed to appeal to Maxwell's personality and sense of humour.* He was delighted with the way that Cavendish had used the pain reaction of his own body as a 'meter' to test the strength of electrostatic energy. Maxwell recreated this approach, using anyone he could persuade to volunteer to try out the technique.

---

* It's easy to think of Maxwell, from his dour photographic portraits, as a typical miserable Victorian, but all his life he displayed an impish (appropriately enough) sense of humour. This often came through, as we have seen, in his letters, which could be distinctly whimsical. For example, he tended to refer to himself as dp/dt – since an equation in his friend Peter Tait's book read dp/dt=JCM, Maxwell's initials. (See notes for detail of the equation.)

Arthur Schuster remembered:

> ... a young American astronomer expressing in severe terms his
> disappointment that, after travelling on purpose to Cambridge
> to make Maxwell's acquaintance and to get some hints on astro-
> nomical subjects, the latter would only talk about Cavendish,
> and almost compelled him to take his coat off, plunge his hands
> into basins of water and submit himself to the sensation of a
> series of electrical shocks.

Maxwell's edition of Cavendish's papers was published in
October 1879, shortly before his death.

Passing fancies

Typically, though Maxwell spent the majority of his time on
his better-known fields of interest, he could not help being
drawn briefly into any passing curiosity that took his fancy. For
example, in 1874 he wrote to his old friend Lewis Campbell
about what would now be called genetics, considerably before
the concept of the gene was widely known.* Admittedly, Maxwell
did so to dismiss the concept – but based on what he thought
was the case, his argument made sense. He wrote:

> If atoms are finite in number, each of them being of a certain
> weight, then it becomes impossible that the germ from which a
> man is developed [i.e. the cell] should contain ... gemmules† of

---

* The word 'gene' was not introduced until 1905. The concept behind the
gene was in Gregor Mendel's work suggesting that a unit of inheritance did
exist, which was published in 1866, but it was not widely known until after
Maxwell's death.
† Darwin's term for a unit of inheritance.

everything which the man is to inherit, and by which he is dif-
ferentiated from other animals and men, – his father's temper,
his mother's memory, his grandfather's way of blowing his nose,
his arboreal ancestor's arrangement of hairs on his arms ... if we
are sure that there are not more than a few million molecules
in [the cell], each molecule being composed of component
molecules, identical with those of carbon, oxygen, nitrogen,
hydrogen etc., there is no room left for the sort of structure
which is required for pangenesis on purely physical principles.

Maxwell proved to be wrong, by attributing too much to genetics, underestimating the number of molecules available, and in
the scale of information that can be stored in a germ cell – but
the fact that he was discussing this concept long before the significance of DNA and its genes was realised shows the range of
his interests and thoughts.

One of the most entertaining of Maxwell's diversions was the
Crookes radiometer, a puzzling device that was introduced to
the world in 1874. William Crookes was an English scientist a
year younger than Maxwell, who would go on to do important
work on electrical effects in vacuum tubes, the precursors of
thermionic valves, which were the first electronic devices.

Crookes' radiometer looked a little like a light bulb. It was a
sealed glass chamber with the air mostly pumped out, but rather
than having a filament inside, it had four paddles suspended from
a central spindle so that they could freely rotate. Each paddle was
white or silver on one side and black on the other – and when
exposed to light, the paddles would rotate at high speed.* At first

* You can see a Crookes radiometer in action at the *Universe Inside You* website
page: http://www.universeinsideyou.com/experiment4.html

sight this might have seemed to be a vindication of Maxwell's idea of radiation pressure – but the radiometer spins in the wrong direction for this (and anyway, Maxwell was aware that the amount of pressure on the paddles from radiation would not be sufficient to turn them).

If the spinning paddles had been pushed by radiation pressure you would expect that the white/silver sides would move away from the light, as they would be reflecting far more of the light than would the black sides. In fact, however, it's the black sides that move away from the light, which caused confusion and delight in equal measures among the scientific intelligentsia.

Maxwell would come to the rescue, aided by some practical information from his friend Peter Tait. With his colleague James Dewar, the inventor of the vacuum flask and an expert on low-pressure work, Tait had discovered that the working of the radiometer was dependent on the amount of air that was left in the bulb. No experimental vacuum was perfect – there would always be some gas molecules present. With too much air or with too little, the radiometer would not work.

Realising that the mechanism must depend on one of his other favourite topics, the kinetic theory of gases, Maxwell seems to have put himself in the frame of mind of one of his demons, able to see the gas molecules in action near to the paddles. When the light was shone onto the paddles, the black sides would absorb more of the light and would heat up. This, he thought, would speed up gas molecules that came into contact with the paddles, producing convection currents around the sides of the paddles, which would effectively suck the paddles round.

With many of his theories, Maxwell wasn't quite there in his first attempt – this turned out to be the case with the radiometer, where his calculations let him down a little. The actual solution turned out, if anything, to be simpler. Gas molecules coming into contact with a warmer black paddle surface would gain a little more energy than those that hit a cooler white side. This meant that, on average, the black sides were being bombarded with more momentum than the white and started to move away from the pressure.

Although Maxwell's convection currents turned out not to be the driving force behind the radiometer, his effort was by no means wasted. In writing up his work for the Royal Society he generalised his mathematics to provide an equation for the behaviour of gases in such rarefied conditions which would prove to be valuable in studies of the upper atmosphere.

A sudden end

The paper inspired by the radiometer would be Maxwell's last contribution to science. In early 1877, he began to have problems with his digestive system. Maxwell suffered frequent heartburn and found swallowing an increasing problem. The discomfort got worse over two years before he consulted his doctor, who took him off meat and replaced it with a milk-based diet.

In the summer of 1879, James Clerk Maxwell was diagnosed with abdominal cancer. He died in Cambridge on 5 November 1879, just 48 years old – the same age as his mother at her death.

Maxwell's funeral was effectively split in two. The first part of the burial service took place in the Trinity College chapel in Cambridge, attended by many of his academic colleagues and friends. His coffin was then taken home to Glenlair, with the closing part of the funeral service held at Parton church, before his burial in the churchyard there.

In which the demon lives on to fight another day

With the sad news of James Clerk Maxwell's demise, it would be easy to consider us both finished. You may recall that many thought I could no longer do my job as a result of the suggestion that it would take energy to erase information, and so, in the overall system including me, entropy would not be reduced. There was one attempt, with a complex argument, to provide a way in which it would be possible to make the erasure process reversible, though many argued that this was cheating, since it required dumping information into an external store – which surely itself was also finite.

The reality of loopholes

Even so, I was thought to be doomed. However, there is a loophole which remains to this day. Although it's true that eventually I would have to wipe some memory, it is also true that in any particular experiment I could still perform my task. There may be many billions of molecules in an experiment, but I can also have an equivalent amount of storage and perform my task without any erasure. Admittedly I wouldn't have enough memory space to run the experiment for ever, but I would be able to reduce entropy to a considerable degree before I ran out of storage – there is no need for me to deal with every single molecule available over all time.

Those who wish me disposed of suggest that this is not an acceptable argument because of the way thermodynamics is practised. Generally, this involves cyclic processes where the components are returned to their initial state, and the argument is that I should be left in the same state after working on a molecule as I was before it – which wouldn't allow me a memory. But that's something of a silly argument, as without any memory I couldn't do my job in the first place – and the whole process is clearly not cyclical when molecules are being irreversibly moved from one side of the box to the other. It seems an argument more from dogma than from physics.

More to the point, other physicists have come up with a concept they call blending which achieves the irreversibility of erasure without any influence over the entropy of the system. To quote one such example from Meir Hemmo and Orly Shenker:

> In particular, the principles of mechanics entail no specific relation between the pre-erasure and post-erasure entropy of the universe. In any case, our analysis of erasure demonstrates that, contrary to the conventional wisdom, classical mechanics does not entail that an erasure is necessarily dissipative [i.e. increasing entropy].

That being the case, I hope you are clear that the challenge JCM set so long ago is, to a degree, still there. I am still a thorn in the side of the second law – almost inevitably, given its statistical nature. Despite the theoretical objections to my existence (rather demonist, if you ask me), a number of recent experiments have shown that demonic action is possible as long as you work on a small enough scale.

In 2016 for example, at the University of Oxford, a team made use of two pulses of light instead of the two sides of my original box. Rather than employ a true demon (apparently, despite the suggestions of Philip Pullman, we are of limited availability in Oxford labs), they made a measurement on two pulses of light and depending on which was stronger, took one pulse in one direction and the other in another. The difference between the voltage produced by photodiodes receiving the two pulses charged up a capacitor. Because the more energetic pulses always go the same way, the result is to be able to produce work from a demon-style interaction.

Closer to the original because it has a thermodynamic element, also in 2016 a Brazilian physicist and his team performed an experiment that seemed to contain its own tiny demon. They managed to produce a situation where the second law was more clearly spontaneously broken, if on a small scale. This was something that must have struck joy into the heart of those who peddle 'free energy' devices. Physicists dismiss perpetual motion machines and free energy devices out of hand. Some consider this a lack of open-mindedness, but in reality, it's just that the physicists understand the second law of thermodynamics.

In the Brazilian experiment, heat moves from a colder to a hotter place. As we've seen, there's nothing odd about heat moving from a colder to a hotter body: it's what a fridge does, after all. But this can only be the case if energy is supplied to make it happen – this is what the 'closed system' bit of the definition of the second law precludes. What is interesting in the described experiment is that heat was transferred spontaneously from 'colder' to 'hotter' (I'll come back to those inverted commas soon), which is what you need for perpetual motion and free energy.

Physicist Roberto Serra of the Federal University of ABC in Santo André, Brazil and the University of York, and his colleagues, got molecules of chloroform – a simple organic compound where a carbon atom has one hydrogen and three chlorine atoms attached – into a special state. The hydrogen atom and the carbon atom in a molecule had one of their properties – spin* – correlated, giving them a kind of linkage. The hydrogen atom was in a higher energy state than the carbon, making the hydrogen technically hotter than the carbon (hence the inverted commas above). And without outside help, as the correlation decayed, heat was transferred from the carbon to the hydrogen. From the colder to the hotter atom.

To understand why this happened requires the alternative definition of the second law, the one involving entropy. As we've seen, entropy is measured by the number of different ways the components of a system can be organised. The more ways, the higher the entropy. In the case of the chloroform experiment, entropy decreases because there are more ways to arrange the quantum states† when they are correlated than when the correlation goes away – it's a bit like there being more different ways to throw a six with two dice together than there are with either of two dice individually.‡ However, free energy

---

* Quantum spin is one of the standard properties of a quantum particle. It was called spin as it loosely corresponds to the familiar rotation of an object, but it does not involve actual spinning around: quantum spin has values of multiples of a half, and can only be either up or down in the direction of measurement.

† The quantum state of a particle (or system) is the collection of the values of its properties such as charge, spin, etc.

‡ With two dice individually, you can only get a six if six comes up on one of the dice. But with two dice together you can get a six with 1+5, 2+4 and 3+3.

enthusiasts don't need to get too excited. Although there does appear to have been a spontaneous reduction in entropy, getting the molecules into the right state to start with would have taken far more energy than could be extracted. This is never going to be a free source of energy.

Dr Serra's chloroform molecules have none of the sophistication of a true demon at work – but they are an actual physical manifestation of that most enduring of laws being, at the very least, cheekily tweaked. Whether or not *my* challenge remains, one thing is certain. Our man, James Clerk Maxwell, has an impressive legacy and I'm proud to be associated with it.

Chapter 10

The legacy

I f you were to ask a scientist around the end of the nineteenth century who – of that century's big names – scientists in the future would regard as the leading British physicist of the era, he would undoubtedly have said Lord Kelvin. Maxwell's old friend William Thomson was feted in his day – a reality that was reflected in his elevation to the House of Lords, the first scientist ever to have received this honour.* Certainly, there is no doubt that Kelvin did essential work in thermodynamics, as well as working on many practical applications of science – his name appears on over 70 patents of the period, and as we have seen, he was a leading figure in the laying of the transatlantic cable.

It was Kelvin, not Maxwell, who ended up alongside Isaac Newton in Westminster Abbey, and Kelvin who has a scientific unit named after him. Not long after Kelvin died in 1907, statues of him would be erected in both his birthplace of Belfast and in the city where he did the majority of his work, Glasgow. By

---

* It's often said, incidentally, that Isaac Newton was the first person to be knighted for his contribution to science. I've even heard this claimed on that unequalled source of knowledge, the BBC's *Pointless* TV quiz programme. But in reality, Newton was knighted for doing a good job at the Royal Mint, where his enthusiasm for catching those who clipped the edges of coins to sell the metal (and for having them hanged, drawn and quartered) was legendary. A man after my own demonic heart, was Newton.

contrast, Maxwell was buried in a little country churchyard and there would not be a statue of him put up in his native Scotland for over 100 years after his death.

By the first half of the twentieth century, though, the picture was transformed. While Kelvin's achievements have not been belittled, they now seem a lot less significant in the grand scheme of things. By comparison, the appreciation of Maxwell's work on electromagnetism, his contribution to statistical mechanics, and his transformation of the way that theoretical physics was undertaken have made him far more of a hero to modern physicists. It's not for nothing that Einstein is reported as saying: 'There would be no modern physics without Maxwell's electromagnetic equations; I owe more to Maxwell than to anyone.'

Those who knew him best were aware from the outset that there was something special about James Clerk Maxwell. His friend since schooldays, Peter Tait, wrote of him in *Nature* at the end of a summary of Maxwell's work, providing a eulogy that would stand up well today in its reference to the resistance to 'vain-babbling' and pseudo-science:

> I cannot adequately express in words the extent of the loss which his early death has inflicted not merely on his personal friends, on the University of Cambridge, on the whole scientific world, but also, and most especially, on the cause of common sense, of true science, and of religion itself, in these days of much vain-babbling, pseudo-science, and materialism.

There was something very special about Maxwell's breadth of contribution. Charles Coulson, who in 1947 took on the same chair that Maxwell had held at King's College London, remarked: 'There is scarcely a single topic that he touched upon which he did not change almost beyond recognition.' This volume of output was matched by Maxwell's insight – his ability to develop models and mathematics to model reality when this approach was a novelty. In a booklet put together to celebrate the centenary of Maxwell's birth in 1931, the English physicist James Jeans gave striking testimony to this intuitive force when describing Maxwell's distribution for the velocities of gas molecules. Jeans wrote:

> Maxwell, by a train of argument which seems to bear no relation at all to molecules, or even to the dynamics of their movements, or to logic, or even to ordinary common sense, reached a formula which according to all precedents and all the rules of scientific philosophy, ought to have been hopelessly wrong. In actual fact, it was subsequently shown to be exactly right … it was this power of profound physical intuition, coupled with adequate, though not outstanding mathematical technique, that lay at the basis of Maxwell's greatness.

It's not uncommon when trying to give Maxwell his rightful place in the pantheon of physics to bracket him with Newton and Einstein (probably throwing in Faraday as well). And though there is no doubt that a considerable amount of Newton's work was concerned with physics, it's arguable that Newton was far less of a physicist than Maxwell.

It's interesting to compare the catalogues of Newton's and Maxwell's libraries. Newton left behind a remarkable* 2,100 books. Of these 109 were on physics and astronomy, 138

---

* Particularly for the period, when books were an expensive rarity.

on alchemy, 126 on maths and 477 on theology.* By comparison, over half Maxwell's books were on physics. Newton was arguably an applied mathematician (when he wasn't occupied as an alchemist, a theologian or working at the Mint). Maxwell, like Einstein, was undoubtedly a physicist.

Looking back from the twenty-first century, Maxwell comes across as unusually unstuffy for his day. Someone far from the stereotypical image of the humourless Victorian scientist. And we are now also in a position to appreciate just how much his work on electromagnetism would help launch a technological revolution.

Considering how fresh much of Maxwell's science was, we can see in some of his later writing and talks a very modern approach to scientific matters. In 1873, he gave a speech entitled 'Discourse on molecules' at the British Association's meeting in Bradford, a small part of which makes an ideal dip into Maxwell's own words.

> In the heavens we discover by their light, and by their light alone, stars so distant from each other that no material thing can have passed from one to the other;† and yet this light, which is to us the sole evidence of the existence of these distant worlds, tells us also that each of them is built up of molecules of the

---

* Which Newton would have considered a science.
† Funnily, in saying this Maxwell was wrong, but for the right reason. At the time, the universe was thought to be much smaller than we now know it to be – but it was also considered so much younger that it was assumed there wasn't time for anything to get from one extreme of the universe to the other. The current Big Bang theory fixes this with a universe that expanded so quickly during the inflation phase that right at the beginning, the extremes of the observable universe could still have been in direct physical contact.

same kinds as those which we find on earth. A molecule of hydrogen, for example, whether in Sirius or in Arcturus, executes its vibrations in precisely the same time.

Each molecule therefore through the universe bears impressed on it the stamp of a metric system* as distinctly as does the metre of the Archives at Paris, or the double royal cubit in the temple of Karnak.

No theory of evolution can be formed to account for the similarity of molecules, for evolution necessarily implies continuous change, and the molecule is incapable of growth or decay, of generation or destruction.

None of the processes of Nature, since the time when Nature began, have produced the slightest difference in the properties of any molecule.

At this point, Maxwell deviates from modern science as he assumes that molecules have to have been made from something, given their identical nature, but that there was no process that could do so that 'we can call natural'. We now know there are perfectly natural processes for the interchange of matter and energy which can account for the creation of matter, but when Maxwell wrote, this science was still 23 years in the future, with Einstein's special theory of relativity. (And the special theory would not have come about without Maxwell's work.) Even so, until that point, and allowing for some change in language, we could have just as easily have been listening to a Brian Cox or Neil deGrasse Tyson expanding on the wonders of the universe as to a Victorian. Maxwell's vision was far removed from the

---

* By 'metric system' Maxwell did not intend the modern usage of a system to base 10, but just a system of measurement.

semi-mystical meandering of earlier physics, a fault that stayed with it even when Newton and his followers had started to include some mathematics.

Throughout this book, a minor part of Maxwell's output – the demon – has played a significant role. I wanted the demon to have his say because he reflects so well Maxwell's ability to challenge the way that his colleagues thought about things, to use interesting new approaches to modelling, and to incorporate a touch of humour into what can be a very po-faced science. More than a great scientist, Maxwell seems to have been a remarkable man, someone with whom it must have been a pleasure to be a friend.

Maxwell and his demon deserve to be remembered as long as science has an impact on our lives.

Notes

Chapter 1

7 – The description of travel to the Maxwell estate is from Lewis Campbell and William Garnett, *The Life of James Clerk Maxwell* (London: Macmillan, 1882), p. 26.

8 – The idea that Maxwell's father undertook scientific experiments is from Lewis Campbell and William Garnett, *The Life of James Clerk Maxwell* (London: Macmillan, 1882), p. 4.

8 – John Clerk Maxwell's paper 'Outline of a plan for combining machinery with the manual printing-press' was published in *The Edinburgh New Philosophical Journal*, 10 (1831): 352–7.

9 – The assertion that the family of Mary Godwin (Shelley) was of 'a very restricted income' despite having a governess is from Kathryn Harkup, *Making the Monster* (London: Bloomsbury Sigma, 2018), p. 11.

9 – The ill treatment of Maxwell by his tutor is recorded in Lewis Campbell and William Garnett, *The Life of James Clerk Maxwell* (London: Macmillan, 1882), p. 43.

10 – Maxwell's early questions about how things worked are recorded in Lewis Campbell and William Garnett, *The Life of James Clerk Maxwell* (London: Macmillan, 1882), p. 12.

11 – Maxwell's arriving back with his tunic in rags after his first day is described in Lewis Campbell and William Garnett, *The Life of James Clerk Maxwell* (London: Macmillan, 1882), p. 50.

12 – Descriptions of contemporary English public schools and their limited curricula are from David Turner, *BBC History*, 'Georgian and Victorian public schools: Schools of hard knocks', June 2015, available at https://www.historyextra.com/period/georgian/georgian-and-victoria n-public-schools-schools-of-hard-knocks/

12 – Baden Powell's concern that the higher classes were not gaining scientific knowledge is quoted in Pietro Corsi, *Science and Religion: Baden Powell and the Anglican Debate, 1800–1860* (Cambridge: Cambridge University Press, 1988), p. 116.

16 – Maxwell's first paper at the age of fourteen is in *Proceedings of the Royal Society of Edinburgh*, Vol. 2 (April 1846) and reproduced in Peter Harman (ed.), *The Scientific Letters and Papers of James Clerk Maxwell*, Vol. 1 (Cambridge: Cambridge University Press, 1990) pp. 35–42.

18 – The description of Maxwell's holiday activities at Glenlair is from

Peter Tait, 'James Clerk Maxwell: Obituary', *Proceedings of the Royal Society of Edinburgh*, Vol. 10 (1878–80): 331–9.

19 – Maxwell's letter to Lewis Campbell detailing his day at university, written from 31 Heriot Row, Edinburgh in November 1847 is reproduced in Peter Harman (ed.), *The Scientific Letters and Papers of James Clerk Maxwell*, Vol. 1 (Cambridge: Cambridge University Press, 1990). p. 69.

20 – Maxwell's maintenance of some odd behaviour at Edinburgh University is described in Lewis Campbell and William Garnett, *The Life of James Clerk Maxwell* (London: Macmillan, 1882), p. 105.

21 – Maxwell's letter to Lewis Campbell on both the barometer experiment and the devil with two sticks, written from Glenlair on 26 April 1848, is quoted in Lewis Campbell and William Garnett, *The Life of James Clerk Maxwell* (London: Macmillan, 1882), p. 116.

22 – Maxwell's letter to Lewis Campbell about his home lab, written on 5 July 1848, is in Peter Harman (ed.), *The Scientific Letters and Papers of James Clerk Maxwell*, Vol. 1 (Cambridge: Cambridge University Press, 1990), p. 71.

23 – The paper 'On the Theory of Rolling Curves' was published in the *Trans. Roy. Soc. Edinb.*, 16 (1849): 519–40 and is reproduced in Peter Harman (ed.), *The Scientific Letters and Papers of James Clerk Maxwell*, Vol. 1 (Cambridge: Cambridge University Press, 1990), pp. 74–95.

23 – Maxwell's change of subject to pursue 'another kind of laws' is noted in Lewis Campbell and William Garnett, *The Life of James Clerk Maxwell* (London: Macmillan, 1882), p. 131.

25 – Maxwell's description of heating shapes of window glass is from a letter to Lewis Campbell, written from Glenlair on 22 September 1848, reproduced in Peter Harman (ed.), *The Scientific Letters and Papers of James Clerk Maxwell*, Vol. 1 (Cambridge: Cambridge University Press, 1990), pp. 96–8.

26 – Maxwell's use of a saucer of treacle to 'observe the latitude' is mentioned in a letter to Lewis Campbell, quoted in Lewis Campbell and William Garnett, *The Life of James Clerk Maxwell* (London: Macmillan, 1882), p. 126.

27 – Forbes' criticism of Maxwell's writing style is quoted in Peter Harman (ed.), *The Scientific Letters and Papers of James Clerk Maxwell*, Vol. 1 (Cambridge: Cambridge University Press, 1990), p. 186.

28 – The draft with revisions of Maxwell's paper 'On the Equilibrium of Elastic Solids' is reproduced in Peter Harman (ed.), *The Scientific Letters and Papers of James Clerk Maxwell*, Vol. 1 (Cambridge: Cambridge University Press, 1990), pp. 133–83.

28 – Maxwell's change of plans from law is from a letter to Lewis

Campbell quoted in Lewis Campbell and William Garnett, *The Life of James Clerk Maxwell* (London: Macmillan, 1882), p. 130.

Demonic Interlude II

38 – Roger Bacon's praise for Peter Peregrinus is from Roger Bacon, *Opus Tertius*, quoted in Brian Clegg, *The First Scientist* (London: Constable & Robinson, 2003), p. 33.

Chapter 2

49 – John Clerk Maxwell's list of suggested places for his son to visit in Birmingham is noted in Lewis Campbell and William Garnett, *The Life of James Clerk Maxwell* (London: Macmillan, 1882), p. 3.

50 – Maxwell's observations on Cambridge colleges before applying, noted by Mrs Morrison, are recorded in Lewis Campbell and William Garnett, *The Life of James Clerk Maxwell* (London: Macmillan, 1882), p. 132.

52 – Maxwell's experiment with working during the night was in a letter to Lewis Campbell dated 11 March 1851, written from his lodgings in King's Parade, Cambridge, quoted in Lewis Campbell and William Garnett, *The Life of James Clerk Maxwell* (London: Macmillan, 1882), p. 155.

53 – Lewis Campbell's note that Maxwell ironically engaged with electro-biology and table-turning is from Lewis Campbell and William Garnett, *The Life of James Clerk Maxwell* (London: Macmillan, 1882), p. 166.

53 – John Clerk Maxwell's letter to his son on the dangers of electro-biology is quoted in Lewis Campbell and William Garnett, *The Life of James Clerk Maxwell* (London: Macmillan, 1882), p. 156.

54 – Maxwell's concern on the way Faraday was treated after explaining the phenomenon of table-turning is from a letter to the Reverend C.B. Tayler, written from Trinity College, Cambridge on 8 July 1853, quoted in Lewis Campbell and William Garnett, *The Life of James Clerk Maxwell* (London: Macmillan, 1882), p. 189.

54 – Lewis Campbell's description of Maxwell's appearance while at Cambridge is from Lewis Campbell and William Garnett, *The Life of James Clerk Maxwell* (London: Macmillan, 1882), p. 162.

55 – Maxwell's letter to his wife from Trinity College, Cambridge, written on 4 January 1870, is quoted in Lewis Campbell and William Garnett, *The Life of James Clerk Maxwell* (London: Macmillan, 1882), p. 499.

56 – Maxwell's poem 'A Vision' is reproduced in Lewis Campbell and William Garnett, *The Life of James Clerk Maxwell* (London: Macmillan, 1882), p. 632.

58 – Maxwell's letter to his aunt, Miss Cay, on using his instrument to see into eyes, particularly of dogs, from Trinity College, dated Whitsun Eve,

1854, is quoted in Lewis Campbell and William Garnett, *The Life of James Clerk Maxwell* (London: Macmillan, 1882), p. 208.

62 – Maxwell's manuscript on the colour top is 'Description of the Chromatic Teetotum as Constructed by Mr J.M. Bryson, Optician Edinburgh', from 27 February 1855, reproduced in Peter Harman (ed.), *The Scientific Letters and Papers of James Clerk Maxwell*, Vol. 1 (Cambridge: Cambridge University Press, 1990) pp. 284–6.

66 – Maxwell's paper 'Experiments on Colour as Perceived by the Eye, with Remarks on Colour-Blindness', based on his initial colour top work, was published in *Trans. Roy. Soc. Edinb.*, 21 (1855): 275–98 and its abstract is reproduced in Peter Harman (ed.), *The Scientific Letters and Papers of James Clerk Maxwell*, Vol. 1 (Cambridge: Cambridge University Press, 1990) pp. 287–9.

66 – Maxwell's letter to William Thomson which mentions the difficulties of reproducing brown colours was written from Trinity College on 15 May 1855 and is reproduced in Peter Harman (ed.), *The Scientific Letters and Papers of James Clerk Maxwell*, Vol. 1 (Cambridge: Cambridge University Press, 1990) pp. 305–13.

67 – Maxwell's letter to Monro on colour theory, written from Glenlair on 6 July 1870, is quoted in Lewis Campbell and William Garnett, *The Life of James Clerk Maxwell* (London: Macmillan, 1882), p. 346.

71 – Maxwell's description of his fluid model of electromagnetism as not containing the shadow of a theory is from Peter Harman (ed.), *The Scientific Letters and Papers of James Clerk Maxwell*, Vol. 1 (Cambridge: Cambridge University Press, 1990), p. 207.

73 – Maxwell's letter to his father about getting shops to shut early for the Working Men's College was written from Trinity College, Cambridge on 12 March 1856 and is reproduced in Peter Harman (ed.), *The Scientific Letters and Papers of James Clerk Maxwell*, Vol. 1 (Cambridge: Cambridge University Press, 1990), p. 404.

74 – Cecil Monro's letter on translating Newton, dated 20 January 1855, and Maxwell's reply from 18 India Street, Edinburgh, dated 7 February 1855, are reproduced in Peter Harman (ed.), *The Scientific Letters and Papers of James Clerk Maxwell*, Vol. 1 (Cambridge: Cambridge University Press, 1990), p. 280.

76 – Professor Forbes' letter to Maxwell recommending he consider the position at Marischal College is quoted in Lewis Campbell and William Garnett, *The Life of James Clerk Maxwell* (London: Macmillan, 1882), p. 250.

76 – Maxwell's letter to his father about applying for the position at Marischal, written at Trinity College, Cambridge on 15 February 1856, is quoted in Lewis Campbell and William Garnett, *The Life of James Clerk Maxwell* (London: Macmillan, 1882), p. 251.

77 – Maxwell's letter to Lewis Campbell about his responsibilities after

his father's death was written from Trinity College, Cambridge on 22 April 1856 and is reproduced in Peter Harman (ed.), *The Scientific Letters and Papers of James Clerk Maxwell*, Vol. 1 (Cambridge: Cambridge University Press, 1990), p. 405.

78 – Maxwell's letter to Richard Litchfield about his unexpected workload at Glenlair was written from Glenlair on 4 July 1856 and is reproduced in Peter Harman (ed.), *The Scientific Letters and Papers of James Clerk Maxwell*, Vol. 1 (Cambridge: Cambridge University Press, 1990), p. 410.

Demonic Interlude III

85 – Tait's observation that Thomson was 'dead against atoms' was in a letter to Thomas Andrews written on 18 December 1861, noted in Crosbie Smith and Norton Wise, *Energy and Empire: A Biographical Study of Lord Kelvin* (Cambridge: Cambridge University Press, 1989), p. 354.

Chapter 3

90 – Information on the Royal Commission from *Universities of Kings College and Marischal College, Aberdeen. First Report of the Commissioners, 1838* (1837–38), cited: http://gdl.cdlr.strath.ac.uk/haynin/haynin0509.htm

91 – Maxwell's letter to his aunt, Miss Cay, from 129 Union Street, Aberdeen, dated 27 February 1857, is quoted in Lewis Campbell and William Garnett, *The Life of James Clerk Maxwell* (London: Macmillan, 1882), p. 263.

92 – Maxwell's letter to Lewis Campbell complaining of the lack of jokes at Marischal is quoted in J.J. Thomson's section of *James Clerk Maxwell: A Commemorative Volume 1831–1931* (Cambridge: Cambridge University Press, 1931), p. 13.

93 Peter Tait's explanation for Maxwell's poor lecturing is from Peter Tait, 'James Clerk Maxwell: Obituary', *Proceedings of the Royal Society of Edinburgh*, Vol. 10 (1878–80), 331–9.

93 – David Gill's opinion that Maxwell's lectures were terrible is taken from George Forbes, *David Gill: Man and Astronomer* (London: John Murray, 1916), p. 14.

95 – Maxwell's emphasis on mathematics and stress on the importance of experiment from his inaugural lecture at Marischal comes from Peter Harman (ed.), *The Scientific Letters and Papers of James Clerk Maxwell*, Vol. 1 (Cambridge: Cambridge University Press, 1990), pp. 419–31.

97 – The wording of the Adams Prize topic is taken from Lewis Campbell and William Garnett, *The Life of James Clerk Maxwell* (London: Macmillan, 1882), p. 505.

98 – The assertion that Maxwell only took on the rings of Saturn as

an exercise in puzzle-solving is from Andrew Whitaker's contribution to Raymond Flood, Mark McCartney and Andrew Whitaker (eds.), *James Clerk Maxwell: Perspectives on his Life and Work* (Oxford: Oxford University Press, 2014), p. 116.

99 – Maxwell's letter describing his move from solid to liquid rings for Saturn is taken from Peter Harman (ed.), *The Scientific Letters and Papers of James Clerk Maxwell*, Vol. 1 (Cambridge: Cambridge University Press, 1990), p. 538.

101 – Maxwell's final statement on the nature of Saturn's rings is from James Clerk Maxwell, *On the Stability of the Motion of Saturn's Rings* (Cambridge: Macmillan and Company, 1859), p. 67.

105 – Maxwell's 'Song of the Atlantic Telegraph Company' is in a letter to Lewis Campbell from Ardhollow, Dunoon, dated 4 September 1857, quoted in Lewis Campbell and William Garnett, *The Life of James Clerk Maxwell* (London: Macmillan, 1882), p. 279.

105 – Faraday's letter to Maxwell on his papers is from Albemarle Street, London (location of the Royal Institution), dated 7 November 1857. It is quoted in Lewis Campbell and William Garnett, *The Life of James Clerk Maxwell* (London: Macmillan, 1882), p. 288.

106 – Rudolf Peierls' remark about waking a physicist in the middle of the night is from Peierls' contribution, 'Field Theory Since Maxwell', to Cyril Dombe (ed.), *Clerk Maxwell and Modern Science* (London: The Athlone Press, 1963), p. 26.

111 – Maxwell's citing of George Stokes' results on gas particles is from 'On the Dynamical Theory of Gases', presented at the 29th meeting of the British Association for the Advancement of Science, held at Aberdeen in September 1859, reproduced in Peter Harman (ed.), *The Scientific Letters and Papers of James Clerk Maxwell*, Vol. 1 (Cambridge: Cambridge University Press, 1990), pp. 615–16.

113 – Faraday's remark that Maxwell should be able to find his way through a crowd is noted in Lewis Campbell and William Garnett, *The Life of James Clerk Maxwell* (London: Macmillan, 1882), p. 319.

113 – Kirchhoff's remark that Maxwell's calculations needed checking is frequently quoted, though I could not find an original source. It is quoted in this form in Robyn Arianrhod, *Einstein's Heroes: Imagining the World through the Language of Mathematics* (Oxford: Oxford University Press, 2006) p. 94.

114 – The Marischal College reaction to the appointment of Daniel Dewar is noted in P.J. Anderson, *Fasti Academiae Marsicallanae Aberdonensis*, Vol. II (Aberdeen: New Spalding Club, 1898), p. 30.

114 – The suggestion that the book *Gaelic Astronomy* brought Maxwell and Principal Dewar of Marischal College together is made by John Read in his chapter 'Maxwell at Aberdeen' for Raymond Flood, Mark McCartney and

Andrew Whitaker (eds.), *James Clerk Maxwell: Perspectives on his Life and Work* (Oxford: Oxford University Press, 2014), p. 236.

115 – Maxwell's letter to Miss Cay from 129 Union Street, Aberdeen, dated 18 February 1858, is quoted in Lewis Campbell and William Garnett, *The Life of James Clerk Maxwell* (London: Macmillan, 1882), p. 303.

120 – The article reflecting on Tait getting the job at Edinburgh over Maxwell is from David Forfar and Chris Pritchard, *The Remarkable Story of Maxwell and Tait*, accessed on the Clerk Maxwell Foundation website available at: www.clerkmaxwellfoundation.org/Maxwell_and_TaitSMC24_1_2002.pdf

Demonic Interlude IV

121 – Maxwell's postcard to Peter Tait from London, dated 23 October 1871, featuring 'O T'! R. U. AT 'OME?' is quoted in Peter Harman (ed.), *The Scientific Letters and Papers of James Clerk Maxwell*, Vol. 2 (Cambridge: Cambridge University Press, 1995), p. 682.

123 – Thomson's statement that the second law made the end of the world inevitable is quoted in Stephen Brush, *The Kind of Motion We Call Heat: A History of Kinetic Theory in the 19th Century* (Amsterdam: North Holland, 1976), p. 569.

125 – Maxwell's comparison of the second law to pouring a tumbler of water into the sea is from a letter to John Strutt (Lord Rayleigh), noted in Robert Strutt, *John William Strutt: Third Baron Rayleigh* (London: Edward Arnold, 1924), pp. 47–8.

126 – Maxwell's first mention of the demon is in a letter to Peter Tait from Glenlair, dated 11 December 1867, reproduced in Peter Harman (ed.), *The Scientific Letters and Papers of James Clerk Maxwell*, Vol. 2 (Cambridge: Cambridge University Press, 1995), pp. 328–33.

127 – Maxwell's description of the demon as intelligent and exceedingly quick was in a letter to John William Strutt from Glenlair, dated 6 December 1870, reproduced in Peter Harman (ed.), *The Scientific Letters and Papers of James Clerk Maxwell*, Vol. 2 (Cambridge: Cambridge University Press, 1995), pp. 582–3.

127 – William Thomson's paper introducing the term 'demon' was William Thomson, 'The Kinetic Theory of the Dissipation of Energy', *Nature*, 9 (1874): 441–4.

Chapter 4

131 – The extract from Maxwell's inaugural lecture is taken from Peter Harman (ed.), *The Scientific Letters and Papers of James Clerk Maxwell*, Vol. 1 (Cambridge: Cambridge University Press, 1990), p. 671.

133 – The description of Maxwell borrowing books from the King's College library for his students is from Lewis Campbell and William Garnett, *The Life of James Clerk Maxwell* (London: Macmillan, 1882), p. 177.

137 – The description of the wet collodion photographic process is taken from Brian Clegg, *The Man Who Stopped Time* (Washington, DC: Joseph Henry Press, 2007), p. 34.

146 – Poincaré's observation of his contemporaries' mistrust for Maxwell's mechanical model is from John Heilbron, 'Lectures on the history of atomic physics 1900–1922', in *History of twentieth century physics: 57th Varenna International School of Physics, 'Enrico Fermi'* (New York: Academic Press, 1977), pp. 40–108.

146 – Maxwell's remarks on the benefit of analogy are taken from James Clerk Maxwell, 'Analogies in nature' (1856), in Peter Harman (ed.), *The Scientific Letters and Papers of James Clerk Maxwell*, Vol. 1 (Cambridge: Cambridge University Press, 1990) pp. 376–83.

148 – William Thomson's assertions that vortices existed in the magnetic field are made in William Thomson, 'Dynamical Illustrations of the Magnetic and Helicoidal Rotatory Effects of Transparent Bodies on Polarized Light', *Proceedings of the Royal Society* (1856), 8: 150–58.

149 – Maxwell's admission that the ball bearings in his model were 'awkward' is from William Davidson Niven (ed.), *The Scientific Papers of James Clerk Maxwell*, Vol. 1 (Cambridge: Cambridge University Press, 1890), p. 486.

Demonic Interlude V

151 – Richard Feynman's citing the development of Maxwell's electromagnetic theory as the most significant event of the nineteenth century is from Richard Feynman, *The Feynman Lectures on Physics, Vol. II – the new millennium edition* (New York: Hachette, 2015), section 1–11.*

152 – Maxwell's answers to Francis Galton's psychological questionnaire were accessed on the Clerk Maxwell Foundation website, available at: www.clerkmaxwellfoundation.org/FrancisGaltonQuestionnaire2007_10_26.pdf

154 – Maxwell's mini-biography for the demon sent to Peter Tait is from Cargill Gilston Knott, *Life and Work of Peter Guthrie Tait* (Cambridge: Cambridge University Press, 1911), pp. 214–15.

155 William Thomson's remark that the demon was, in effect, living is from William Thomson, 'The Kinetic Theory of the Dissipation of Energy', *Nature*, 9 (1874): 441–3.

155 – Richard Feynman's demonstration that a mechanical valve could

---

* The Feynman lectures don't have page numbers.

not do the work of a demon is in Richard Feynman, *The Feynman Lectures on Physics, Vol. II – the new millennium edition* (New York: Hachette, 2015) sections 46.1 to 46.9.

156 – The comment by James Jeans that the demon merely achieved what could happen over a longer timescale anyway is from James Jeans, *The Dynamical Theory of Gases* (Cambridge: Cambridge University Press, 1921), p. 183.

Chapter 5

162 – Albert Einstein's observation that the most fascinating subject as a student was Maxwell's theory is quoted in Albert Einstein, *Autobiographical Notes* (Illinois: Open Court, 1996), pp. 31–3.

165 – Maxwell's words on the inference that light was an electromagnetic wave are from his paper 'On Physical Lines of Force' in Peter Harman (ed.), *The Scientific Letters and Papers of James Clerk Maxwell*, Vol. 1 (Cambridge: Cambridge University Press, 1990), pp. 499–500.

166 – Maxwell's comment about his teaching duties being too heavy is from King's College London Archives, King's College Council, Vol. I, minute 42, 11 October 1861.

166 – Maxwell's reference for Smalley's application as Astronomer Royal for New South Wales is reproduced in Peter Harman (ed.), *The Scientific Letters and Papers of James Clerk Maxwell*, Vol. 2 (Cambridge: Cambridge University Press, 1995), p. 87.

167 – Maxwell's letter mentioning the noise of curling to Henry Droop, written from Glenlair on 28 December 1861, is reproduced in Peter Harman (ed.), *The Scientific Letters and Papers of James Clerk Maxwell*, Vol. 1 (Cambridge: Cambridge University Press, 1990), p. 703.

167 – Maxwell's letter to J.W. Cunningham on printing the examination papers in mechanics is reproduced in Peter Harman (ed.), *The Scientific Letters and Papers of James Clerk Maxwell*, Vol. 2 (Cambridge: Cambridge University Press, 1995), p. 61.

Chapter 6

171 – Maxwell's letter including his speculation on comets and the mechanism of gravity to John Phillips Bond from Glenlair and dated 25 August 1863 is reproduced in Peter Harman (ed.), *The Scientific Letters and Papers of James Clerk Maxwell*, Vol. 2 (Cambridge: Cambridge University Press, 1995), pp. 104–9.

174 – The details of Maxwell's viscosity experiment are from James Clerk Maxwell, 'The Bakerian Lecture: On the Viscosity or Internal Friction of Air and other Gases', *Proceedings of the Royal Society of London*, Vol. 15 (1866–67), pp. 14–17.

175 – The efforts the Maxwells went to in order to increase and decrease the temperature of the attic for the viscosity experiments are noted in Lewis Campbell and William Garnett, *The Life of James Clerk Maxwell* (London: Macmillan, 1882), p. 318.

178 – Maxwell's letter to Lewis Campbell describing seeing Wheatstone's Stereoscope, written from Edinburgh in October 1849, is reproduced in Peter Harman (ed.), *The Scientific Letters and Papers of James Clerk Maxwell*, Vol. 1 (Cambridge: Cambridge University Press, 1990), p. 119.

185 – Maxwell's notes on an electromagnetic explanation of reflection and refraction based on Jamin's theory are reproduced in Peter Harman (ed.), *The Scientific Letters and Papers of James Clerk Maxwell*, Vol. 2 (Cambridge: Cambridge University Press, 1995), p. 182.

185 – Maxwell's admission that he did not attempt to deal with reflection is quoted in Peter Harman (ed.), *The Scientific Letters and Papers of James Clerk Maxwell*, Vol. 3 (Cambridge: Cambridge University Press, 2002), p. 752.

188 – Maxwell's belfry analogy for the use of 'black box' mathematical models is in William Davidson Niven (ed.), *The Scientific Papers of James Clerk Maxwell*, Vol. 2 (Cambridge: Cambridge University Press, 1890), pp. 783–4.

191 – Michael Faraday's request that mathematical physicists also give a lay summary of their work is in a letter from Albemarle Street, London, dated 13 November 1857 and quoted in Lewis Campbell and William Garnett, *The Life of James Clerk Maxwell* (London: Macmillan, 1882), p. 290.

192 – Details of Michael Pupin's attempt to understand Maxwell's mathematical explanation of electromagnetism are from Freeman Dyson, 'Why is Maxwell's Theory so hard to understand?', *James Clerk Maxwell Commemorative Booklet* (Edinburgh: James Clerk Maxwell Foundation, 1999), pp. 6–11.

194 – Maxwell's humorous comment about Nabla and Nablody is in a letter to Lewis Campbell written on 19 October 1872, found in William Davidson Niven (ed.), *The Scientific Papers of James Clerk Maxwell*, Vol. 2 (Cambridge: Cambridge University Press, 1890), p. 760.

194 – Maxwell's suggestion of using 'space-variation' as a name for Nabla is in a letter to Peter Tait written on 1 December 1873, found in William Davidson Niven (ed.), *The Scientific Papers of James Clerk Maxwell*, Vol. 2 (Cambridge: Cambridge University Press, 1890), p. 945.

Demonic Interlude VI

199 – Einstein's assertion that Maxwell made the most profound change in the perception of reality since Newton is taken from Albert Einstein, 'Maxwell's influence on the development of the conception of physical

reality', in *James Clerk Maxwell: A Commemorative Volume 1831–1931* (Cambridge: Cambridge University Press, 1931), pp. 66–73.

201 – Leo Szilard's analysis of the demon's measurement and memory storage process is in his paper 'On the Decrease in Entropy in a Thermodynamic System by the Intervention of Intelligent Beings', in Bernard Feld and Gertrud Weiss (eds.), *The Collected Works of Leo Szilard: Scientific Papers* (Cambridge, MA: MIT Press, 1972), pp. 103–29.

201 – The suggestion that Szilard's consideration of the demon as part of the experiment was influenced by quantum theory comes from Andrew Whitaker's contribution to Raymond Flood, Mark McCartney and Andrew Whitaker (eds.), *James Clerk Maxwell: Perspectives on his Life and Work* (Oxford: Oxford University Press, 2014), p. 183.

201 – Bragg's remark about God running electromagnetics by the wave theory and the devil by quantum mechanics is quoted in Daniel Kevles, *The Physicists* (Harvard: Harvard University Press, 1977), p. 159.

Chapter 7

207 – Maxwell's fondness for entertaining children is described in Lewis Campbell and William Garnett, *The Life of James Clerk Maxwell* (London: Macmillan, 1882), p. 40.

215 – Norbert Wiener's reference to Maxwell as the father of modern automatic control is noted in Rodolphe Sepulchre, *Governors and Feedback Control*, available at: www.clerkmaxwellfoundation.org/Governors.pdf

217 – Maxwell's letter to Peter Tait, written from Glenlair on 7 November 1870, on the naming of the mathematical term that would be called 'curl', is reproduced in Peter Harman (ed.), *The Scientific Letters and Papers of James Clerk Maxwell*, Vol. 2 (Cambridge: Cambridge University Press, 1995), pp. 568–9.

217 – Maxwell's introduction of the mathematical term 'curl' is from James Clerk Maxwell, 'Remarks on the Mathematical Classification of Physical Quantities', *Proceedings of the London Mathematical Society*, s1–3 (1871), pp. 224–33.

218 – Maxwell's letter to Thomson from Glenlair dated 30 October 1868, citing the east wind as a reason not to apply for a post at St Andrews, is reproduced in Peter Harman (ed.), *The Scientific Letters and Papers of James Clerk Maxwell*, Vol. 2 (Cambridge: Cambridge University Press, 1995), pp. 457–9.

218 – Maxwell's letter to Campbell from Glenlair dated 3 November 1868, saying he would not stand, is reproduced in Peter Harman (ed.), *The Scientific Letters and Papers of James Clerk Maxwell*, Vol. 2 (Cambridge: Cambridge University Press, 1995), p. 460.

219 – Maxwell's letter to Thomson from Glenlair dated 9 November 1868

saying he would stand is reproduced in Peter Harman (ed.), *The Scientific Letters and Papers of James Clerk Maxwell*, Vol. 2 (Cambridge: Cambridge University Press, 1995), p. 463.

Chapter 8

222 – The details of the Grace of the Senate of the University of Cambridge of 9 February 1871 are reproduced in Lewis Campbell and William Garnett, *The Life of James Clerk Maxwell* (London: Macmillan, 1882), p. 350.

222 – Blore's letter to Maxwell telling him about the Cavendish Professorship is reproduced in Peter Harman (ed.), *The Scientific Letters and Papers of James Clerk Maxwell*, Vol. 2 (Cambridge: Cambridge University Press, 1995), p. 611.

223 – Maxwell's reply to Blore about the Cavendish Professorship is reproduced in Peter Harman (ed.), *The Scientific Letters and Papers of James Clerk Maxwell*, Vol. 2 (Cambridge: Cambridge University Press, 1995), p. 611.

225 – Lord Rayleigh's letter to Maxwell hoping that Maxwell will take the new professorship was written from Cambridge on 14 February 1871 and is quoted in Lewis Campbell and William Garnett, *The Life of James Clerk Maxwell* (London: Macmillan, 1882), p. 349.

225 – The initial quote from Maxwell's inaugural Cambridge lecture is taken from William Davidson Niven (ed.), *The Scientific Papers of James Clerk Maxwell*, Vol. 2 (Cambridge: Cambridge University Press, 1890), p. 250.

226 – The later quote from Maxwell's inaugural Cambridge lecture is taken from Lewis Campbell and William Garnett, *The Life of James Clerk Maxwell* (London: Macmillan, 1882), p. 356.

228 – George Stokes' comment about the difficulty of getting a house in Cambridge is reproduced in Peter Harman (ed.), *The Scientific Letters and Papers of James Clerk Maxwell*, Vol. 2 (Cambridge: Cambridge University Press, 1995), p. 615.

230 – Maxwell's comments on John Hunter in a letter to Peter Tait are recorded in Peter Harman (ed.), *The Scientific Letters and Papers of James Clerk Maxwell*, Vol. 2 (Cambridge: Cambridge University Press, 1995), p. 836.

231 – George Bettany's complaint of the difficulty imposed by the Mathematics Tripos on Natural Science at Cambridge is from George Bettany, 'Practical Science at Cambridge', *Nature*, 11 (1874): 132–3.

232 – Isaac Todhunter's 1873 dismissal of the need for students to see experiments is from Isaac Todhunter, *The Conflict of Studies* (Cambridge: Cambridge University Press, 2014), p. 17.

233 – The attempt by Corpus Christi College to sue the university because the Cavendish Laboratory blocked its light is described in R. Wills and J.W. Clerk, *The Architectural History of the University of Cambridge*, Vol. 3 (Cambridge: Cambridge University Press, 1886), p. 183.

234 – Maxwell's comment about moving around like a cuckoo without a fixed location to lecture is from William Davidson Niven (ed.), *The Scientific Papers of James Clerk Maxwell*, Vol. 2 (Cambridge: Cambridge University Press, 1890), p. 760.

234 – Maxwell's letter to Katherine written in London and dated 20 March 1871 about the targets of the laboratory is reproduced in Lewis Campbell and William Garnett, *The Life of James Clerk Maxwell* (London: Macmillan, 1882), p. 381.

234 – Maxwell's list of requirements for the Cavendish Laboratory is from a letter to William Thomson written at the Athenaeum Club on 21 March 1871 and reproduced in Peter Harman (ed.), *The Scientific Letters and Papers of James Clerk Maxwell*, Vol. 2 (Cambridge: Cambridge University Press, 1995), pp. 624–8.

236 – Trotter's letter to Maxwell on the benefits of giving researchers their own space is quoted in Isobel Falconer's chapter on 'Cambridge and the Building of the Cavendish Laboratory' in Raymond Flood, Mark McCartney and Andrew Whitaker (eds.), *James Clerk Maxwell: Perspectives on his Life and Work* (Oxford: Oxford University Press, 2014), p. 84.

237 – The *Nature* editorial comparing Cambridge unfavourably with continental universities was in 'A Voice from Cambridge', *Nature*, Vol. 8 (1873): 21.

240 – Maxwell's undated poem 'Lectures to Women on Physical Science I' is quoted in Lewis Campbell and William Garnett, *The Life of James Clerk Maxwell* (London: Macmillan, 1882), p. 631.

240 – Garnett's observation on Maxwell's reluctant acceptance of women on a Long Vacation course is quoted in Isobel Falconer's chapter on 'Cambridge and the Building of the Cavendish Laboratory' in Raymond Flood, Mark McCartney and Andrew Whitaker (eds.), *James Clerk Maxwell: Perspectives on his Life and Work* (Oxford: Oxford University Press, 2014), p. 86.

240 (footnote) – The *Telegraph*'s reaction to a woman coming top in the Mathematics Tripos is quoted in Caroline Series, 'And what became of the women?', *Mathematical Spectrum*, Vol. 30 (1997/8), pp. 49–52.

Demonic Interlude VII

246 – The paper apparently probing a quantum demon's mind is Nathanaël Cottet, Sébastien Jezouin, Landry Bretheau, Philippe Campagne-Ibarcq, Quentin Ficheux, Janet Anders, Alexia Auffèves, Rémi

Azouit, Pierre Rouchon and Benjamin Huard, 'Observing a quantum Maxwell demon at work', *Proceedings of the National Academy of Sciences*, 114(29) (2017), pp. 7561–64.

Chapter 9

247 – Arthur Schuster's quote from Maxwell on never dissuading a man from trying an experiment is from Arthur Schuster, *A History of the Cavendish Laboratory* (London: Longmans, Green and Co., 1910), p. 39.

247 – The review from *The Ironmonger* on Maxwell's *Theory of Heat* is reproduced in Isobel Falconer's chapter on 'Cambridge and the Building of the Cavendish Laboratory' in Raymond Flood, Mark McCartney and Andrew Whitaker (eds.), *James Clerk Maxwell: Perspectives on his Life and Work* (Oxford: Oxford University Press, 2014), p. 76.

251 – Maxwell's comment that if Cavendish's papers had been published earlier, electrical measurement would have been developed much earlier is in William Davidson Niven (ed.), *The Scientific Papers of James Clerk Maxwell*, Vol. 2 (Cambridge: Cambridge University Press, 1890), p. 539.

251 (footnote) – Maxwell's description of himself as dp/dt comes from the equation dp/dt = JCM which shows the change of pressure as temperature changes, with J being Joule's equivalent, C Carnot's function and M the rate at which heat must be supplied per unit increase of volume, the temperature being constant.

252 – Arthur Schuster's recollection about an American complaining of being subject to electrical shocks by Maxwell is from Arthur Schuster, *A History of the Cavendish Laboratory* (London: Longmans, Green and Co., 1910), p. 33.

252 – Maxwell's comments on units of inheritance are quoted in Lewis Campbell and William Garnett, *The Life of James Clerk Maxwell* (London: Macmillan, 1882), p. 390.

Demonic Interlude VIII

258 – The assertion that the erasure of information need not result in the increase of entropy comes from Meir Hemmo and Orly Shenker, *Maxwell's Demon* (Oxford: Oxford Handbooks Online, 2016), accessed at: http://www.oxfordhandbooks.com/view/10.1093/oxfordhb/9780199935314.001.0001/oxfordhb-9780199935314-e-63

259 – The Oxford University paper describing a photonic demon is Mihai Vidrighin, et al., 'Photonic Maxwell's Demon', *Physical Review Letters*, 116 (2016), p. 050401.

259 – The chloroform-based quantum demon is described in Patrice Camati, et al., 'Experimental rectification of entropy production by Maxwell's demon in a quantum system', *Physical Review Letters*, V. 117 (2016), p. 240502.

Chapter 10

264 – Einstein's claim to owe more to Maxwell than anyone is from Esther Salaman, 'A Talk with Einstein', *The Listener*, 8 September 1955, Vol. 54, pp. 370–71.

264 – Peter Tait's eulogy to Maxwell is from *Nature*, Vol. 21 (1880): 317–21.

264 – Charles Coulson's remark about Maxwell is from C. Dombe (ed.), *Clerk Maxwell and Modern Science: Six Commemorative Lectures* (London: Athlone Press, 1963), pp. 43–4.

265 – James Jeans' remarks about Maxwell's intuitive ability are from his entry, 'Clerk Maxwell's Method', in *James Clerk Maxwell: A Commemorative Volume 1831-1931* (Cambridge: Cambridge University Press, 1931), pp. 97–8.

266 – The quote from Maxwell's lecture on molecules is from Lewis Campbell and William Garnett, *The Life of James Clerk Maxwell* (London: Macmillan, 1882), p. 358.

Index

A

Aberdeen
    as city divided 89–92
    evening education 103–4
    JCM move to 75–9
    see also Marischal College
Aberdeen School of Science and Art
    103–4
Aberdeen, University of 119
absolute zero 174
action at a distance 48, 71, 162
Adams, John Couch 96
Adams, William Grylls 166, 198
Adams Prize 96–103, 215
aether *see* ether
Airy, George 96, 102, 166
Albert, Prince 117–18
amp 182, 184
analogy, power of 146–50
analytical engine 147
antigravity 139
antimatter 139
Apostles, the 52–3
Archimedes 180, 186
argon 224
Aristotle 82–3
atoms 81–5, 237
attraction
    electrostatic 158–9
    modelling 140–1

B

B.A. standard resistor 184
B.A. unit 183
Babbage, Charles 147
Bacon, Roger 38–9, 186
Baden-Powell, Robert 12
balloons 32, 35
Balmoral 118
Bartholin, Erasmus 23
belfry 188–9

Bernoulli, Daniel 108
Bettany, George 231–2
Big Bang theory 149, 266
billiard balls 208–9
binary systems 111
Birmingham 49
Blackburn, Hugh 10
blending 258
Blore, Edward 222–3, 224
Boltzmann, Ludwig 249
Bond, George Phillips 171–2, 173–4
books by JCM
    Elementary Treatise on Electricity
        238
    Theory of Heat 127, 206, 247
    Treatise on Electricity and
        Magnetism 206, 215, 226, 247–8
    A Treatise on Natural Philosophy
        205
Bragg, William 201, 241
Bragg, William Lawrence 241
Brewster, David 178
British Association for the
    Advancement of Science (BA)
    116–18, 120, 179–80, 266–7
Byers, Mrs 91

C

cable car 181
calculus 103
caloric theory 70, 85–6, 108
Calotype 178
Cambridge, New Museums Site 233
Cambridge, University of
    Department of Applied
        Mathematics and Theoretical
        Physics 231
    JCM inaugural lecture on return
        225
    leading university for physics 237
    religious requirements 90

"Yoga has benefited the lives of people for millennia. Its power to promote human health has now been verified scientifically. In this book, Suza Francina shows how anyone can reap the healing benefits of this ancient yet modern therapy."

Larry Dossey, M.D.
author, *Be Careful What You Pray For, Prayer Is Good Medicine* and *Healing Words*
executive editor, *Alternative Therapies in Health and Medicine*

"Few people are as qualified as Suza Francina to write about the life-enhancing benefits of yoga for those over 50. Ms. Francina's book confirms my own experience as well as inspires me to continue yoga with enthusiasm and confidence. What better gift can we give ourselves than the ability to move, breath and enjoy life, whatever our age."

Judith H. Lasater, Ph.D.
physical therapist, author, *Relax & Renew: Restful Yoga for Stressful Times*

"This is the book we've been waiting for! At last, we latecomers to yoga have a commonsense guide that initiates us in a way that is eminently "do-able." We're recommending it to our patients as a means of freeing both mind and body from the inner tension and outward rigidity that foster anxiety and depression."

John E. Nelson, M.D., and Andrea Nelson, Psy.D.
editors, *Sacred Sorrows, Embracing and Transforming Depression*

"The *New Yoga for People Over 50* will be my constant companion. In this crazy world that loves control, where most people are trying desperately just to cope, *The New Yoga for People Over 50* gives me the energy and flexibility to swim the laps, work at home and in the garden, be a grandma and laugh with my friends and family."

Malchia Olshan
swim champion, National Masters and Senior Olympics

"Studying with Vanda Scaravelli through her late 70s and 80s gave me a remarkable vision of aging. I hope Suza Francina's new book, filled with success stories about older teachers and students, will inspire many people of all ages to embark on the wonderful path of growth and transformation we call yoga."

Esther Myers
author, *Yoga and You: Energizing and Relaxing Yoga for New and Experienced Students*

"Master Iyengar yoga instructor Suza Francina has distilled a lifetime of teaching into a practical handbook that anyone can use to improve their health. I've strengthened my back muscles and increased my flexibility using her methods. I recommend this book to all, especially those who want to improve the quality of their life and health in midlife and beyond."

George Jaidar
author, *The Soul: An Owner's Manual, Discovering the Life of Fullness*

"I loved the people in this book! Their stories show how yoga can expand an older person's world, help them to improve their posture and health and bring them joy. They are an inspiration."

Noreen Morris, R.N., S.C.M.
hospice nurse and yoga student

The New Yoga
for People
Over 50

A Comprehensive
Guide for Midlife and
Older Beginners

Suza Francina

Health Communications, Inc.
Deerfield Beach, Florida

www.hci-online.com

We would like to acknowledge the publishers and individuals below for permission to reprint the following material:

Photographs on pages xix, 15, 30, 35, 36, 37, 40, 41, 43, 47, 55, 57, 69, 71, 75, 76, 79, 80, 86, 102, 104, 109, 125, 133, 151, 172, 174, 176, 177, 211, 215, 221, 227, 231, 234 and 235 reprinted with permission of Jim Jacobs. ©1997 Jim Jacobs.

Photographs on pages 54, 122, 123, 128, 129, 135, 138, 140, 143, 144 and 246 reprinted with permission of Paul Del Signore Photography. ©1996 Paul Del Signore.

Photographs on pages vii, 22 and 216 reprinted with permission of John Henebry. ©1997 John Henebry.

Photographs on pages xvii, xxii, xxvii, 17, 29, 38, 45, 49, 52, 66, 68, 72, 73, 101, 170, 171 and 222 reprinted with permission of Ron Seba. ©1993 Ron Seba.

Photographs on pages xxvii, 64 and 238 reprinted with permission of Benette Rottman. ©1993 Benette Rottman.

Photograph on page 207 reprinted with permission of Nils Larsen. ©1993 Nils Larsen.

(Continued on page 281)

Library of Congress Cataloging-in-Publication Data

Francina, Suza, (date)
    The new yoga for people over 50: a comprehensive guide for midlife and older beginners / Suza Francina.
        p.   cm.
        Includes bibliographical references.
        ISBN 1-55874-453-3 (pbk.: alk. paper)
        1. Yoga, Hatha. 2. Middle aged persons—Health and hygiene. 3. Aged—Health and hygiene. I. Title.
RA781.7.F7   1997   97-20984
613.7'046—dc21   CIP

Publisher: Health Communications, Inc.
        3201 S.W. 15th Street
        Deerfield Beach, Florida 33442-8190

Cover design by Andrea Perrine Brower
Cover photo of Betty Eiler by Jim Jacobs

*For my students
at the Ojai Yoga Center
and my children,
Bo and Monica Hebenstreit*

B.K.S. Iyengar at age 75.

*Yoga is an immortal art, science and philosophy.
It is the best subjective psycho-anatomy of mankind ever
conceived for the experience of physical, mental, intellectual, and
spiritual well-being. It has stood the test of time from the
beginning of civilization and it will remain supreme as a precise
psycho-physical science for centuries to come.*

—B.K.S. Iyengar, author of *Light on Yoga,*
Light on Pranayama and other treatises on yoga.
Born on December 14, 1918, Iyengar still performs a
rigorous daily yoga practice that would challenge
anyone at any age.

Contents

~

• Sitting Comfortably on the Floor • How Chairs Support Your Yoga
Practice • Yoga Sitting on a Chair • Chair Twist • Chair Forward Bend
• The Wall: Your Best At-Home Teacher • How to Practice Triangle Pose
at a Wall • Props Help You to Balance • Tree Pose • Using Doorways
to Improve Posture and Stretch Shoulders • Stretching Stiff Shoulders
with Straps or Belts • Shoulder Stretch Practice • Stretching Stiff Leg
Muscles with Straps or Belts • Lying-Down Leg Stretch with a Strap
• Improving Your Posture and Breathing Using Bolsters and Blankets
• How to Use a Bolster or Folded Blanket to Relax • Covering Your Eyes
Enhances Relaxation • Props to Relieve Stress on the Head and Neck
• The Backbending Bench • Back and Neck Pain • Exercising After Back
Injury or Surgery • B.K.S. Iyengar on Using Props • *Eric Small: Yoga for
Multiple Sclerosis and Other Problems Affecting Balance and Mobility*

4. Key Yoga Postures for Reversing the Aging Process 63
• Downward- and Upward-Facing Dog Pose • Downward-Facing
Dog Pose with a Chair • Upward-Facing Dog Pose with a Chair
• Downward-Facing Dog Pose from the Floor • Upward-Facing
Dog Pose • Other Variations of Downward-Facing Dog:
Downward-Facing Dog Hanging from a Rope or Strap • Restful
Downward-Facing Dog with Chair Variation • Cautions for Downward
Dog Pose Hanging from a Rope or Strap • Right-Angle Handstand with
a Wall or Other Yoga Prop • How to Practice Right-Angle Handstand
• Handstand in the Hallway or Doorway • *Ruth Barati: Feeling Vital
and Vibrant in Your 70s • Elizabeth T. Shuey, a student of Ruth Barati*

5. Yoga for Feet and Knees Over 50 85
• Healthy Feet and Knees Can Make the Difference Between
Walking and a Wheelchair • Revitalizing and Stretching Your Feet
• Fingers and Toes Entwind • Fingers and Toes Entwined While Seated
• Fingers and Toes Entwined Lying Down • More Ways to Stretch
the Toes • Foot Reflexology • Your Knees Need Yoga, Too • Practice the
Hero (or Heroine) Pose for Healthy Knees • How to Practice Hero Pose
• *Ruth K. Lain at 82: Yoga Opens Doors to New Friends and Interests*

Acknowledgments

First and foremost I wish to thank my students in the Yoga-Over-60 program at the Ojai Yoga Center, my living laboratory. Their sense of adventure and growth inspires me daily. I am also grateful to the students and teachers from all over the world who responded to my questionnaire and shared their stories, interviews and photographs. Their letters of support and proddings about when the book would be done have helped me to persevere. My special thanks go to Betty Eiler and her students for their enthusiastic contributions.

I also want to thank my partner at the Ojai Yoga Center, Judi Flannery-Lukas, who taught for me endlessly so I could finish this manuscript. Judi's trips to India to study at the Ramamani Iyengar Memorial Institute have illuminated our Yoga Center with "Light on Yoga."

Likewise, I want to express my appreciation to the teachers at The Iyengar Yoga Institute in San Francisco and to other teachers I have studied with over the years, especially Diana Clifton and Felicity Green. You opened my eyes and dispelled all my notions about aging, both by your personal example and the confidence with which you took students more than twice my age into Handstands at a time when this pose put the fear of God into me.

Thanks to B.K.S. Iyengar for permission to use his photos and quotes on yoga, health and aging. I was almost as impressed by his prompt response to my inquiry (living proof that he does not procrastinate like the rest of us mortals!) as I was with his inspiring demonstration of the yoga poses at age 75 during his visit to the U.S.

I hardly know how to thank photographer Jim Jacobs who appeared on the horizon at exactly the right moment. A serious yoga-over-50 practitioner himself, Jim has a special passion for photographing older people exhibiting beauty, grace and character. The photographs of his teacher, Ramanand Patel, and others enliven the pages of this book.

And then there are the many people who have helped me directly and indirectly over the years to complete this project—my agent, Barbara Neighbors Deal, for bringing the manuscript to the attention of my publisher; Karen Marita, for secretarial assistance and keeping my house and family afloat; Malchia Olshan and George Jaidar, surrogate parents who dried my tears and helped me to grow up; Barbara Uniker, Michael Callahan, Lyn Hebenstreit, Dale Hanson, Robert Houston, Chip Lukas, George and Jean Kalogridis, Julie Cross Hebenstreit and my parents, Mary and René Diets, for their invaluable support.

Thank you also, Judith Gustafson, for casting light on my computer, searching out typos and extra spaces, and for endless hours of all kinds of help. I deeply appreciate your friendship through thick and thin, for better and for worse.

I want to express my appreciation to everyone at Health Communications Inc., who helped assure that this book would see the light of day. Peter Vegso and Gary Seidler, HCI's president and vice president and avid yoga students, for believing in the project. Christine Belleris for discovering this book in her mountain of prospective manuscripts and guiding it through the editorial process. Andrea Perrine Brower for her cover design. Lawna Oldfield, for the text layout and design. And the rest of the talented HCI staff: Allison Janse, Kim Weiss, Ronni O'Brien, Kelly Johnson Maragni, Kathryn Butterfield and Randee Goldsmith.

Last, but not least, Karen McAuley, the Mr. Iyengar of editors—merciless, rigorous, unrelenting. Every day for many months as I sat chained bleary-eyed to the computer, pecking away with one hopeless finger, she would arrive on her bicycle, take one cheerful look at my rambling efforts, and insist then and there that I streamline and clarify my thoughts. Her daily presence as she scrutinized each page has been both a curse and a blessing. This book has been a true test of our friendship.

Introduction

Welcome to the New Yoga
for People Over 50!

Yoga is a gift for older people. One who studies yoga in the later years gains not only health and happiness, but also freshness of mind since yoga gives one a bright outlook on life. One can look forward to a satisfying, more healthful future rather than looking back into the past. With yoga, a new life begins, even if started later. Yoga is a rebirth which teaches one to face the rest of one's life happily, peacefully and courageously.

—Geeta S. Iyengar, *Yoga, A Gem for Women*

As preventive medicine,
yoga is unequaled.

—KAREEM ABDUL-JABBAR

Twenty years have passed since I wrote the first edition of *Yoga for People Over 50*. The time has gone by in the twinkling of an eye, and many of my students who began yoga after age 60 are now in their 80s. Recently, when one of my octogenarians remarked how challenging the "Over-60 Class" was for newcomers, they laughingly decided to call themselves the "Advanced Old Age Class." The members of this class are the most dedicated and have the best attendance record of all my students because, they've told me, "We don't dare miss! We don't dare quit!"

When new students drop in on the class, thinking it's going to be easy, the experienced students seem to get a thrill out of staying in Downward-Facing Dog long after the gaping novice collapses. The octogenarians especially enjoy demonstrating the Right-Angle Handstand, a pose requiring upper-body strength, flexibility and a little bit of daring. It's all in good fun, of course, and generally inspires new people to start practicing.

You are as young as
your spine is flexible.
—*Source Unknown*

I see the biggest changes among my students who come regularly two or three times a week for several years. It is a source of inspiration and encouragement to see someone over 60 practicing poses that people of any age may find difficult when starting out. I know how impressed I was by my teachers who were more than 40 years older and far more supple than I was.

It's one thing to be under 20 and practicing all these wonderful, exotic-looking poses, and people can just shrug and say, "Well, she's still young. Her body can do that naturally." But when you see older students and teachers practicing difficult poses with relative ease, poses that your poor stiff body perhaps half their age can only dream about, then you start to wake up and realize that *yoga really works!*

Right-Angle Handstand.

Right-Angle Handstand at the wall with extra support under the hands.

At every stage of spiritual growth, the greatest ally you have is your body.
—*Deepak Chopra, M.D., Ageless Body, Timeless Mind: The Quantum Alternative to Growing Old*

You're never too old to do yoga. On the contrary, you're too old *not* to do yoga! No segment of our population can benefit more from yoga than people over 50. In fact, the older you are, the more you will benefit from practicing this ancient science and healing art. Yoga goes against the grain by removing the stiffness and inertia from the body. "It takes time to take care of your self," I remind my students, "but it takes even more time if you don't take care of your self." Yoga improves the quality of life for people of all ages, but most especially for those who are older.

My teaching experience confirms what Dr. Deepak Chopra and other experts on aging have written: that older people in particular need a conscious, non-mechanical, intelligent, expansive approach to exercise that involves the whole person—body, mind and spirit. The fragmented geriatric exercise and rehabilitation programs commonly prescribed are of limited benefit to seniors. Yoga's preventive and rehabilitative gifts are becoming more widely known; even more important is yoga's timeless wisdom that aging can bring greater perspective and illumination, expanded awareness and continuing growth, rather than deterioration. This underlying philosophy can help an aging population bring balance to our culture's obsession with the superficial trappings of youth.

How I Began Teaching Yoga

Like many of my generation, I began practicing yoga from books in the late 1960s and early 1970s. My first teachers were older men and women who were considered, back then, eccentric senior citizen types—

that is, they were pioneers far ahead of their time. In the first edition of my book, published in 1977, I described the pleasant agony my body experienced while attempting to mimic instructors more than twice my age. During the first yoga class I ever attended, I was situated between two students in their mid-60s. They were bending and stretching like youngsters while I, not half their age, felt as if my poor, stiff back was stuck in cement. I didn't feel too kindly toward the flexible 60-year-old teacher who told me outright that "if we grow old and stiff, we have no one to blame but ourselves."

While I was starting yoga, I was working as a home health-care provider for convalescents and the elderly, something I had done since my teens. I befriended and cared for many of these same people until they died, some of them for as long as 25 years. This gave me the opportunity to observe firsthand the mental and physical changes that often occur in the later years. The contrast between the elderly people I cared for and the seemingly ageless yoga practitioners I met was striking. The people who practiced yoga showed almost no evidence of such common problems as arthritis, bone fractures caused by osteoporosis, heart conditions, breathing problems, incontinence, confusion, memory loss and other problems mistakenly blamed on aging.

When I first began teaching yoga at a local senior center, like most young, enthusiastic but ignorant exercise instructors, I thought that older people were very stiff and brittle and that they would break if I asked them to bend. I automatically relegated people with white hair to armchair exercises or gentle stretches on the floor. Gradually, as I gained experience and confidence, I saw that as with

Yoga is for everyone, at any age. You can begin at any time of your life. If you are over 65 and an absolute beginner, welcome! There is no better time to start than this moment!
—*Lilias Folan, author of Lilias, Yoga and Your Life*

any age group, older people come into a yoga class with various levels of ability and medical histories.

While both the frail elderly and late-life yoga students with severe balance problems may initially benefit and gain confidence by practicing modified yoga postures sitting in a chair, practicing in this way can be counterproductive to the goal of keeping older students independent and out of a wheelchair. In 20 years of teaching yoga to older beginners, I've learned that most

Variation.

Sandy Yost, in her 70s, enjoys a Handstand.

can benefit from the same vital weight-bearing postures that are taught in my regular classes, while those with medical problems such as heart disease, high blood pressure, osteoporosis, arthritis and other concerns discussed in this book can practice at a slower pace and with modified poses.

People of all ages generally start yoga to stretch the "kinks" out of their body, to strengthen their bones and muscles, to improve their posture, to breathe better, to relax and improve their overall health and vitality. Older students who attend class regularly for at least six months report that their increased strength and range of movement enables them to return to physical activities they thought they had lost forever: gardening, climbing uphill or climbing stairs, biking, dancing, reaching and bending without strain, being able to sit comfortably on the floor in various positions, and getting up and off the floor with confidence.

My older students and teachers are pioneers—positive role models who demonstrate that yoga reverses the aging process and allows us to enjoy our bodies well into our real old age.

What You Will Find in This Book

Throughout this book you will find inspiring photographs, quotes and personal stories of teachers and students from all over the world, many in their 70s and older, who demonstrate that age is irrelevant when it comes to practicing and enjoying the benefits of yoga. B.K.S. Iyengar, born in 1918, still performs a rigorous daily yoga practice. Also featured are Diana Clifton, a highly respected teacher in her mid-70s; Vanda Scaravelli, born in 1908, and author of *Awakening the Spine;* and Indra Devi, who, when I interviewed her at age 94, was still traveling the world, lecturing on the benefits of yoga. Many other practitioners who began yoga after midlife, and who overcame various health problems associated with aging, appear throughout the text.

The beginning chapters of this book explore the changing view of aging and yoga's well-known reputation for slowing down and reversing the premature aging process. You will learn how the health of your spine and posture affects every system of the body and how yoga postures and breathing exercises affect the circulatory system, the heart and other vital organs. Physicians and physiologists have long recognized that reversing gravity, by inverting parts or all of the body, has a beneficial effect on the circulation, lungs and brain. After 50 it becomes increasingly important to reverse the downward pull of gravity, and the benefits and precautions of turning the body halfway and completely upside down are explained.

The New Yoga for People Over 50 also explores yoga's special benefits for women during the menopausal years, how yoga strengthens our bones and helps prevent osteoporosis and arthritis, yoga's role in preventing and reversing heart disease, and other common health concerns of those of us at midlife and older.

The practice guides that appear throughout the book demonstrate how yoga props—bolsters, straps, chairs, backbending benches, walls and wall ropes—enable older students to experience the benefits of difficult or challenging poses safely. Yoga props help teach the principles of correct body alignment and correct movement, and work to improve strength, flexibility, balance and endurance. Props also provide support for the practice of many deeply relaxing, rejuvenating, restorative postures.

The practice guides focus on the special needs of middle-aged and older people who have begun to feel the "wearing down" effect of other forms of exercise, or who may be coming to yoga with various health problems. Directions are geared to beginners well over 50 and demonstrate how the more challenging yoga postures, which are whole-body movements, can more easily be learned in their component parts. In the beginning, many of the more difficult positions can be practiced lying on the floor or with the support of a wall or chair, and stretching each part of the body individually, removing the stiffness from the body one joint at a time.

Despite the fact that people over 50 can be in better physical condition than those half their age, professionally prescribed home exercise routines, specifically designed for the "geriatric adult" or "geriatric segment of the population," still tend to relegate the older practitioner to easy generic exercises done on the floor, in a chair or wheelchair, bed or bathtub. This book describes and illustrates how older beginners can adapt the practice of yoga to their special needs, including medical considerations such as cardiovascular disease, arthritis, hip surgery and osteoporosis, and also how they can safely progress.

The practice guides were tested in class on students who started practicing yoga after the age of 70, 80 and older, and demonstrate how even those beginners with difficulty getting up from the floor can progress from gentle floor and chair stretching to vital weight-bearing standing and inverted postures. Beginners will gain confidence practicing at home and in a class setting.

Use Common Sense!

My purpose in featuring older yoga practitioners is to encourage readers of all ages to exercise in a way that will keep their bodies healthy for a lifetime. I wish to remind readers that the more advanced postures, all demonstrated by people well over 50, are best learned under the guidance of an experienced and qualified instructor.

I especially want to caution readers not to strain their knees by attempting prematurely to sit in the Lotus posture or other more advanced bent-knee seated positions. Over the years, many overzealous Westerners have learned the hard way that the knee is a most unforgiving joint. If your knees have become stiff from years of sitting in a chair, refer to the practice guides in chapters 3, 5 and 8 for safe stretches to remove the stiffness from your hips and knees.

The directions for practicing yoga in this book are based on an approach developed by one of the most influential yoga teachers of this century, B.K.S. Iyengar. Iyengar yoga is known for its emphasis on developing strength, flexibility, endurance and correct body alignment. Standing poses unique to Iyengar yoga strengthen the whole body and are known to relieve many common back and neck problems. They also help to correct foot, ankle and knee problems, which are almost epidemic in our culture in the over-50 population. You will find further information in chapter 5 on yoga for feet and knees.

I hope that this book will bring yoga to the attention of those who are ready to reject the notion that aging is a process of decline and who will encourage many other people to explore yoga's ascending path to physical and spiritual transformation.

Suza Francina helps a student stretch her legs.

No book can ever replace a teacher who directs, adjusts, encourages and inspires you. It is my dream that *The New Yoga for People Over 50* will motivate many more people to attend their local yoga classes.

Frank White:
Yoga, A Better Way to Spend the
Autumn and Winter Years

Frank White, 76, demonstrates an advanced balance pose. He was honored in 1993 as Yoga Teacher of the Year.

I am in better shape now than when I was 40. I can move better. I'm more flexible. I'm stronger, with more stamina. Yoga absolutely not only retards, but reverses the aging process.

I was in my 66th year of life, and it was becoming increasingly difficult to get out of bed in the morning. Stiffness and pain in my spine, legs and joints were keeping me from doing all the wonderful things that my body was capable of doing as a younger person. Something was gradually taking away my freedom of movement.

The doctors called it "severe degenerative osteoarthritis"— a very fancy name for a crippling disease. I just called it "plain misery." Depressed and anxious, the spark and zest for life that I once had were rapidly diminishing.

I had been under the care of a cardiologist for about 15 years for hypertensive cardiovascular disease (high blood pressure) and arteriosclerosis. I took medication daily. Not a pleasant way to spend the autumn and winter years of my life.

Somehow I found myself in a yoga class. Not only did I know that I was home, but I realized that a miraculous, almost magical transformation was possible. When I finished the class, I was weeping. Something released inside my body that said, "YES!"

From that day until the present—and I am now 76 years old—I have been doing yoga on a daily basis. I became a vegetarian and, as my life transformed, so did my health. I lost 50 pounds. My blood pressure returned to normal without medication. My cholesterol dropped from 400 to 150.

I became a yoga instructor and am devoting my life to teaching this ancient healing art to everyone. I now teach about 12 to 15 classes a week, to people of all ages: youngsters and senior citizens, people with arthritis, osteoporosis, back problems, etc.

The physical benefits of yoga are indeed remarkable. People who are young stay younger longer, older people become younger. To see this happening before my very eyes is a joy to behold.

I am no longer using any medication for past physical problems. Arthritis no longer holds fear in me: arteriosclerosis is no longer a threat. My flexibility for a man of my years is, I believe, remarkable. I can glide through a vigorous yoga class. My whole life has changed. I function more and more as I did when I was a much younger person.

One

Our Changing View of Aging: How Expectations Determine Outcome

We need to change our idea of what aging is. If I know my biological potential is 130 years, then I don't consider myself middle-aged until I'm 65. . . . One of the great principles of mind/body medicine is that expectancies determine outcome. If you expect to remain strong in old age, you will.

—Deepak Chopra M.D., *Ageless Body, Timeless Mind*

As a society and as individuals, we can expect that our notions of aging will continue to change dramatically in the years ahead. Leading pioneers in the field of mind/body medicine such as Deepak Chopra, M.D., an endocrinologist, bestselling author and internationally recognized authority on how our consciousness affects our health, urge us to consider the power that our beliefs about aging have over us. The latest research shows that how we age has more to do with our belief system and mindset about aging than any other factor.

In the last several decades, gerontologists have proved that remaining active throughout life halts the loss of muscle and skeletal tissue. The news is spreading among older people that they should continue all the activities they enjoyed in earlier years—walking, hiking, bicycling, gardening, golf, tennis, karate, swimming, dancing, yoga—you name it. Not long ago, a wild, 100-year-old daredevil named S.L. Potter, defying age, common sense and the fears of his physician and children, made his first bungee jump from a 210-foot tower. Further evidence that we are redefining what is appropriate in old age were photographs in the news of two of America's oldest sisters, Sarah and Elizabeth Delany, then ages 102 and 104—one practicing Shoulderstand, the other stretching in a yoga pose with one foot behind her head.

What happens when we change our old expectations about aging? Gerontologists from Tufts University found out when they put a group of the frailest nursing home residents, ages 87 to 96, on a weight-training regimen. Traditionally doctors believed that this type of elderly person belonged in bed, in a rocker or wheelchair out on the porch or in front of the television. Exercise would exhaust or kill these fragile people. Instead, they thrived. Within eight weeks muscle tone improved by 300 percent. Coordination and balance improved as well. Most important, these elderly people's confidence in being active returned. Some of them who had not been able to walk unassisted could now get up and go to the bathroom

Advanced age is not a static, irreversible biological condition of unwavering decrepitude. Rather, it's a dynamic state that, in most people, can be changed for the better no matter how many years they've lived or neglected their body in the past—Yes, you do have a second chance to right the wrongs you've committed against your body. Your body can be rejuvenated. You can regain vigor, vitality, muscular strength, and aerobic endurance you thought were gone forever . . . this is possible, whether you're middle-aged or pushing 80. The 'markers' of biological aging can be more than altered: in the case of specific physiological functions, they can actually be reversed.

—*William Evans,*
Biomarkers, The 10 Keys
to Prolonging Vitality

by themselves—an act of reclaimed dignity and independence that cannot be underestimated.

With Yoga, the Body Remains Open and Flexible

The accepted view of the aging process has been one
of stiffening, rigidity and closing down. Without proper
exercise, the body contracts and we lose height, strength
and flexibility. As a result, our natural free range of
motion is restricted so daily activities become difficult
and in some cases impossible. Yoga exercises reverse the
aging process by moving each joint in the body through
its full range of motion—stretching, strengthening and
balancing each part. Most popular forms of weight bearing exercise contract muscles and tighten the musculoskeletal system, adding to the stiffness that normally
settles into the body with the passage of time. In our
youth-oriented culture, obsessed with thinness, we
tighten the muscles to make the body look firmer. What
is much more important, however, especially as we grow
older, is opening and expanding the body so that the
aging process is tempered. Insurance companies, gerontologists, cardiologists, senior exercise physiologists and
other health care professionals interested in preventing
chronic illness and disability in the older population are
becoming increasingly interested in what yoga has to
offer older people.

Chronological vs. Biological Age

According to *Age Wave* authors Ken Dychtwald and
Joe Flowers, 100 million Americans are changing what

There is no age limit, one can start yoga when 70 or 80 years old and no damage will occur if the movement comes from the spine. People feel elated and it gives them comfort and encouragement to discover that it is possible for them to control and modify their bodies. To talk about old age as an impediment is an excuse to be lazy.
—*Vanda Scaravelli, born in 1908, author of* Awakening the Spine

If we had a culture that nurtured the intelligence implicit in the blood and bone from infancy on, we wouldn't need remedial efforts and there's no telling how far we might advance. Many people have made big changes in later life by learning how to honor the wisdom of their bodies.
—*Gloria Steinem, who took up weight-training and yoga after age 50, author of* Revolution from Within: A Book of Self-Esteem

How do you define normal? There are people who are biologically 40 and chronologically 80. There are other people who are 20 and their biology reflects the age of 50 or 60. . . . What we call normal aging is the psychopathology of the average. We shouldn't confuse average with normal because there are so many exceptions.

—*Deepak Chopra, M.D., Ageless Body, Timeless Mind*

we think about aging. Today, there are more older people alive than ever before, and increasing numbers of them are over 85. As the population grows older, we are discovering that chronological age is only one way to measure aging, and perhaps the least accurate one. Chronological age matches what physiologists refer to as our "biological age" only when we are very young. As the years pass by, biological time slows down; the older you are the more slowly you age. By the time we enter midlife, biological age becomes less significant than qualities like zest, energy, enthusiasm, confidence and contentment. In over 20 years of teaching yoga to people of all ages, shapes, sizes and temperaments, I've seen wide variations in the ways human beings change.

In the 1970s, the American Medical Association's Committee on Aging concluded a 10-year study by declaring that it had not found a single physical or mental condition that could be directly attributed to the passage of time. Stress- and diet-related conditions such as high blood pressure, heart disease, arthritis, osteoporosis, loss of muscle strength, reduction in motor fitness (i.e., balance, flexibility, agility, power and reaction time), reduction in respiratory reserves (breathing capacity), constipation and diseases associated with elimination, diabetes, sleep disorders and depression are experienced by young and old alike.

We now know that many of the classic symptoms of aging are caused by inactivity or the wrong type of activity (i.e., mechanical, imbalanced forms of exercise that strain the body), inadequate nutrition and accumulated stress and tension. Even such common outer symptoms of aging—poor posture, rounded shoulders, dowager's hump, closed chest, stiffness and loss of mobility—originate when we are younger and become increasingly pronounced as the years go by.

Health and Aging

Our aging population is increasingly concerned about the availability and affordability of adequate health care. A growing number of people are realizing that what we refer to as our "Health-Care System," which consists primarily of medical doctors, surgeons, drugs and hospitals, should be more accurately labeled as our "Ill-Health-Care System."

Medical services have become very efficient at diagnosing what is wrong with older people. Physician services and the advent of antibiotics have given doctors excellent control of infectious diseases. Hospital and surgical services for illness and injuries have become so proficient that many who would have died at 50, 60 or 70 now live longer.

Many people assume that because the elderly are living longer, they are in good health. Certainly a small percentage of them are, but many aging people suffer from serious and disabling health problems. A closer look at older people who have been saved by modern medicine finds that a high percentage of them are suffering from degenerative diseases. Arthritis, osteoporosis, heart disease, chronic fatigue, cataracts, macular (visual) degeneration, diabetes and cancer are common. While traditional medicine offers drugs, surgery and other methods to treat degenerative diseases, it does not offer the means to prevent them.

Statistics from the U.S. Department of Health and Human Services demonstrate that Americans are not the healthiest people in the world. For example:

Instead of going about the business of healing this sick society, we're lulling it to sleep with services. We need to see services for the aged for what they really are: Novocain. They're not really changing anything. They are simply dulling the pain of loss and deprivation, alienation and frustration and despair, much of what we call senility, and confusion is not organic brain damage but induced by frustration, despair, a sense of loss of purpose and one's role in life.
—Maggie Kuhn, 1970s activist for the elderly

The National Academy of Sciences has reported that if institutionalization could be postponed by just one month, it would save $3 billion in Medicare and Medicaid. And that doesn't include the savings in dignity and independence for elderly people.

—*William Evans, Biomarkers. The 10 Keys to Prolonging Vitality.*

- Men in 22 countries have a longer life expectancy
- American women rank seventh in longevity
- Our adult mortality rate is eleventh in the world
- Dental disease affects 98 percent of people of all ages
- Cancer strikes 930,000 Americans annually
- There are 440,000 cancer deaths per year
- Each year 19 million people get heart disease and 750,000 die from it (37 percent of all deaths)
- High blood pressure affects 28 million people
- Arthritis affects 30 million; bursitis affects another 30 million
- Five to 8 million people have asthma, cataracts, diabetes or migraines
- One million people have kidney stones

Eight prescriptions are written each year for every person in the United States, and even more for the elderly. Older people have been known to take as many as 30 different medications during the last year of their life. Each year in the U.S., medication problems are the cause of more than 250,000 hospitalizations for people 65 and older.

People are no longer content to be shuffled from doctor to doctor, only to be told that their illness, pain or degeneration is normal for their age, and to go home, take their medication and get plenty of rest. Our present primary health-care delivery system, which until recently has ignored prevention and health, is failing us at a cost of billions. This is the reason "alternative" health systems are now becoming mainstream. Chiropractic, homeopathy, herbs, acupuncture, massage, nutrition, Ayurvedic medicine, yoga and others are coming into their own as people share positive experiences with these health systems. These more holistic therapies are

Vanda Scaravelli, in her 80s, is a testament to the benefit of yoga.

safe and generally offer a less expensive means of help-ing people suffering from many acute and chronic con-ditions. They are restorative in nature, help to recharge and revitalize the body's precious stores of energy, and thus help to prevent illness, disease and degeneration.

Yoga offers a uniquely holistic approach to health. This ancient science, with its deep roots in Ayurvedic medi-cine, is truly the most complete system of self-health care that exists. The best part of it is that it is something you do to yourself, for yourself. It is an active, rather than a passive, approach to keeping yourself healthy and fit.

As I grow older, it is liberating and exhilarating to feel more freedom of movement, openness and exten-sion in my joints and muscles than back in my younger years. As a teacher, I find it a continual revelation to see how the bodies of people of all ages respond to yoga and proper exercise.

Ayurveda, "the science of life" or "knowledge of life span" in Sanskrit, is a 5,000 year-old system of mind/body medicine.

Betty Eiler: After 50, Include Yoga in Your Lifestyle

Betty Eiler is a yoga teacher who began practicing yoga when she was almost 50.

My general health, well-being and joy in life have increased immeasurably. It seems to me that the most important thing that seniors can do to enhance their lives is to change their lifestyle to include yoga with experienced and qualified instructors. To have it work for each of us, we must DO IT frequently.

I have had many inspiring experiences working with students over the age of 50. They involve all aspects of healing, including reduction of rheumatoid arthritis; recovery from frequent headaches, numbness in hands; backaches; and remission of multiple sclerosis.

Those who have the discipline to stay for six months or more report their increased feeling of well-being and enthusiasm for life. Their range of movement increases dramatically, enabling many to return to physical activities they thought were lost forever: gardening, climbing uphill, biking, reaching and bending, and simply sitting comfortably on the floor.

The wonderful progress people in this age range make has inspired me to start taking "before" pictures of the ones I believe will make a commitment to yoga. After six months or so, we can document their progress with "after" pictures. The changes might be dramatic or subtle. People often forget how they felt or looked pre-yoga.

As for myself, I began yoga approximately 13 years ago, and some of the changes I have experienced are:

- My mild scoliosis disappeared.
- My shoe size slowly moved from a 6½ N to now an 8½ N or B, and the flat footprints of my childhood are now footprints with arches.
- My seated poses, such as sitting with the soles of my feet together *(Baddha Konasana)*, were once with rounded back and knees very elevated. My knees now touch the floor without wall support (on a stretched-out day).
- I am comfortable doing the Lotus in Headstand *(Pindasana)*.
- At age 52, for the first time in my life, I did the Splits *(Hanumasana)*, and at age 55 I did a mid-room, Full Arm Balance *(Adho Mukha Vrksasana)*, dropping back to the Upward Bow *(Urdhva Dhanurasana)*.

For people over 50 thinking about starting yoga I say, "Try it, you'll like it." Live life with excitement and enthusiasm and do something different. You will be amazed how the body responds. I find a new awareness in every class I attend —a subtle opening that I didn't have the last time I did a certain posture, a reaching for the ceiling and feeling that I'm getting there. I have become more centered and have a greater amount of stamina and physical strength. This, too, is because of the guidance of the Inner Spirit.

—Marleen Burrow, a student of Betty Eiler

Two

How Yoga Slows Down and Reverses the Aging Process

Yoga deals with the most profound of mysteries, the essential nature of the human being in relation to the universe. The meaning of Yoga is union or yoking, from the Sanskrit "yug," to unite. In the context of yoga philosophy, the union is between the individual soul and universal soul. The individual has to search for the divine within, and Yoga provides the systematic steps to achieve this.

—SILVA MEHTA, *YOGA: THE IYENGAR WAY*

Yoga: The Ideal Health System for People Over 50

Jim Jacobs, in his 50s, practices Lotus Pose in Shoulderstand.

It's interesting to me how yoga is becoming incredibly popular. . . . We are really living in a very complex time—a time of great turmoil and change. The more irrational of us are worried about the millennium ending—as if a date would really matter. Yoga is a good antidote for all that. Yoga will take us out of all this historical paranoia. It's a long haul we're in.

—*Sting (Interview with Ganga White, Yoga Journal, November/December 1995)*

Yoga philosophy has appealed to great thinkers for centuries and today, as we head into the 21st century, it is evident that yoga is becoming more and more a part of our lifestyle. Millions of Americans from all walks of life practice yoga, and the popularity of this ageless, timeless, holistic health system will flourish even more as modern medicine rediscovers and documents its value.

Dean Ornish, M.D., author of *Dr. Dean Ornish's Program for Reversing Heart Disease*, and other renowned medical specialists recommend yoga as a key part of a program for preventing and reversing heart disease. Most of today's stress management techniques have their roots in yoga. Hospitals throughout the country are using yoga and meditation to help patients suffering from chronic pain and stress-related medical disorders. Doctors at Cedars-Sinai Medical Center in Los Angeles are so certain of yoga's health benefits that it is a key part of their program for people who have had heart attacks. C. Noel Bairey Merz, a cardiologist there, states emphatically, "The bottom line is that yoga is an exercise that is good for the heart."

If you are new to yoga, let me assure you that whatever your age or physical condition and whatever your religious belief or cultural heritage, the practice of yoga's stretching and strengthening exercises and the breathing and relaxation techniques can help you to vastly improve the quality of your life and health.

The word "healing" comes from the root, "to make whole." The word "yoga" comes from the Sanskrit, meaning to yoke or discipline, to unite, to make whole. Because true health involves body, mind and spirit, yoga incorporates physiological, psychological and spiritual processes. There are various systems of yoga, and each provides different ways of unifying the physical, mental, emotional and spiritual aspects of a human being.

Yoga is not a religion. It is a nonsectarian method for promoting a healthy and harmonious lifestyle. Any person of any faith can practice yoga and find his or her religion enhanced as a result.

What Is Hatha Yoga?

This book focuses on hatha yoga—a physical discipline that explores the intimate connection between the body, mind and spirit. The Sanskrit word *ha* means "sun" and *tha* means "moon." The goal of hatha yoga is

When I go into gyms and health clubs, I see men with rounded shoulders use a weight machine and do 50 lifts that tend to round the shoulders and three lifts that tend to open the chest. The lifts that open the chest are hard for them to do and, hence, are not as pleasing psychologically; so they don't do many. But what they don't know is that the pleasing lifts are reinforcing all the negative patterns about their bodies: rounding their shoulders, collapsing their lungs, curtailing their breathing. What they need is somebody to show them how to open their bodies to oxygen, work with alignment, and exercise for health.

—*Sue Luby, fitness expert, yoga teacher and author of* Bodysense: The Hazard-Free Fitness Program for Men and Women to balance and unify the positive and negative energy flows (life forces) within the body. Using the flow of the breath and the internal flow of these energies, yoga helps us to realize our potential for health and self-healing.

All of the approaches to hatha yoga involve the practice of various movements and postures. In general, they consist of forward bends, backbends, twists, inversions, standing and balancing poses as well as relaxation and breathing techniques. These movements and postures, along with conscious use of the breath, remove stiffness and tension from the body, restore vitality, strength and stamina, and improve balance and coordination; they also promote the efficiency of the body processes of digestion, assimilation and detoxification.

Older people, in particular, need a conscious, intelligent, expansive and non-mechanical approach to exercise that involves the whole person—body, mind and spirit. At the physical level, the practice of yoga strengthens and balances all the systems of the body; yoga also offers relief from the various ailments mistakenly attributed to aging. From the mental or psychological viewpoint, yoga improves concentration, calms and steadies the emotions and helps the practitioner to see life with greater perspective. In the realm of the spiritual, yoga expands awareness and teaches the mind and body to be quiet and at peace. Yoga has a unique place in our search for health because it offers a practical technique for experiencing unity and harmony of every aspect of our being.

According to yogic tradition, the years after 50 are the ideal time for psychological and spiritual growth. The practice of yoga not only restores the health and vitality of the body, but the philosophy behind yoga aims to open and expand a human being on all levels so that aging can become a time of greater perspective and illumination, rather than deterioration.

Yoga Prevents and Corrects the Most Visible Symptom of Aging— the Rounding of the Spine

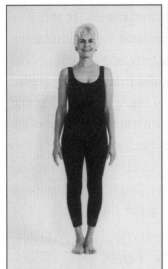  

Yoga prevents and can even reverse the most visible and obvious symptom of aging—one that cannot be disguised or transformed cosmetically—the shortening and rounding of the spine. In our culture, where people spend years hunched over a desk or steering wheel, or engaged in other activities that tend to pull our upper body forward, a rounded upper back, forward head and collapsed chest are so prevalent among people over the age of 50 that we almost consider it a normal part of aging.

Over the course of a "normal" lifetime, the spine degenerates and the body becomes shorter. When the back becomes rounded, it compresses the chest, which

Women in particular need to understand how something so simple as physical posture can undermine, or enhance, self-confidence. How we stand says something not just about us, but to us. . . . Body-mind communication is a two-way street; mind/change can begin with body-change. Wherever negative physical imagery has been part of low self-esteem, a counterpoint of positive imagery can be part of raising it.

—*Gloria Steinem, Revolution From Within: A Book of Self-Esteem*

causes shallow breathing. By limiting the amount of oxygen the cells receive, this 'old age posture' contributes to cardiovascular and other health problems. Yoga counteracts and reverses all of this.

Poor posture and the degeneration of the spinal column affect the health of every system of the body. Not only do a rounded back and collapsed chest restrict breathing, but they interfere with the vital flow of blood and of nerve impulses to the internal organs. In this way poor posture interferes with digestion and elimination. Maintaining the health and integrity of the spine is the central theme of yoga. With regular practice, yoga helps restore the strength and agility of the spine, slowing and even reversing the common degenerative changes often found in people over 50. The practice guides in this book will teach you specific exercises for preventing and decreasing upper back roundness, elongating the spine and expanding and opening the chest.

Yoga: A Superior Form of Weight-Bearing Exercise for Preventing Osteoporosis

In the last 20 years, scientists have learned how important exercise is to maintaining the strength of the bones. During weight-bearing exercise, muscles transmit mechanical and bioelectrical signals to the bone, causing it to thicken. Research has found that daily weight-bearing exercise is crucial to helping both women and men avoid osteoporosis, or porous, brittle bones. In a person with severe osteoporosis, a fall, blow or lifting action that would not affect the average person can easily cause one or more bones to break.

Medical textbooks and articles about exercise frequently claim that yoga is not a weight-bearing exercise and is therefore not considered effective for strengthening bones and preventing osteoporosis. Yoga's stretching exercises are wrongfully relegated to the role of warming up the body for other physical activities or helping to cool it down afterwards.

In fact, yoga is a superior form of weight-bearing exercise because it stimulates bones throughout the

Posture affects every system of the body— not only the neuro-muscular system (joints, ligaments, bones, muscles and the nerves that move them), but the endocrine system (pituitary, thyroid, adrenal, etc.) and the cardiovascular, circulatory and respiratory system. All of these systems can be directly correlated and related to problems with posture.

—Nikolaas Tinbergen, 1973 Nobel laureate in Physiology/Medicine

The author and Sandy Yost, age 72, demonstrating the Half-Moon Pose. This standing position stimulates the bones of the legs, arms and spine to retain calcium. Beginners can practice this pose with their hand on a block or chair, their back against a wall or counter, or with the help of a teacher or friend, as illustrated.

body. In yoga's flowing, graceful movements, weight is systematically applied to bones in the hands, arms, upper body, neck and even the head, as well as the feet and legs. Inverted weight-bearing yoga postures such as Handstands and Headstands, where the bones in the arms, wrists and hands are strengthened by supporting the weight of one's own body, all work to prevent osteoporosis and other problems related to a weak skeletal structure. Yoga is one of the few exercise systems in which weight is borne through the entire body. Because yoga postures are learned gradually, the weight applied to the bones increases safely, incrementally, as the student becomes stronger and can hold postures for longer periods of time. This is especially important after the age of 50. (See chapter 7 on osteoporosis.)

Keeping Your Joints Young

While yoga is already known for restoring our natural, youthful flexibility, most people are not aware of yoga's therapeutic effect on the joints. When a joint is injured, one of the first rehabilitation techniques in physical therapy is called passive range of motion. In passive range of motion, the therapist moves the joint as far as possible without pain. Through many repetitions of the movement, the range of motion is gradually increased. The patient then graduates to active range of motion exercise, doing the same maneuvers without assistance.

Western medicine has long recognized the value of this kind of therapy for injured joints. More recently, however, physicians are realizing that the same principle, which is the basis of much of the work done in yoga, helps prevent the degeneration of healthy joints.

Health is reflected in lifestyle, and if we ignore our bodies we may, even by our mid-thirties, stiffen far more than necessary. Movement lubricates the muscles, ligaments, and joints. If we become sedentary, the muscles lose tone and the muscle groups that hold us upright become unevenly matched. The joints then feel strain, lose space in which to articulate, and we start to suffer from wear and tear.

At this point we become less inclined to move because it is not comfortable to do so. This is premature aging. Stretching daily can reduce this stiffness, allowing us eventually to dance, play tennis, and enjoy our bodies well into our real old age.

—*Maxine Tobias,*
Complete Stretching

Inverted Poses: The Fountain of Youth

Newcomers to yoga are usually fascinated—and sometimes put off—by the sight of their more experienced classmates hanging upside down from wall ropes like bats or going upside down in Handstand, Headstand or another inverted position. Naturally, they want to know about the benefits of inverting the body.

Physicians have long recognized the beneficial effects of reversing the downward pull of gravity on the circulation, lungs and brain. They frequently advise their patients to put their feet up to compensate for the weakening of veins in the lower legs of many older people. Doctors have also reversed gravity in the treatment of respiratory problems. For example, a method called "posture drainage" predates antibiotics for treating certain lung infections. Using this technique, the patient assumes various positions to turn various parts of the lungs (which are arranged like the branches of a tree) upside down. However, the most effective form of postural drainage is one not yet commonly described by Western medicine: that of turning the body upside down, as in the inverted yoga postures.

The gravitational force of the earth is among the most powerful physical influences on the human body. Just as plants and trees are shaped by the direction of sunlight and wind, our bodies are shaped by the pull of gravity. As time goes by, the body has a tendency to narrow at the top and settle toward the bottom. Reversing the downward pull of gravity helps the body retain its balance and symmetry.

Gravity also tends to cause the compression and flattening of the cells and blood vessels of the brain, and the collapse of the top lobes of the lungs. It also tends to create stress and strain on the ligaments, blood vessels and organs in the abdominal cavity. Being upside down helps to reverse all these tendencies. The large intestines are especially responsive to the relief from gravitational compression, and inverted poses are an effective treatment for constipation.

To further appreciate the benefits of reversing gravity, consider the obvious fact that water always flows downhill. The fluids that comprise 75 to 80 percent of our body are generally at the mercy of gravity, and the circulatory system must constantly adjust to gravity's effects. One of the most vital functions of the cardiovascular system is to deliver fresh blood to the brain. When we are in a horizontal position, the heart is at the same level as the brain, and gravity has little effect on the flow of blood between the brain and the heart. When we are standing or sitting, the brain is above the heart and the cardiovascular system must compensate by working harder to pump blood to the brain. In a partially or fully inverted position, the brain is below the level of the heart and gravity pulls blood into the brain; the brain receives a proper blood supply with minimum effort by the heart.

When the body is completely inverted, venous blood flows from the legs and abdomen to the heart without strain. According to yoga experts and doctors studying yoga, the regular and long-term practice of forward bends and inversions can reduce arterial blood pressure by helping to reset the pressure regulating reflexes. The Headstand helps to increase venous return to the heart, bringing the deoxygenated blood toward the heart and relieving pressure in the passive venous system caused by the pooling of blood in the legs during standing.

During the course of a typical day, most people spend 16 or more hours with the head (brain) above the heart and the legs and pelvic area below the heart. I always advise students who are not yet ready to practice more difficult upside-down positions to at least put their legs up against a wall for five minutes. Inverted positions, especially when combined with proper nutritional support, can alleviate problems with varicose veins and, if begun in time, can prevent them.

Reversing Gravity Reverses the Aging Process

It is well known among yoga practitioners that the inverted yoga positions slow down and even reverse the common physical changes that come with the passage of time.

Turning the body even halfway upside down by bending forward from a standing position increases the circulation to the entire upper body, including the brain. The revitalizing and relaxing effect of standing forward bends and completely inverted positions is related, in part, to this change. When the body is inverted, blood circulates easily around the neck, chest and head. This increased circulation stimulates the thyroid and parathyroid glands and helps the lungs, throat and sinuses become resistant to infection. It is also possible that inversions have a positive effect on the brain, as we come out of these postures feeling invigorated and clear-headed.

My early interest in yoga was kindled by a 60-year-old-teacher who told me that she discovered yoga about the time she put her mother and mother-in-law in a rest home. She explained, "The agony of taking them to a

B.K.S. Iyengar at age 75. Inverted poses revitalize the whole body.

rest home after they were senile made a profound impression on me. I began to consider that perhaps I could hold back or maybe even prevent senility by increasing the flow of blood to my brain through the inverted postures."

The ravages of senility are apparent in every nursing home in the country. While Western medicine accepts the fact that this is a degenerative disease associated with inadequate circulation to the brain, researchers have found few effective ways of preventing or treating it. The blood flow to the brain gradually reduces as one grows older. Dr. Krishna Raman, in his writings on yoga and the circulatory system, states that by the time a person is 65, the blood flow to the brain may be a third of what it was at age 25. Yoga teaches that the most effective way of increasing blood to the brain is to allow gravity to do the work for you by bringing the brain below the level of the heart, permitting circulation to the upper body to increase without putting strain on the heart.

After age 50, it becomes increasingly important to reverse the downward pull of gravity on the body.

Precautions on Practicing Inversions

Inverted positions, whether done with special equipment, pelvic swings, antigravity devices or your own upper body strength, are only as safe as people are educated to make them. Inverted positions like the Headstand should be learned under the guidance of an experienced teacher and should not be practiced by individuals with the following conditions: high blood pressure; glaucoma; detached retina; heart problems or stroke; epilepsy, seizures, or other brain disorders; acute infections of the ear, throat or sinuses; osteoporosis; obesity; conditions requiring aspirin therapy; chronic neck problems or whiplash. These conditions may improve with gentle supported forward bends and supported inversions.

Chapter 11 describes simple Legs-Up-the-Wall Pose and other inverted positions that most people new to yoga can safely practice.

Joyce Rudduck: No One Is Ever Too Old or Too Stiff to Do Yoga

Joyce Rudduck in the rejuvenating Shoulderstand.

Joyce Rudduck, who is in her 60s, is co-director and teacher at the Australian School of Yoga. She began practicing yoga at 54 as a last resort to ease the constant pain she still suffered from two accidents years ago. In one accident, a crowbar fell on her head, injuring her neck. A year later she was thrown from a turning truck and jammed her lower back.

The doctor told me, "You will just have to learn to live with it," but my interior self said, "No, I will not!" I had always been a very active person, but the growing pain throughout my body was gradually slowing me down. In every position, I was always shifting to get out of pain. Because of the accidents, I was always suffering and in pain.

At last I had the good fortune to be referred to an excellent physical therapist who asked, "Joyce, have you ever done yoga?" I explained that I'd gone to a few classes a few years ago and that it felt fine but nothing to rave about. However, she persisted, saying that this was a different approach to yoga. She described it as "therapeutic yoga," and told me to think about it.

After mulling it over for a week I decided to give it a whirl, telling myself that I had nothing to lose. At that point I was desperate and willing to try anything.

Thus I arrived at the Australian School of Yoga. I remember well the first things we did in that class. We lay on our back, feet to the wall, and stretched our legs this way and that with straps. Our instructor was the late Martyn Jackson, one of B.K.S. Iyengar's advanced senior teachers. His big cheery voice boomed out over everyone. Every moment was agony! Our shoulders were opened by lying on our stomachs with our hands raised up on chairs. His instructions were "five minutes before breakfast—five minutes before your supper every day."

Obediently I complied and months later reaped the reward. That was in November 1985.

After the first six months I began coming to class twice a week. After a year, four times a week—and so I grew into yoga.

Then I developed an ardent desire to ease the pain of others. No one in my opinion was too old, too stiff or too anything to do yoga because I had done it and was still doing it and improving all the time.

In 1986 Martyn began a weekday 6 A.M. intensive class, which I attended every day before work. It was hard, but rewarding. Fear was dissolving, my whole being was changing physically and mentally. Slowly I began to feel a new me. The constant pain was disappearing.

In 1988 I took a teacher's training course. For a "stiffy over 50," it is not an easy road, but with determination and perseverance and doing something quietly every day no matter how small, one grows and blossoms and learns that, "Your body is the teacher," and teachers become guiding lights to assist you.

Three

In Praise of Props

How Common Household Objects Can Help You Improve Your Posture, Maintain Your Balance, and Stretch, Strengthen and Relax

Props are silent instructors that teach directly to the intelligence of the body, and through that direct perception the harmony of mind, body and spirit can be experienced.

—MARY DUNN, YOGA TEACHER

Most of us think we don't have enough time to exercise. What a distorted paradigm! We don't have time not to. We're talking about three to six hours a week or a minimum of 30 minutes a day, every other day. That hardly seems like an inordinate amount of time considering the tremendous benefits in terms of the impact on the other 162-165 hours of the week.

—STEPHEN R. COVEY, *THE 7 HABITS OF HIGHLY EFFECTIVE PEOPLE*

B ack in the 1960s, when Westerners first began turning to yoga, the use of a prop to practice yoga was practically unheard of. Potential yogis simply came to class with an exercise mat, a blanket or beach towel. Everyone mimicked the teacher's demonstration of the yoga positions as best he or she could. You were not expected to sweat or be perfect. As I recall, emphasis was on deep relaxation in the corpse pose. Classes were often held by candlelight, with incense and Indian sitar music to evoke a mystical atmosphere.

Nowadays, at the turn of the 21st century, when you walk into a yoga classroom, you will likely see folding chairs and benches stacked in one corner of the room and shelves filled with blankets, nonskid mats, pillows and bolsters of various shapes and sizes, wooden blocks, dowels and 10-pound sandbags. There may also be ropes, straps, pelvic slings or other anti-gravity equipment hanging from the walls or suspended from the ceiling. You may see a whale-shaped piece of furniture—a unique wooden object known as a backbending bench.

If you arrive early, some students will probably be sitting or lying down quietly, but others are using these props, also known as yoga tools or yoga aids, to warm

This is the single most powerful investment we can ever make in life—investment in ourselves, in the only instrument we have with which to deal with life and to contribute.

—*Stephen R. Covey, The 7 Habits of Highly Effective People*

Sandy Yost shows how invigorating using ropes can be.

up or to relax before class starts. People of all ages are hanging halfway or completely upside down from ropes, and those that aren't reversing gravity are lying on the floor, stretching their legs with a strap around one foot. Other students are lying over bolsters, benches and chairs. Some are sitting on top of a block or bolsters with the soles of their feet together and sandbags weighing down their thighs.

According to Ruth Steiger and Kay Eskenazi, cofounders of the company Yoga Props and authors of the *Yoga Prop Usage Guides* (see Resources), "A prop is any object that helps you to stretch, strengthen, relax or improve your body alignment. By providing more height, weight or support, props help you to extend beyond habitual limitations and teach you that your body is capable of doing much more than you think it can!"

Props are adapted to a student's body type and flexibility.

The backbending bench allows you to stretch with minimum effort.

Using yoga props makes postures safer and more accessible. Since many people are already stiff by the time they start yoga, props allow them to practice poses they would not ordinarily be able to do.

People over 50 also frequently come to yoga with problems, ranging from back and neck pain or knee problems to old injuries. The more problems a student has, the more useful yoga props are. Props allow you to hold poses longer, so you can experience their healing effects. By supporting the body in the yoga posture, muscles can lengthen in a passive, non-strenuous way. The use of props also helps to improve blood circulation and breathing capacity.

For example, if you cannot bend forward and bring your hands to the floor without straining or bending your knees, try placing your hands on a desk, table or chair. As you become more flexible you will find that you can put your hands on a lower prop like a bench, a stack of books or a block. With practice, most people's hands will touch the floor and the prop will no longer be necessary.

The creative use of props expands the help a teacher can give, especially when teaching classes of various levels of ability. In a group situation, for example, those students who are not strong enough to practice inversions on their own can safely receive their benefits by being supported by ropes suspended from the wall or ceiling. The ropes absorb some or all of a student's weight, inversions can be performed without strain, and the student can receive the benefits of the pose.

Props are also used to teach students how a pose done correctly should feel. A rope hanging from a wall hook or doorknob and placed at the top of the legs in *Adho Mukha Svanasana,* the Downward-Facing Dog Pose, allows the student to stretch the torso and arms as far forward as possible. Because the rope pulls the student's weight back into the legs, it helps the student experience the elongation of the abdomen and the deep muscles of the torso in the pose. The head can rest on a bolster or pillow. In this way, a wonderful, passive stretch is experienced. The student gets a taste of what it feels like to "let go" in a pose, to relax and enjoy the pose. Use of the props facilitates imprinting of the correct action in the pose so that the student understands it when the prop has been removed.

By using props, students who need to conserve their energy can practice more strenuous poses without overexerting themselves. People with chronic illness can use props to practice without undue strain and fatigue.

Props are adapted to a student's body type and flexibility. They are especially helpful to anyone who may avoid trying certain poses because of fear, problems with balance such as loss of hearing and eyesight, pain and various other limitations. In therapeutic situations, props are invaluable. People who have scoliosis (curvature of the spine), rounded back or other chronic postural problems can significantly improve their posture by stretching with the help of a prop.

Props give tremendous encouragement, create confidence, reduce pain, support the body and guide students to practice the poses correctly. In addition to using common objects around the house as props, a wide range of professionally designed yoga props are available and well worth the cost.

Sitting Comfortably on the Floor

A lifetime of sitting in chairs causes stiffness in the hips and knees. As such, most people unaccustomed to sitting on the floor tend to slump backwards and collapse their chest. By sitting on two firmly folded blankets, a bolster, a big book or other height, older adults can quickly learn what it feels like to sit on the floor with their back straight and chest open.

The stiffer you are in the hips, the thicker the height under your bottom should be. If you feel strain in your knees, try placing a folded blanket or other support under your knees.

How Chairs Support Your Yoga Practice

The prop I use most frequently in classes for beginners over 50 is an ordinary, metal folding chair. The best folding chair to use is very stable and sturdy, with a level seat that does not collapse when you put weight on it, and it has a wide space in between the backrest and the seat. If you can find one with a rung between the back legs, that would be helpful, but if this type is not readily available, just find a chair that matches the first part of this description as closely as possible.

While other armless chairs can also be used for yoga, standard metal folding chairs are more versatile. They may be used for supported restorative postures and backbends. If you don't have access to a folding chair at the moment, use the plainest, sturdiest armless chair in your house. Avoid cushioned or upholstered chairs, since they do not provide a firm, even base of support. An inexpensive folding chair can be used in a multitude of ways to stretch and strengthen your body in ways that challenge the most sophisticated gym equipment. You can bend forward, backward and sideways, do push-ups, relieve backaches and shoulder aches, and go upside down safely with the assistance of this simple piece of furniture.

Chairs can be used in dozens of innovative ways. They allow new students to practice poses that are often too challenging for beginners of any age—but chairs are especially useful for students over 50. Many of my older students initially practice yoga standing poses with the help of a chair or wall. Safety is another important reason for using walls or chairs, especially for beginners over 60 or 70. These props help older people with balance problems practice yoga without slipping or falling. In this way, props help build confidence and stamina. The energy that you might normally divert into struggling to keep your balance can be channeled into practicing the poses correctly and holding them longer. As the poses become more familiar, most students develop the balance to practice without a wall or chair.

Yoga Sitting on a Chair

In cases where sitting or lying on the floor is really not possible or practical, the use of one or two chairs to sit on can be helpful. A folded sticky mat prevents the student from being uncomfortable or, in cases where muscles are very weak, from sliding out of the chair. The surface of the chair becomes a floor, without the hassle of getting up and down off the floor, for someone who does not have sufficient strength in the stomach and back muscles to keep the body erect. A blanket on top of the sticky mat allows the person to sit forward on the sitting bones, keeping the spine in an elongated position.

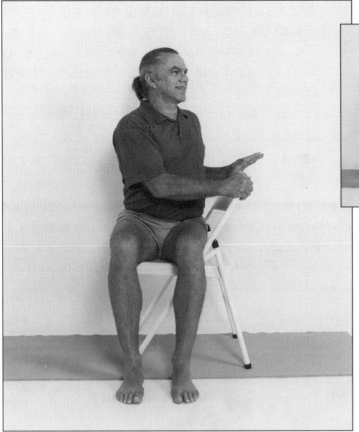

Variation with block.

Jay Myers, a new student in his 50s, feels the benefit of this simple Chair Twist.

Chair Twist

1. Sit sideways on the whole seat of a chair, with the right hip against the back of the chair.
2. Pressing your feet firmly into the floor, lift your rib cage, stretch your shoulders back and down, and lengthen your spine upward. Keep your knees straight in line with your feet.
3. Continue pressing your feet firmly into the floor as you turn to hold the back of the chair. Pull with the left hand to bring the left side of the body toward

the back of the chair, and push with the right hand to turn the right side away from the chair. Keep your upper body stretching toward the ceiling as you turn to the maximum. Turn your head slowly and look over your right shoulder.

4. Continue twisting for several more breaths; exhale as you come back to the center. Pause for a moment, sitting tall. Move to the opposite side of the chair to repeat on your left side.

Note: If your feet do not reach the floor when seated in the chair, place a block, stool or other height under your feet. If your feet and knees tend to splay out when twisting, hold a block or book between your knees as illustrated on page 35.

Yoga teacher Betty Eiler helps her student turn more in Chair Twist.

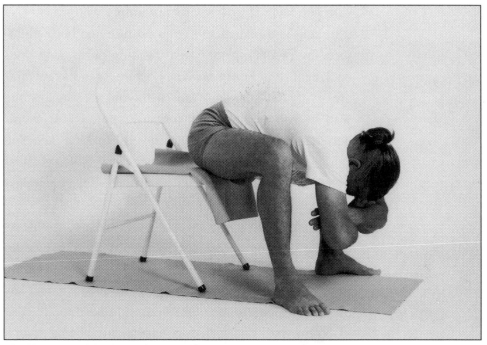

The Chair Forward Bend is a refreshing pick-me-up.

Chair Forward Bend

After the Chair Twist, it is very restful to sit on the front of the chair. Spread the legs wide apart, press firmly into the floor with your feet and stretch your trunk upward. Slowly bend forward, relaxing the back and the head. You can hold your elbows as illustrated, or place your hands on a stool, bolster or the floor.

The Wall: Your Best At-Home Teacher

One of the best yet most overlooked props for stretching is the wall. "The wall is your best guru—it doesn't lie!" my teachers told me when I first began yoga. I recommend that you start your at-home practice by making

sure you have at least six feet of empty wall space. Use
it to help you understand good posture and body alignment by practicing the yoga standing poses with your
back against it. For improving posture, spend a few minutes every day standing "tall" with your back against the
wall. If you tend to stoop forward, stretch your shoulders back and down the wall several times a day. You
can also use a counter, railing or sturdy table, as illustrated on page 39, to remind you how healthy, open posture feels.

Students who come to class with balance problems
can gain strength and confidence by first practicing
Triangle Pose and other vital weight-bearing standing
poses with their back against a wall. In cases where it has
become difficult to maintain an upright position, the use
of both a wall and wall ropes (available at many yoga
centers or from a yoga prop source) are invaluable for
rebuilding strength and stamina. Standing poses can also
be practiced by bracing the back foot against a wall and
holding wall ropes, as illustrated on page 39. I especially
encourage my older beginners who find it difficult to
maintain their balance in the middle of the room to practice standing poses at home with the support of a wall,
sturdy table, railing or kitchen counter.

How to Practice Triangle
Pose at a Wall (Trikonasana)

1. Stand tall with your posture open, shoulders relaxed
   away from your ears, near a wall. Step your feet
   about three-and-a-half to four feet apart (depending
   on the length of your legs), keeping your feet in
   line, facing forward, heels close to the wall.

Sandy Yost and a student practicing the Triangle Pose with wall ropes. Walls and wall ropes help students gain strength and confidence.

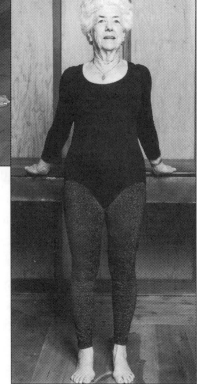

This simple stance immediately gives you a more youthful posture.

Barbara Wiechmann, in her 50s, is grateful for the gift of yoga. The wall allows her to feel more open and relaxed in the Triangle Pose.

2. Breathe normally. Anchor and root your feet to the earth by pressing the soles of your feet deep into the floor. Activate your legs by pulling up the thigh muscles. Allow your body to become taller and taller, lengthening your spine upward. Raise your arms to shoulder level, palms facing down, and stretch out through your fingertips. Feel the center of your body expand and open.

3. When you feel stable and centered in this position, turn the left foot about 15 degrees in, and the right foot 90 degrees out. Line up the right heel directly in line with the center of the left arch.

Triangle Pose. The use of a block and strap opens the shoulders and hips.

4. Inhale, and on exhalation stretch to the right from the hip joint, so your torso bends sideways as a unit toward your right leg. In the beginning you may need to place your right hand on your leg, a chair or block. Extend your left arm up, in line with the right arm, palm facing forward. If you feel unusual strain in your shoulder, try placing your left hand on your hip.

5. Stay in the pose for several breaths, keeping your legs active, shoulders and neck relaxed. Come out of the pose on an inhalation, keeping your body close to the wall. Turn your feet to face forward. Relax back into the wall and pause for a moment to feel the effects of the pose. Repeat on the other side.

The use of a block and strap, as illustrated on the previous page, is invaluable for learning to bend from the hip joint with the chest open and the spine lengthening.

Standing poses are refreshing and invigorating for all ages and levels of ability. They remove the aches and pains of "old age." They also improve circulation and breathing, stimulate digestion, regulate the kidneys and relieve constipation. The back, hips, knees, neck and shoulders all gain strength and mobility with regular practice.

Props Help You to Balance

Walls are also helpful for learning balance when standing upright on one leg or turning halfway or completely upside down, as described in chapters 4 and 11.

My older students report that practicing balance poses such as the Tree Pose in class helps them improve their balance in daily life. To build strength and confidence, Tree Pose can first be practiced standing near a wall for support. The wall reminds students to maintain good posture while balancing in the pose. Beginners tend to hunch their shoulders forward when attempting to place their foot against their inner thigh. Practicing with a strap around the ankle of the lifted leg, as illustrated on the next page, is useful for keeping the foot from slipping and for maintaining healthy, open posture in the pose. In addition to giving a sense of balance, Tree Pose strengthens the feet and legs.

Tree Pose

1. The ability to balance on one foot starts by learning to stand firm on both feet, rooted to the earth.

A strap can keep your foot from slipping in this balance pose.

Stand tall with your feet and knees facing forward. Be aware that the feet are your foundation when you stand. Create stability while you have both feet on the ground. Feel the contact of the soles of your feet on the floor. As awareness of your body deepens, the soles of the feet feel increasingly sensitive and alive, like the palms of your hands on the floor.

2. Standing firm on the left leg, bend the right leg to the side. Catch the ankle (or wrap a strap around your ankle) and place the foot at the top of the left

inner thigh. Bring your foot as high up the leg as possible. Press the right knee back, in line with your hip.

3. Tree Pose can be practiced with the arms and hands in various positions. Whether you practice with your hands in prayer position as illustrated, or with your arms stretching upward above your head, allow your chest to lift and open, your spine to lengthen upward. Stay steady by continuously firming and anchoring the standing leg. Soften your eyes, focus your gaze. Find your balance. To release out of the pose, bring the arms and leg down. Pause for a moment, standing steady on both feet. Repeat on the other side. *Note: If you feel unusual strain bringing the sole of your foot against your inner thigh, try practicing with your foot lower down the leg.*

Using Doorways to Improve Posture and Stretch Shoulders

I never tire of showing my students the many uses of a doorway and the door frame. The side of the door itself, the frame around the doorway and even the door handles (see chapter 4) are all useful yoga props.

Doors make an excellent prop for lengthening the spine and improving posture. Remember that the spine has four natural, gentle curves, but many people over 50 have developed too much rounding of the midback and upper back, and too much of a sway in the lower back. To understand and improve your posture, stand so that your heels press securely against either side of a door. Position your back so you can gently press your spine against the edge of the door, then grasp the top edge of

A door can help you discover your true height.

the door as high above your head as you can. The pressure of your back against the door lets you feel your usual alignment and what happens when you stretch upward. The midback may move slightly away from the door while the lower back presses toward the door. Gently draw your navel back toward your spine as you stretch upward and make yourself taller and taller.

This door stretch is also an excellent exercise for removing stiffness in the shoulder joints. In many older people the shoulders are so stiff that when they stretch their arms above their head or try to stretch up the doorway, their elbows bend. With regular practice, shoulders stretch and arms straighten.

Depending on the design of the door or doorway, this exercise can also be done in the doorway itself (taller people will get a better stretch in the doorway). Line up your spine with the center of the doorway and follow the directions for lengthening the spine. As your shoulders regain their natural flexibility, walk your hands farther back, while gently drawing your navel back toward your spine to avoid over-arching or swaying your back.

Stretching Stiff Shoulders with Straps or Belts

To avoid pulled muscles, overstretching and joint strain, never force or rush your body into a yoga position. Use straps and belts, instead, to help you achieve a healthier, more balanced stretch.

The following shoulder stretch, known in Sanskrit as *Gomukhasana* or Cow Face Pose, works on the muscles that control the shoulder joint. Regular practice develops greater freedom of movement so that rounded shoulders can return to healthy posture. If you cannot clasp your hands together, use a strap as a bridge between your two hands as illustrated. This shoulder rotator exercise is one of the most basic corrective poses for removing stiffness from the shoulder joints (the collarbones at the front, arm bones and shoulder blades at the back). It is especially valuable for those who play tennis and other racket sports.

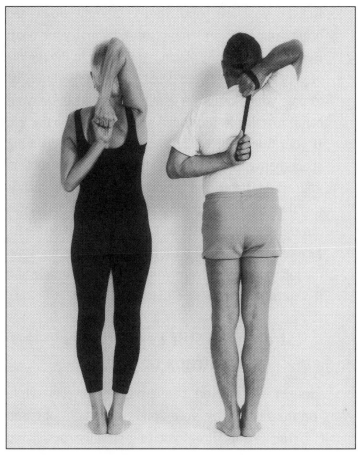

If your hands do not touch, use a strap to help you stretch your shoulders.

Shoulder Stretch Practice

1. Stand (or sit) in your best, tallest posture. Pause for a moment to observe your breath. Allow yourself to smile. This will naturally relax your jaw and face muscles.

2. Stretch your right arm straight up over your head and then bend your elbow so that your palm touches your back between the shoulder blades. Reach across with your left hand to move your elbow closer to your head.

3. Release your left hand from your right elbow and bring your left arm straight back behind your body. Bend the left elbow, placing your hand in the middle of your back above your waist, palm out. Without distorting your posture or straining, try to clasp your hands together as illustrated by the teacher on the left.

4. If your fingers just barely touch, or if there is a big space between your hands, hold a strap or sock in your right hand and gradually work your hands together. Stretch up through the top elbow and down through the bottom elbow. Keep your head centered, face relaxed. Hold at least half a minute. Repeat on the left side. Learn from your more flexible side and repeat or hold longer on the tighter side.

Stretching Stiff Leg Muscles with Straps or Belts

A strap, towel or soft belt around your feet while lying on the floor helps your spine to remain long and stable. Using a strap allows you to gradually stretch and lengthen stiff leg muscles without straining your back.

Lying-Down Leg Stretch with a Strap

1. Lie on your back with your knees bent, feet flat on the floor. If your head tilts back with your chin higher then your forehead, place a folded blanket under your head and neck. Check to see that your upper body is in line with your legs. Allow your back to relax into the floor.

2. Bend your right knee in toward your chest and wrap a strap, towel, soft belt or necktie around the

This student in her 80s uses a strap to stretch her legs while the more experienced student next to her practices the Lying Down Big Toe Pose.

ball of your foot. Hold the strap with your right hand. Stretch your left arm out in line with your shoulder on the floor, palm facing up.

3. Slowly straighten your right leg and stretch your toes toward your face. Walk your hand higher up the strap, toward your foot, till your arm is straight. Keep your shoulders and rest of your back relaxed on the floor.

4. If your right hand is quite far away from your foot, keep your left knee bent. If you find it easy to hold your big toe, or if your right hand is high up the strap close to your foot, you can deepen the stretch by practicing this pose with the bottom leg straight, extending through both heels, as illustrated. Stretch your toes toward your face to lengthen your calves and Achilles tendons.

5. Smile and allow your face muscles to relax. Let your breath flow freely, stretching deeper as you exhale. Hold the strap firmly without creating tension in your hand. Enjoy the feeling of the back of your leg lengthening. Hold for about half a minute, longer as you learn to relax and cooperate with the pose. Repeat on the opposite side. If you are practicing with both legs straight, it is helpful to extend the lower heel into the wall.

Improving Your Posture and Breathing Using Bolsters and Blankets

Bolsters, folded blankets and towels are used to support the back of the body in various lying-down positions. Placed along the length of the spinal column and back of the head, these props help correct poor postural and breathing habits, which are often magnified after age 50. The shape of the blanket or pillow supports and encourages the natural curves of the spine. Supported lying-down positions open and expand the chest to improve your posture and breathing.

Most people respond with a delightful sigh of relief when they are positioned on the folded blankets or firm pillows. I consider the use of these props absolutely essential in a yoga program for people over 50.

Because the body tends to sink and collapse into soft materials such as thermal or fluffy blankets, quilts, comforters, sleeping bags or soft pillows, these do not work as yoga props. Heavy cotton or wool blankets that form a firm pad when folded, or bolsters (both of which can be ordered from the addresses in the Resources section or are available through your local yoga center) are best for practicing yoga postures.

The *pranayama,* or breathing bolster, is firm and narrow and supports your spinal column from the lower back to your head when you are lying down. By supporting your spine and opening and expanding the chest, the muscles of your abdomen, chest and back release their tension, lengthen and relax deeply. This bolster is specifically designed so that the sides of your rib cage open and expand over the bolster and move downward toward the floor. When your rib cage expands laterally in this manner, your breathing capacity naturally deepens. The bolster leaves a lasting vital impression on the body of what it feels like to have the chest open and free. Using it enhances your awareness of the breath and your ability to regulate and deepen both your inhalations and exhalations by encouraging you to relax.

You can achieve a similar effect using two or three firm blankets, folded lengthwise.

How to Use a Bolster or Folded Blanket to Relax

1. Sit on the floor with your bolster or folded blankets behind you—the edge of the bolster touching your lower spine, and its length extending behind you. (Your bottom should be completely off the bolster.) Place one or two neatly folded blankets at the other end of the bolster to support your head. Look to see that you are centered with the bolster or blankets.
2. Lower your body backwards, supporting your weight on your elbows, so that your back and head rest on the bolster.

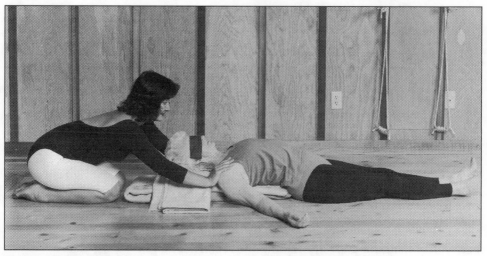

Lying back on blankets or bolsters opens your chest and allows your breath to flow freely.

3. Before lying all the way down, look down the front of your body and see if your chin is in line with the center of your chest, navel and pubic bone. If necessary, adjust the position of your torso so that the straight line formed by your nose, chin, center of your chest, navel and pubic bone extends directly toward a center point between your heels.

4. Continue lowering your spine slowly on the bolster or blankets, without shifting your torso to the left or the right. Before resting the back of your head on the bolster or folded blanket, tuck the additional folded blanket under your head, to bring your forehead slightly higher than your chin. Your forehead should not drop back.

This relaxing position can be practiced with the legs extended, as illustrated, or with a rolled blanket under the knees. Your body should feel supremely comfortable, supported by the bolster, blankets and the floor. (See chapter 12 for more information on the art of conscious relaxation.)

Covering Your Eyes
Enhances Relaxation

Placing an eyebag over your eyes will help you relax. An eye covering helps quiet the mind by creating darkness and removing visual stimuli, by relaxing the muscles around your eyes and by calming the involuntary movements of your eyes. After you become familiar with placing your body on the bolster or folded blanket, hold the eye covering in both hands. Eyebags are usually filled with seeds or rice, so distribute the contents evenly. Place the eyebag over your closed eyelids. The eyebag's gentle pressure should feel soothing. Once you've positioned the eyebag, relax your arms on the floor with your hands about a foot away from your hips. Turn your palms up and stretch your fingers outward, turning your thumbs toward the floor. Let your hands soften and relax. (See photo on previous page.)

If an eyebag is not available, use a folded damp washcloth or similar material, placed neatly like a band over your eyes. Eyebags are easy to make by sewing cotton or soft, silky fabric into an 8- or 10-inch bag, about $3^1/2$-inches wide (to cover eyes but not nose), filled with about 8 ounces of rice or flax seeds. Or you can even fill a soft cotton sock with rice or flax seed and sew up one end.

Props to Relieve Stress on the
Head and Neck

Certain yoga postures such as Headstand or Shoulderstand put unaccustomed weight on the head, neck and shoulders. Yoga props are available to relieve the stress of weight-bearing by supporting most or all of

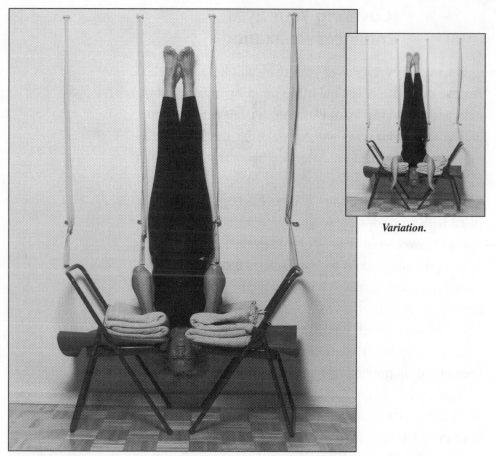

Variation.

Headstand practiced with the shoulders supported by two chairs prevents neck strain and allows beginners or people with neck problems to practice without injuring the neck.

your body weight as you practice these poses. A prop known as the Headstander, for example, supports nearly 90 percent of your weight, enabling you to practice the Headstand without bearing weight on your neck. Similarly, the Halasana bench supports most of your weight and frees your neck as you practice Shoulderstand and Plow Pose. Both of these props are available through the sources listed in the Resources section. (Refer to chapter 11 on Restful Inversions.)

The Backbending Bench

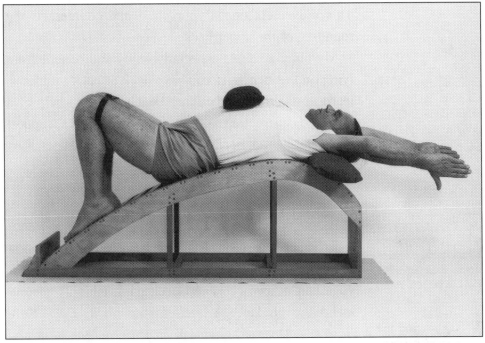

With the help of a backbending bench and other yoga props, older students can safely experience the rejuvenating effects of a backbend.

Yoga postures give a sense of delight, lightness, buoyancy, vitality, balance and health. When our muscles are too tight or weak, or we are afraid of putting our bodies in positions or shapes that they have not been in for many years, props are a way of removing the obstacles between you and a pose that seems out of reach. With a backbending bench, available for use at many yoga centers, almost anyone can begin to practice a safe, supported backbend. This bench supports and lengthens your spine, stretches your arms and shoulders, opens your rib cage and lungs, deepens your breathing, stretches your groin, abdomen and the front of your thighs.

Lying on the backbending bench presses the vertebrae of your upper back toward the center of your body. This movement opens your chest and counteracts the rounding of the upper back.

My older students especially enjoy and benefit from lying on this bench. If you have neck or back problems, learn to use the backbending bench with the help of a qualified instructor, as you may need extra support under your head, neck or back.

Back and Neck Pain

Yoga differs from other types of rehabilitative exercise in that it engages the whole person. Yoga-based relaxation techniques and stretching and strengthening exercises are effective because the mind is focused in a meditative way on your movements, skin and muscle sensations, and relaxed breathing. Mind and body work together, creating a physiological and psychological environment that optimizes the potential for healing.
—Mary Pullig Schatz, M.D., Back Care Basics: A Doctor's Gentle Yoga Program for Back and Neck Pain Relief

People with back and neck pain generally experience relief practicing yoga, with the help of the props presented in this chapter. For in-depth instruction on how yoga can relieve back and neck pain, I highly recommend the guidance offered in the book, *Back Care Basics: A Doctor's Gentle Yoga Program for Back and Neck Pain Relief*, by Mary Pullig Schatz, M.D.

In addition to a gentle yoga program, other therapeutic techniques such as chiropractic and massage can be helpful. Many people with back and neck problems experience both temporary and long-term relief from chiropractic adjustments. A chiropractor familiar with yoga is ideal. Yoga will allow you to progress beyond symptomatic relief by correcting the weaknesses and misalignments in your body, changing your postural habits, and teaching you to relax and cope constructively with stress.

Exercising After Back Injury or Surgery

When the back or neck hurts, it is instinctive to seek relief from pain by resting and avoiding further strain.

Standing poses like this Forward Stretch with Hands High on the Wall help to keep your back healthy for a lifetime.

People with back problems often fear that any move-ment or exercise might cause re-injury. But without an intelligent exercise program that addresses the underly-ing causes of back pain—unhealthy posture, muscles that are too tight or weak and poor body mechanics in daily life—re-injury is practically inevitable.

If the body remains immobilized beyond a reasonable period of rest, the muscles that support the back become weaker and increasingly vulnerable to re-injury. To break the cycle of further deterioration, inactivity and pain, we need to learn how to practice yoga, and which type of exercise rehabilitates our back and which type of exercise or yoga practice hurts it.

Dr. Schatz points out that some people feel so good after surgery or in between "back attacks" that they mistakenly imagine that their injuries have healed. Without pain as a reminder, the need to care for the back tends to recede in our consciousness. However, if surgery has taken place, post-surgical structural weaknesses and muscular weakness, aggravated by prolonged bed rest and combined with underlying problems contribute to the likelihood of more back problems in the future.

B.K.S. Iyengar On Using Props

B.K.S. Iyengar asserts that very few people make use of the last phase of life in a fruitful way.

It is an art to check the aging process. To stop its ascendance, one should learn to make old age a useful weapon. In this stage of life, one becomes negative. Courage starts declining and intelligence becomes dull. Anxiety encircles the older person. Laziness becomes a part of old age.

Some in old age realize the importance of yoga and come for help. They have not done any yogic practice before and want to learn and do something; yet they are unable to do the yoga postures. At that state, the profound utility of props and their values are realized. Even incapable persons will find hope of doing something that keeps life flowing with joy.

Iyengar believes that students who come to yoga late in life get the advantage of keeping themselves fit physically and mentally using props. His experience has been that bolsters, backbending and other benches, ropes and other props are useful in old age, when people may not be able to do the posture independently. He believes that props free the older student from anxiety. "Practice on props leads one toward non-attachment of the body. The brain calms down and sound sleep, a dream for many old people, comes naturally through the use of props," says Iyengar.

More ways to use yoga props are explained throughout this book.

Eric Small:
Yoga for Multiple Sclerosis and Other Problems Affecting Balance and Movement

Eric Small, in his 60s, demonstrates the Lotus Pose in Headstand.

Eric Small, a Los Angeles yoga teacher in his 60s, keeps multiple sclerosis in remission by practicing yoga.

Multiple sclerosis (MS) is a disease that prevents proper functioning of the body's electrical circuitry. It attacks the insulation (the myelin sheath) around the electrical wiring (the nerves) of the body, causing the myelin sheath to unravel and deteriorate, leaving behind gaps and scar tissue. Consequently, electrical signals do not flow effectively along the nerve pathways, and the system short-circuits. Symptoms of MS include loss of balance, loss of sight, speech impairment, organ dysfunction, motor impairment, numbness, pain and loss of energy. Stress and/or fatigue can severely exacerbate these conditions.

Over 40 years ago I found that my body was no longer responding to signals my brain sent it. I felt like a telephone that always dialed a wrong number, got a busy signal, or found no one at home. The diagnosis: MS.

After I was diagnosed with MS, I underwent every known therapy at the time. Then I discovered hatha yoga. At first the discipline was very hard to come to. As I gained confidence and stopped falling down so much, I started to respond to yoga and gained strength and flexibility. My nervous system was relieved of stress. Little by little, I learned to use my mind to communicate with non-responsive areas in my body. Sometimes I even developed alternative solutions to execute normal functions.

I found that by using props, I could maintain the *asanas* (yoga postures) better. One of the most important tools was water. In the buoyancy of the pool, gravity and muscle strength were no longer negative factors. The water was also very relaxing, and helped moderate my body temperature. (Overheating is a "no-no" for anyone with MS).

When I took my first Iyengar-style class, bells went off. The alignment, the concentration and, of course, the props were exciting and effective. I still use many props in my own daily practice, as well as when I teach yoga to other people who have MS. I find that the Iyengar system allows the MS student to achieve the beneficial effects of the asanas without too much frustration or fatigue.

I have been teaching hatha yoga to other MS students for many years, and find the best place to start is with the breath, to set aside time both in and out of class to develop a smooth breathing pattern. This helps students to calm down and center themselves. It also soothes the nervous system. Working with this general breathing pattern is time well spent. It builds a confidence that the MS student carries into the adventure of learning asanas.

Each case of MS is unique. The symptoms, the degree of nerve degradation, and thus the general health of each individual can vary tremendously. It is of vital importance to ascertain each student's abilities to discover which asanas are most important. The key to asanas for the MS student is to be gentle, not over-ambitious. No pose should be held too long, as this causes fatigue and overheating. Poses need to be supported in whatever way possible to accomplish this. I use walls, chairs, belts, benches, pillows, bolsters, the horse, and even lie flat on the floor for certain poses.

For MS students who use wheelchairs, I have developed an amazing series of asanas that require a folding chair and a belt. Some of my students have even been able to get up into supported Shoulderstand with my help. Though they may still need it, bit by bit these students become less dependent on the wheelchair. Then their whole attitude toward themselves and life improves.

Although hatha yoga is not a cure for MS, it can slow some of its negative effects. I have seen so many positive results over the years: increased blood circulation, improvement in digestive and eliminative processes, more effective muscle performance and increased security in balance, among others. Yoga improves the use of secondary and alternative ways for the body to func-tion. Self-esteem and confidence increase. Students learn to incorporate the knowledge and skills developed through the practice of yoga into their everyday lives. This reduces stress and fatigue, and dramatically improves their lives. Using yoga tech-niques, I have remained symptom-free.

We make a disclaimer that yoga will not cure MS, but that it is a way of teaching the physically limited. I get people out of a wheel-chair and into a regular chair right away—by the second or third class, often the first. One man came in like a crab, as if he were palsied instead of suffering from MS—all contorted in on himself and angry. Now he smiles. His feet and legs were all swollen. Now they're normal. He now sits on a regular chair and gets on the floor to do Shoulderstands, with help. Another man was 84 when he first came to me. He didn't have MS, he was an old man dying. His leg was blue-black from lack of circulation. In three months he had a pink and healthy leg. This man just turned 90.

Hatha yoga is effective where physical therapy isn't, in that you become your own teacher. In yoga, one's breath, mind and body work together to control physical movement.

Four

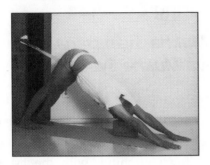

Key Yoga Postures for Reversing the Aging Process

In the mornings, Monday through Friday, we do our yoga exercises. I started doing yoga exercises with Mama about 40 years ago. Mama was starting to shrink up and get bent down, and I started exercising with her to straighten her up again. . . . When Bessie turned 80 she decided that I looked better than her so she decided she would start doing yoga too . . .

—SARAH AND ELIZABETH DELANY, AT AGES 102 AND 104, AUTHORS OF *HAVING OUR SAY, THE DELANY SISTER'S FIRST 100 YEARS*

> The spine is of greatest importance. If the movements you do during the day originate from the spine, then the action is correct. . . . Babies' spines are extremely soft and light and remain so for a long time. Instead, the adult spine is rigid and heavy and yoga, as intended here, consists in breaking bad habits and in re-educating the spine so as to bring back its original suppleness.
>
> —*Vanda Scaravelli, Awakening the Spine*

Downward- and Upward-Facing Dog Pose

Adho Mukha Svanasana and *Urdhva Mukha Svanasana*

The Downward-Facing Dog Pose is named for the way dogs instinctively stretch their bodies. When practiced with the hands on the floor, the shape of the pose resembles a dog stretching, with the arms and hands stretched out like a dog's forepaws, the shoulders, spine and chest stretching and the pelvis and tailbone high, stretching back away from the hands. When dogs stretch they do so with great enjoyment—with all their heart and soul. Naturally, we humans should stretch in a similar way.

Frank White enjoys stretching in the Downward-Facing Dog Pose.

A panacea for people over 50, this ingenious whole-body stretching and strengthening pose combines the benefits of going upside down and bending forward. It is actually like an entire yoga session rolled into one. My experience has been that even octogenarians who may not have stretched for many years—and who may initially have trouble kneeling and getting down and back up from the floor—can begin to enjoy this pose very early in their practice.

Many of my students who started yoga in their 70s or 80s could barely go back and forth from Downward- to Upward-Facing Dog once before their arms collapsed. Now these same experienced students enjoy practicing Upward- and Downward-Facing Dog 10 times or more, both with a chair (as illustrated on the next page) and from the floor.

Downward-Facing Dog Pose is a halfway inverted position that almost everyone can safely practice. This pose inverts the internal organs and increases blood flow to the head. A weight-bearing pose, it strengthens the hands, wrists, arms and shoulders and stimulates bones to retain calcium, thus helping to prevent osteoporosis. Downward- and Upward-Facing Dog poses work together to remove a lifetime of stiffness from the shoulder joints, wrists, hands and fingers. The whole spinal column is lengthened, abdominal muscles are strengthened and neck tension is released. This pose helps prevent and decrease the roundness of the upper back so common among older people in our culture. It corrects rounded shoulders by stretching the pectoral muscles on the front of the chest and·eradicates the obstacles to good posture.

Older students often report that they regain lost height

Those who are afraid to do Headstand *(Sirsasana)* can conveniently practice this position. As the trunk is lowered in this posture (asana) it is fully stretched and healthy blood is brought to this region without any strain on the heart. It rejuvenates the brain cells and invigorates the brain by relieving fatigue. Persons suffering with high blood pressure can do this pose.
—*B.K.S. Iyengar,*
Light on Yoga

after consistently practicing the Downward-Facing Dog and other yoga poses. My students of all ages practice it almost every class, from the floor and with a chair or hanging from ropes or straps.

When practiced from the floor, the Dog Poses require a fair amount of flexibility in the back of the legs, and strength and flexibility in the wrists, arms and shoulders. You can prepare your body by first practicing with the hands on the seat of a chair.

Downward-Facing Dog Pose with a Chair

For people who have back problems or a stiff body, practicing the Downward-Facing Dog Pose with a Chair can help you stretch in a healthy way.

1. Put a sturdy, level chair against a wall. Place your hands shoulder-width apart on the front edge of the chair seat. Keeping your hands on the chair, take a giant step back until you are a full arm's-length away, your heels slightly behind your hips, feet hip-width apart.

2. Press your hands firmly into the chair seat and come up high on your toes, lifting your bottom as high up as possible. Stay up on your toes for several breaths as you push the chair away from you and lengthen your spinal column. Push the chair toward the wall, stretch your fingertips as far away as possible and your buttock bones as far back as possible to lengthen your spine to the maximum. Continue pushing the chair away as you slowly lower your heels to the floor. While you press your heels firmly down, pick up all 10 toes and spread them wide apart until you can see daylight between them! Breathe calmly and freely. Smile so that your face muscles relax.

3. To come out of the pose, bring your body forward toward the chair, bend one knee, and stand up, nice and tall. Sit down in the chair for a few moments if you feel light-headed or need to rest.

Upward-Facing Dog Pose with a Chair

1. Begin in Downward-Facing Dog with hands on the chair seat. To move into Upward-Facing Dog, change the positioning of your hands slightly so that they firmly grip the edges of the chair seat.

2. Keeping your hands firmly gripping the seat of the chair, bring the tops of your thighs and pubic bone

Upward-Facing Dog Pose with a Chair

toward the chair. Continue firmly pressing down into the chair seat, straightening your arms, rolling your shoulders back, lifting your sternum and opening your chest. Look up by taking your head gently back, without constricting your neck.

3. Keep your hands gripping the chair seat as you stretch back into Dog Pose. If you feel unusual strain in your wrists, try padding the chair with a folded sticky mat. As your strength improves, repeat Upward- and Downward-Facing Dog several times. To come out of the pose bring your body forward from Downward Dog back toward the chair, bend one knee and stand up slowly. Sit

down in the chair for a few moments if you feel light-headed or need to rest.

Note: Be careful not to hold your breath. Keep your breath flowing. Stay in the Upward and Downward Dog Poses long enough so you feel your body strengthening and lengthening, but not so long that your arms collapse. Observe the positioning of your toes as you move from Downward into Upward Dog. As your thighs move closer to the chair, you will feel your toes turning.

Downward-Facing Dog Pose from the Floor

1. Kneel on all fours on a non-slippery floor, so that your hands do not slide. Position your knees slightly behind your hips, toes curled under as shown, your feet and knees hip-width (about 18 inches) apart. Place your hands slightly in front of your shoulders, shoulder-distance apart. Spread all

Variation with blocks.

Downward-Facing Dog Pose stretches and strengthens the whole body.

10 fingers wide apart and press both hands down onto the floor.

2. On an exhale, straighten your knees and lift your bottom toward the ceiling, so that your body forms a high upside-down V or pyramid shape. Raise your heels high off the floor and try to lift your bottom higher and higher. Press your hands deep into the floor, as if you are pushing the floor away from you. After stretching for a few breaths with your heels lifted, try pressing your heels down toward the floor, as illustrated on pages 64 and 69.

3. Breathe smoothly, naturally. Keep your face and neck relaxed and soft. Imagine roots pulling your hands and feet into the earth while the top of your buttocks, your tailbone, extends toward the sky. Release, come back to kneeling on all fours. Slowly lower your bottom back toward your heels and lower your torso and forehead to the floor in Child's Pose. (See chapter 6, page 140.)

A common complaint in this pose is pressure on the wrists. If your wrists are extremely sensitive, place a folded sticky mat (or folded blanket on the sticky mat to keep it from slipping) under the heel of your hands, so that the wrist part of your hand is slightly elevated and supported by the extra cushioning.

Do not stay in the pose if your back hurts, if you feel unusual pressure in the head or dizziness, or if your wrists and shoulders ache. Ask a yoga teacher for help.

Upward-Facing Dog Pose

In the Upward-Facing Dog Pose, the body is supported by the hands and feet. The legs and pubic bone

Variation with hands on blocks helps to open the chest.

Upward-Facing Dog Pose.

are off the floor. Older beginners generally find it easier to practice the pose by shifting their weight forward and then lowering their body into Upward-Facing Dog Pose.

1. Lie face down on a non-slippery surface. Position your hands next to your chest, keeping your elbows close to your body. Place your feet hip-width apart. Stretch your legs back and pull your kneecaps up so that your thigh muscles feel firm.

2. On an inhalation, lift your upper body off the floor, straightening your arms. Lift and open your chest. Look forward or slightly up, keeping your neck long and relaxed, your shoulders stretching back and down. Try to bring your legs and pubic bone off the floor. Keep your buttock muscles firm and your legs stretching back. Press down with your hands, straighten the arms more and lift your trunk away from the floor. Keep breathing! Do not sink the weight of your body toward the floor.

A student in her 80s practices Upward-Facing Dog Pose from the floor.

Lower yourself to the floor and relax in Child's Pose.

As strength improves, stretch from Upward-Facing Dog back into Downward-Facing Dog. Rest by bending your knees back to the floor and relax in Child's Pose.

The human body contains approximately 650 muscles, all imprinted with the same ageless message: Use it or lose it. Downward- and Upward-Facing Dog poses develop the strength and shape of long-forgotten muscles and stimulate bone growth, preventing brittle bones and osteoporosis. Most problems in these poses can be overcome by experimenting on your own or with the help of a qualified teacher.

Other Variations of Downward-Facing Dog: Downward-Facing Dog Hanging from a Rope or Strap

Variation with hands on chair.

Ropes lengthen the spine and soothe back tension.

In Downward-Facing Dog Hanging from a Rope, the rope supports most of your weight, allowing you to let go and lengthen your body even more. Hanging from a rope allows you to hold the pose longer, thereby giving the back of the body a deeper stretch. The gentle traction of this pose eases many types of back pain, tension or spasm, while at the same time restoring the natural, youthful curves of the spine.

For people who come to yoga with a rounded back, Dog Pose Hanging from a Rope is invaluable, especially followed by a session with wall ropes going back and forth. You can purchase professional yoga wall ropes from the prop sources listed in the back of this book. Or make a loop out of a strap or soft rope, looped securely over a doorknob or through an eye hook securely attached to a wall stud. Various types of ropes or straps will serve the purpose, but a soft, unbreakable cotton rope or mountain-climbing webbing about six to seven feet long works best. If the strap or rope cuts into your legs, it may not be wide enough. Try either a loop made out of different material, or use a thick towel or smooth blanket to pad the strap if it cuts into your legs.

Look at the photos on pages 75 and 76 and follow these directions carefully.

1. Hook the rope or strap securely around both sides of a sturdy doorknob (the doorknob must be sturdy enough to hold your weight).
2. Step into the loop and place it at the crease at the top of your legs—right where your hips bend at the groin.
3. Lean forward into the rope until your weight feels supported and place your hands on the chair for stability. Make sure the rope is taut—the trick in the beginning is to walk the feet backward while stretching the arms farther forward so that the rope remains taut and does not slip down the legs.
4. Step your feet backwards as you stretch your hands out in front of you on the chair. When you feel secure, move the chair farther away from you, but still keep your hands on the chair.
5. More flexible people can take their hands all the way to the floor so that the body forms an upside-down V

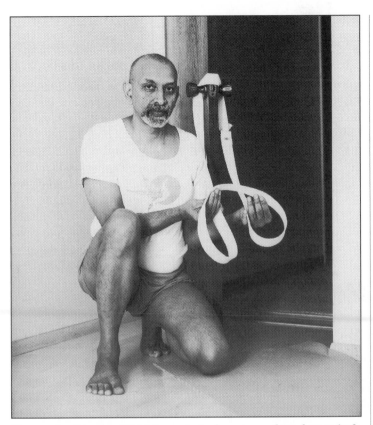

Yoga teacher Ramanand Patel demonstrates how you can hang from a single strap placed at the tops of your legs or with two loops as shown here.

or pyramid. If you find yourself bending your knees to relieve pain in the hamstrings, then place your hands back on the chair. Don't hurry. Eventually you will enjoy stretching in the pose with your hands on the floor.

6. To further lengthen your spine, continue to walk your hands farther away on the chair seat or floor. Keep your legs actively stretching back toward the door behind you, heels stretching down. Feel your spine lengthening (from the top of your tailbone down to the top of your head). Increase the stretch

Downward-Facing Dog Pose, with rope around a door knob and a block under the head.

by lifting all 10 toes up off the floor and spreading them wide apart.

7. Hang like this for several breaths, gradually increasing the length of time to about three minutes. Keep your leg muscles active by stretching your heels to the floor and periodically lifting and spreading your toes.

8. To come out, bend your knees and walk your feet toward your hands. As you become more confident and flexible, you can hang forward a little longer with your heels more underneath your hips. If your legs are stiff or your back feels strain, place your hands on a chair.

9. Stand back up on an inhalation. Step back and lean your body back into the wall or door. Stand quietly and notice how you feel.

Restful Downward-Facing Dog with Chair Variation

Instead of lowering your palms to the chair or floor, clasp each elbow in the opposite hand and rest the elbows on the seat of the chair, lengthening your upper body as much as possible. Rest your forehead on your forearms, or place a bolster, stack of folded blankets or towels on the seat of the chair. If you're still uncomfortable and unable to lengthen your spine, turn the chair around and rest your head on your forearms on the back of the chair.

Cautions for Downward Dog Pose Hanging from a Rope or Strap

Be sure to practice on a non-slippery surface without socks. Be careful not to lose your balance as you step in and out of the rope. If you tend to feel dizzy or light-headed after forward-bending poses, stay in them for shorter periods and increase the time gradually. If you feel unsteady, practice with a sturdy chair within reach. Hanging from ropes secured to a wall makes the pose more stable; plus, you can lean back into the wall when you step out of the rope.

Yoga Master B.K.S. Iyengar designed this way of practicing the Downward-Facing Dog Pose. At his institute in India and yoga centers all over the world, the ropes are attached to hooks in the wall, as illustrated on page 73.

Right-Angle Handstand with a Wall or Other Yoga Prop

Turn page 57 sideways to view Right-Angle Handstand from an upright position. Note that if you were upside down stretching at the wall, your hands would be on the floor, and your feet on the wall approximately at the height where your hands are when you are standing upright. To prepare for Right-Angle Handstand, practice stretching with your hands flat on the wall, shoulder distance apart, feet hip-width apart, your body approximately at a right angle (see page 57).

If you are unsure of your upper body strength, I recommend you first learn Right-Angle Handstand with the help of a teacher before practicing on your own. Some yoga centers have a prop called the "horse," a braced wooden frame similar to the trestle used in gymnastics. This yoga prop is extremely useful for learning the dynamics of Right-Angle Handstand and other important poses.

Before proceeding, review the cautions on inverted poses in chapters 2 and 11.

Right-Angle Handstand is a good example of how yoga strengthens your upper body without sacrificing flexibility. As I mentioned in the introduction, this pose requires quite a bit of strength and daring. Regularly stretching with your hands on a wall, and practicing the Downward- and Upward-Facing Dog Poses, will give you the strength and flexibility you need to safely practice this Upside-Down Pose.

How to Practice
Right-Angle Handstand

1. Practice Right-Angle Handstand by positioning your body as if you are doing the Downward-Facing Dog with your heels on the wall. Kneel on the floor, place your hands about four feet away

Betty Eiler helps a student learn Right-Angle Handstand.

from the wall, shoulder distance apart, and come up into Dog Pose with heels secure on the baseboard.

2. When you feel ready, walk your feet up the wall in a backward Handstand until they are as high (or a little higher) than your hips, forming a right angle as illustrated. Hold as long as your arms feel strong and secure. Walk down the wall. Move your bottom back toward your heels and lower your torso and forehead to the floor (Child's Pose).

If your wrists feel sensitive, place a folded sticky mat, a wedge or a block (made for this purpose, available at most yoga centers) under the heel of your hands as illustrated, so that the wrist part of your hand is slightly elevated and supported by the extra lift. Be sure to place any extra support for your hands on a non-slippery sticky mat.

Handstand in the Hallway or Doorway

Handstands are also known as Full Arm Balance or *Adho Mukha Vrksasana*—meaning Upside-Down Tree Pose in Sanskrit.

Full Arm Balance is energizing and strengthening.

For me as a teacher, it is especially fun and rewarding to help students over 50 learn to kick up into Handstands with confidence. Whenever I feel lethargic or grumpy, I find a wall or a tree and kick up into a Handstand. This pose immediately changes your perspective and you feel like a kid again. If you are a grandparent, you can experience the thrill of your grandchild proudly bringing you to school to demonstrate Handstands in front of the whole class for "show and tell"— as several of my students have reported.

A hallway or doorway is an excellent prop for students who don't have enough momentum or confidence to kick up into a Handstand. Start by practicing the Right-Angle Handstand in a hallway. Start in Downward-Facing Dog with your heels on the wall. Walk the feet up the wall behind you as high as you can. While one foot secures the position by pressing into the wall, the other leg swings up to the vertical position, bringing you into a Handstand. A wide doorway can be used in a similar way.

For safety's sake, be sure to have an experienced teacher guide you the first few times you try this. Your teacher can help you establish the right distance for your hands from the wall or door frame. The exact distance you place your hands from the wall depends on your height, flexibility and the size of the hall or doorway.

Ruth Barati:
Feeling Vital and Vibrant in Your 70s

This pose shows the strength and grace that is the heart of yoga.

For over 40 years, yoga has been my constant companion, my source, my solace, my gateway to joy. What began as a pleasant diversion became a discipline for growth, an ongoing apprenticeship in awareness, attention, mindful observation and commitment.

Its physical, ethical and philosophic foundation allows me in my 70s to wake up each day feeling vital and vibrant, supple and strong, eager to confront the challenges of the world around me and the world within. I continue to delight in the passions of the body and the inquiring journeys of the mind.

Yoga has taught me that the body and mind are partners, not adversaries or strangers to each other, and that there is no paradox in embodied spirituality. This interaction has guided me to a calm center from which I can summon at will serenity and peace, a sense of being part of the mystery and order of the universe and an intuition of union with others. In addition, sharing yoga with my husband and two daughters has given us a refuge and source of strength in adversity and a loving fountain of well-being when the sun has smiled upon us.

The crown has been the humbling and awesome privilege of teaching. The purpose of teaching yoga is for our mutual opening and heart healing. The men and women of all ages (teenagers to people in their 90s) in class journey together into the frontiers of our bodies, minds and spirits. Their insights, wisdom, responsiveness and love are priceless gifts as moment by moment they are teaching me.

The ultimate reward is to watch my students throw off tension and stress and awaken to the messages of the body, as conscious breathing creates a bridge to an enlivened mind and the asanas change their bodies and alter their reality. Divisions within them heal, allowing them to act with wisdom. Centered, they experience in their very tissues a consciousness that makes all that the organism is and does meaningful.

Elizabeth T. Shuey, a student of Ruth Barati

All my life I had been intrigued with yoga, but was never really exposed to it or learned very much about it. That was to change.

After six months on crutches following a hip injury, someone suggested I take a yoga class. As a young person, I had been active in sports, in school and out, was a swimmer and hiker. The long months of living on crutches resulted in stiffness and considerable frustration with my enforced inactivity.

The teacher, Ruth Barati, welcomed me to my first yoga class with such genuine warmth and understanding. She was willing to guide me through activities appropriate to my limited ability. Very soon I realized my upper torso and leg were responding to stretching and relaxation and my spirits were lifted. When I was finally able to bear weight on both legs, she continued to teach and encourage me as I grew stronger. Her quiet and beautiful presence brought a serenity into my life, even when I was grappling with grief over the deaths of both my husband and mother. Ten years later I am still attending her class, and she continues to renew and refresh my spirit and body. I am in better touch with my inner self.

What does yoga mean to me? It means adding a dimension to my life that wasn't there before. When I walk into that room I am aware of a deep spirit, an area of quietness, of loveliness, where there is caring, respect, gentleness; where images and directed thought lead me away from the noise and clutter of the day to an inner awareness of body, mind and spirit.

Yoga has given me a greater respect for the human body and its amazing potential. I am no longer discouraged by its weaknesses, many of which I now see are unconsciously self-induced. Through yoga I have learned the value of stretching and strengthening postures, which have greatly increased my flexibility and feelings of well-being. The awareness of the importance of breathing in its varied forms is a continual theme throughout yoga, the benefits of which are enormous.

I welcome the quiet meditative aspects of yoga, centering myself and the feeling of continuity with the wisdom of ages past. When I am tense and tired and feeling perhaps somewhat older than I'd like to feel, yoga gives me a wonderful feeling of peace, perspective and renewed strength.

Five

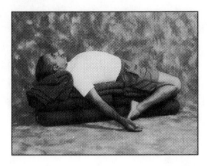

Yoga for Feet and Knees Over 50

Healthy Feet and Knees Can Make the Difference Between Walking and a Wheelchair

Taking up yoga again at age 50, after a hiatus of many years, has been both a revelation and confirmation to me. On the one hand, yoga now has revealed that strength and flexibility can still be mine. On the other hand, somehow I knew that this was true, given my active and healthy lifestyle. As my body increases in flexibility with each week of class, my dedication to the practice increases as well. With the practice comes a new peace of mind through the process of yoga.

—MINDY LORENZ, YOGA STUDENT

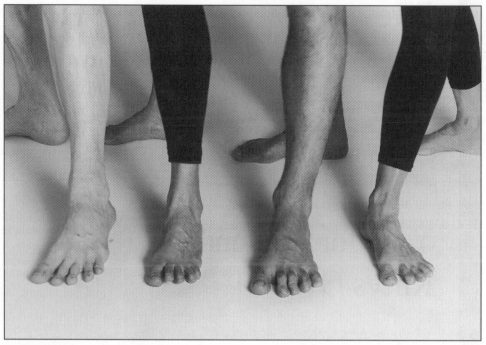

Healthy yoga feet ready for action!

You have to learn to listen to your body, going with it and not against it. . . . You will be amazed to discover that, if you are kind to your body, it will respond in an incredible way.

—*Vanda Scaravelli,*
Awakening the Spine

The feet are our foundation. They connect us to the earth. Yet an awareness of the relationship between our feet and the health of the whole body is often literally the farthest thing from our minds. Let's take a fresh look at these two faithful servants that, in the course of a lifetime, will walk an estimated 115,000 miles—the equivalent of four-and-a-half strolls around the planet.

One of the first features that impressed me about my older yoga teachers was their feet. Their toes could spread wide apart like fingers. In contrast, most of the elderly people I took care of had toes so jammed together from a lifetime imprisoned in too-tight shoes that walking was painful. It was practically impossible to clean the spaces between their toes. Their ankles were swollen and dark from poor circulation. I realized that stretching and strengthening the feet, ankles and toes are absolutely imperative for maintaining one's mobility and independence.

Every architect realizes that the structure of a building depends on a solid foundation for its strength. When the foundation is weak or flawed, problems arise throughout the building. Likewise, many aches and pains—backaches, headaches, leg cramps, even a cranky disposition—can be traced to the body's foundation, the feet. There is a lot of truth to the adage, "When your feet hurt, you hurt all over."

On average, a person takes 10,000 steps every day. Healthy feet are crucial for an active and independent lifestyle, yet four out of five adults experience foot pain—and most of these people are over 50. Surveys indicate that most people think that foot pain is normal. This may be one reason why foot problems that tend to worsen with age are too often ignored.

There are many common ailments of the feet. These include weak, fallen arches; displaced, deformed, overlapping toes; claw toes; hammer toes; calluses and corns; bunions and big toes that seem to be glued to their neighbors. Who or what is the real culprit behind these problems? The shoe industry! Most difficulties people have with their toes and feet are caused or aggravated by shoes that place too much pressure on them. With all the wide, comfortable and attractive shoes available now, it is amazing that shoe manufacturers persist in creating insidious styles that cramp, crowd and squash the feet until the toes are stiff and all but the faintest hint of life has been pressed out of them. Surveys by the University of Southern California School of Medicine and the American Orthopedic Foot

Healing people and healing the planet are part of the same enterprise. People have a deep psychological need for contact with nature; the planet needs the reverential care of humans.

—*Theodore Roszak,*
The Voice of the Earth

and Ankle Society found that 88 percent of women were wearing shoes smaller than their feet and 80 percent complained that their feet hurt.

If the shoe industry stopped manufacturing their tight, narrow, toe-crunching products, we would not lose the spaces between our toes that we had as young children. I am an unabashed fanatic when it comes to wearing shoes that are big enough to spread all 10 toes, just as if I were barefoot. I also find that most people actually believe that their shoes have plenty of room, but the shape of their feet from a lifetime jammed in cramped quarters tells a different story.

The 26 bones of the feet need ample space to distribute and balance the weight of the body. You may not think your shoes are too tight, but if you cannot spread all 10 toes wide apart, then your feet need wider shoes, more time barefoot and yoga. Lack of space results in the over-use or under-use of bones, muscles, tendons and ligaments, and the entire body must compensate. High heels throw off the ankles, knees and hips, the sacroiliac joint, the lower and upper back, the neck, and may even affect the jaw.

If you figure that your parents probably put you in shoes prematurely—often before you began walking—then add the years of adulthood in closed shoes, it is no wonder that feet begin to lose their resilience quite early in life, whether you are athletic or sedentary. Aching feet are one of the first reasons that many older people stop walking—just at a time in their life when the body most needs some healthy, stimulating exercise.

Left untreated, foot ailments alter our body mechanics. Our whole posture is affected by foot problems, which can gradually progress to pain in the ankles, knees, hips, back, and neck and head. Counteracting the effects of aging on the feet requires more than proper hygiene and visits to a local podiatrist. The feet need regular stretching and strengthening exercise such as you experience in yoga.

When I ask my beginning yoga students to lift and stretch all 10 toes wide apart until you can see daylight between them, they usually look at me with utter disbelief. However, as one of my teachers, Mary Dunn, proclaims, "The ability to stretch our toes like fingers and to create a wide, healthy, open space between each and every toe is not some vestigial ability available only to a chosen few."

Revitalizing and Stretching Your Feet

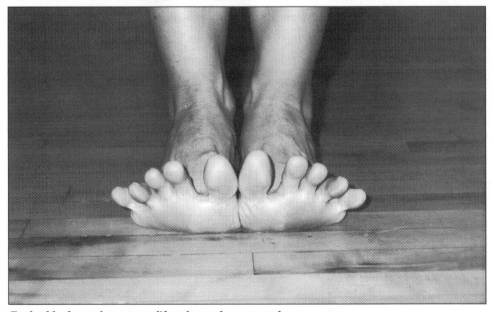

For healthy feet and toes, try to lift and spread your toes whenever you can.

One of the best ways to strengthen your feet is to walk barefoot on natural terrain, just as our ancestors did for the first million years of human history. The more uncomfortable walking barefoot is for you, the more you probably need it! Use common sense. If your feet are very sensitive, start on a smoother, grassy or sandy surface. Then as your feet toughen up, walk on more uneven, pebbly or rocky terrain. This gives the feet a stimulating natural foot massage. (Creek beds are great for this, plus you can slosh through mud or ice cold water, which is very invigorating!)

I'm going to give you the same advice that I give myself: Unless I'm hiking, biking or running, my feet are either bare (in socks around the house on cold days) or in open-toed, comfortable sandal-type shoes. Any chiropractor or podiatrist will be glad to write you a doctor's excuse should your job require heels. The healthier your feet become, the less they will tolerate being crammed into unhealthy shoes.

The combination of walking and yoga is the supreme way to rehabilitate your feet. Walking (in wide shoes, of course!) maintains the overall health of the foot—improves circulation and encourages bone and muscle development as well as improves strength and flexibility in the supporting muscles of the shins, calves and quadriceps.

Try to spread your toes wide apart like fingers. If necessary, reach down with your hands and try to make your two big toes touch each other, as illustrated. Stretch all your toes as far apart as possible; separate the little toe away from its neighbor especially. If your big toe seems unable to move independently of the second toe, pull your toes apart with your fingers every chance you have and practice the toe stretching exercises described at the end of this chapter. Stretching the two big toes toward each other and the little toes away from each other strengthens and lifts the three arches of the foot. The arches hold the bones of the feet and ankles in their correct position, encourage foot muscles to function efficiently, and act as shock absorbers. Fallen arches are often the cause of tired, aching feet.

To evaluate your ankle and foot alignment, have a friend or your yoga teacher stand behind you and check the relationship of your heels and ankles. If the heels and ankles lean inward or outward, this misalignment affects the feet, knees and hips. Check to see if one foot leans in or out more than the other. Are your arches lifted or flattened?

Fingers and Toes Entwined

Separate your toes with your fingers and make them free and independent!

By diligently practicing the Fingers and Toes Entwined exercise described on the following page, you can begin to undo a lifetime of damage to your feet. You will be surprised to discover that all 10 toes will come back to life if you commune with them through stretching and massage.

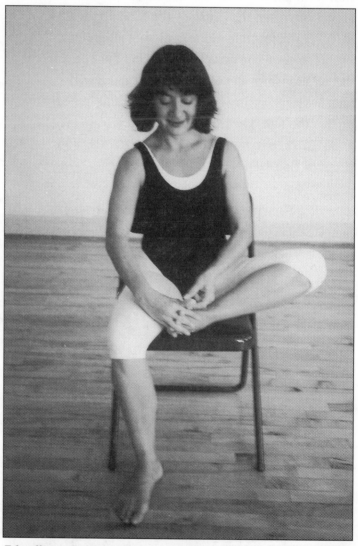

Take off your shoes and stretch your toes.

Fingers and Toes Entwined While Seated

1. Sit in a chair or on the floor. Bend your left knee and take hold of the left foot with both hands.
2. Interlace the fingers of the right hand with the toes of the left foot, as illustrated. Slide the base of your little finger into the base of the little toe, your next finger into the base of the next toe until all your fingers and toes are firmly connected. Then spread your fingers wide apart.
3. Extend through your heel and stretch the toes toward your left knee. Hold for at least one minute and repeat on the opposite side. When you release your fingers, notice how the color of your feet has improved from increased circulation.

If you can't get the fingers between the toes, try placing the heel of your hand on the ball of the foot and wrapping your fingers over the top of the toes. Press the toes toward the knee and stretch them apart as best you can.

As you become adept at stretching your feet, you can simultaneously press the various reflex points on the big toe with your thumb. (For specific points, check a foot reflexology chart.) Use the thumb on your opposite hand to give the big toe an extra stretch by pulling it even farther away from its neighbor. Stretch the little toe and big toe in opposite directions. Pretend that you are "sawing" your fingers in between your toes. Dig deep into the base of any toe that is especially stiff and stubborn, to show your feet that you really mean business and that you are sorry for abusing them all these years.

Keep the base of the fingers and base of the toes connected long enough so that the space between the toes expands. Increase the length of time daily.

Note position of foot on seat of chair. More flexible people can increase hip flexibility by positioning the foot on top of the thigh.

Fingers and Toes Entwined Lying Down

Lie on your back, a folded blanket or book under your head. Bend both knees and place both feet on the floor close to your bottom. Bring your right knee knee toward your chest and place the side of your right foot on your left thigh near your knee. Hold the right foot with your left hand. Interlace the fingers of the left hand with the toes of the right foot, connecting the base of your fingers with the base of the toes. Keeping the fingers and toes entwined, extend through your right heel and stretch your toes toward your right knee.

To increase the stretch in your hips while stretching your toes, lift your left foot off the floor and bring your left knee closer to your chest. Your right hand can be used to support the right knee or to keep your right foot from slipping off your left thigh. (If you have trouble reaching your toes, place a big book or bolster under your left foot and add more height under your head.) Release the right leg. Repeat these instructions with your left leg.

The more painful it is to stretch your toes, the more, I assure you, that you need it! Smile, breathe slowly, and let go of the pain and stiffness. Facing the little pains in the feet will help you to avoid the greater pains in your body.

More Ways to Stretch the Toes

Many of my students ask me what I think about using pedicure spacers, cotton balls or other spacers in between the toes. While they may help a little bit, they are not a substitute for stretching and exercising the feet and toes. Most problems with the feet are mechanical. While it is helpful to walk around with toe separators

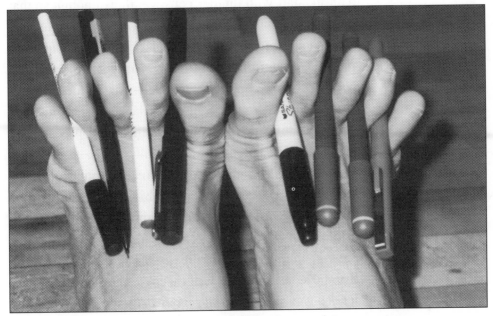

This may look funny, but it is a useful way of expanding the spaces between your toes.

and other orthopedic paraphernalia, nothing takes the place of proper exercise to restore balance to the feet, ankles and rest of the body.

Here is another yoga foot therapy suggestion that has helped many people. In medical classes at the world-renowned Ramamani Iyengar Memorial Yoga Institute in India, teachers insert slender sticks and twigs in between the students' toes. Our Westernized alternative is to replace twigs with pens and pencils of various thicknesses. A practical way to insert them into the base of the toes is to sit on the floor with your feet straight out in front of you, bend the knees to bring the feet within reach, and then experiment with inserting the pens in between the base of each toe. Save the thicker pens for the big toes. If your toes are very tight and painful, start with thinner pens, yet thick enough so that they force you to stretch your toes wider apart. Use your common sense. This usually feels from slightly uncomfortable to painful, but if done sensibly it is stimulating and effective. Keep the pens in between the toes for several minutes. (You can sit with the legs straight, toes stretched toward your face as illustrated.) Gradually increase the length of time.

Yoga places great emphasis on the importance of stretching the feet while stretching the rest of your body. Proper movement and alignment of the feet are often the most effective remedy for foot problems. By paying attention to your feet, spreading the toes and lifting the arches, you will discover that your shoes no longer wear out unevenly, your calluses soften and the spaces between your toes expand as you walk down the road of life.

Foot Reflexology

Many people react with pain when I press certain parts of their feet, a sure sign that their feet need more stimulation. A healing art known as Foot Reflexology teaches that the 72,000 nerve endings on each foot connect to different body areas, and that the toes and their bases contain nerves that extend to glands and sensory organs within the head. Although each toe relates to specific organs within the head, the big toe is said to relate to all glands and organs within the entire head.

Reflexology has its roots in ancient Chinese acupressure and theorizes that by massaging the nerves in the feet, we stimulate corresponding body areas. Ill-fitting shoes place uneven pressure on these nerve endings, whereas walking barefoot, massaging and stretching the feet, improves circulation and benefits the health of the whole body.

According to yoga philosophy, the connection of hands and feet completes the circuits of energy that flow through fingers and soles, and through the right and left sides. In forward-bending postures, the hands often hold the ankles or soles of the feet. By pressing the hands on the ankles or feet, you create resistance and expand the space in joints and spinal vertebrae. This pressure also promotes healthy circulation in the arteries and veins in the bottom of the feet. According to Reflexology, compression massage, even simply placing the hands on the feet, helps the blood on its upward journey to the heart and brain.

According to the Bible and other ancient teachings, the feet are symbols of humility and peace. In many yoga postures the feet are brought above the level of the head (as in upside-down poses), or the head is brought down toward the feet, (as in standing and seated forward bends). In more advanced back-bending postures the feet and head actually meet. If it is true that the head is indeed the seat of the ego, this helps to explain yoga's theory that bringing the head and feet together may help to cultivate more humble, introspective and thoughtful characteristics in our personalities. Yoga also

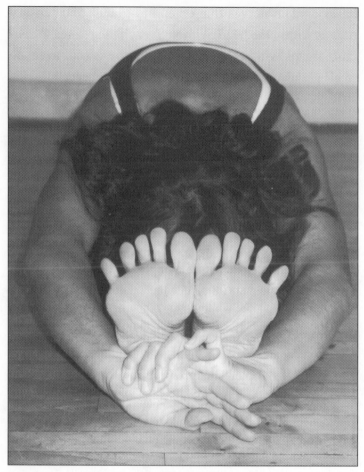

Stretch your toes in all yoga poses. This forward bend stretches the toes and backs of the knees.

emphasizes, however, that before you practice standing on your head or place your body in other unusual positions, you must first "learn to stand on your own two feet."

Your Knees Need Yoga, Too

I can still hear my anatomy instructor rattling on about the knees being the biggest and most complicated joint in the body, while he pointed to the legs of "Miss Bones," the shapely skeleton who dangled faithfully beside his desk.

The knees are between the two longest bones of the body: the femur, or thighbone, and the tibia, or shin bone. The bones of the knee are held together by criss-crossing ligaments, gristle-like structures that join the femur and the tibia to make the knee sturdy and keep it from wobbling. Since almost all the weight of the body passes through the leg bones, thick pads of cartilage called the meniscus act as shock absorbers between the bones when the knee joint is flexed or extended.

These cartilage pads are easily injured. Because of their poor supply of blood, they heal very slowly. Besides torn cartilage, there are at least four other types of knee injuries: sprained ligaments, fractures of the bone joints, wear and tear on the knee cap, (a small, flat triangular bone at the front of the knee) and wear and tear on the muscle tendons.

A quick glance in an anatomy text will help you appreciate the complexities of the knee and show why knees are prone to injury. Exercise that takes the knee joint through its full range of motion is of utmost importance to both prevention of knee problems and repair of injuries.

As in all joints in the body, function is a trade-off between mobility and stability. For example, the shoulder is extremely mobile, but quite unstable and easy to dislocate. The hip is not very mobile, but it is stable. When compared with the hip and shoulder joints for stability, the knee is somewhere in the middle. It is actually one of the most functional joints in the body, allowing us to sit, stand, walk, run, kick, squat and move in hundreds of activities.

As anyone with a knee injury knows, the knees can be mighty unforgiving. When I first began practicing yoga, probably like millions of other Westerners, I strained my right knee attempting to sit cross-legged in the Lotus Pose. When I finally realized that my knee problems were caused by tight hips, I began practicing a series of stretching exercises that gradually gave me enough hip flexibility to sit comfortably on the floor without any complaints from my knees.

Whenever there is a problem in one part of the body, it is always advisable to see how the rest of the body may be causing or contributing to the problem. If the knee joint has been traumatized, it is a good idea to stretch and strengthen the ankle and hip joints, since knee problems are often a reflection of problems at the foot or hip joint. You may remember from the previous section on the feet that if the base of the body is out of balance, strain can occur at the knee. If the hip joints are tight, as they were for me when I first attempted to sit cross-legged, then the rotation needed to sit in this position is forced at the more delicate knee joint instead of coming from the more stable hip joint.

Inflexible joints combined with tight leg muscles force the knees to move in ways inappropriate for their design. Because in our culture the average person chronically tightens muscles, and because we feel this is normal, many people do not realize how restricted their musculoskeletal system is.

The strength and flexibility of the muscles at the front of the thighs (quadriceps), the backs of the legs (hamstrings) and the inner thigh muscles (adductors) all influence the health of the knees. Since most athletic activities involve keeping the knee flexed most of the time, these muscles become tight. Running tightens the hamstrings because the knee is never fully extended and the hamstrings are used as decelerators for the forward-swinging leg, to keep it from moving too far forward. Such repetitive activity tends to tighten both hamstrings and the quadriceps. As the muscles around the knee joint become tighter, they restrict the length of the stride and the freedom of the knee joint. This causes soreness and eventually leads to knee injuries.

A person with an injured knee, or someone with very stiff leg muscles, is generally advised to begin stretching in positions where the knee is kept straight without bearing weight, as illustrated in chapter 3. The knee is most protected when the leg is fully extended. The following leg stretch, described in chapter 3, is recommended for people of all ages, especially those over 50!

This hamstring stretch helps keep your knees healthy.

Practice the Hero (or Heroine) Pose for Healthy Knees (*Virasana*)

In our chair- and car-oriented culture, most older adults lose the natural ability to sit comfortably between their feet. The Hero Pose is especially difficult for runners and cyclists, who often come to yoga with knee problems and extremely tight front thigh muscles. It is not unusual for new students who have not sat in this position since childhood to first attempt the pose by sitting on a high bolster or stack of blankets, folded lengthwise.

Study the photos below, carefully noting the placement of the blocks and folded sticky mat. If you experience cramping in your feet or pain in your ankles, place

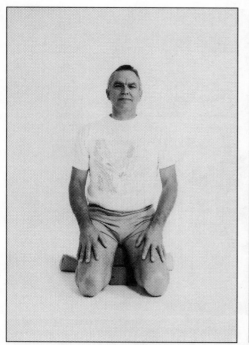

Hero Pose. Note how the props raise the entire torso and cushion the tops of the feet.

a rolled towel, folded blanket or sticky mat under the ankles and feet as illustrated, with the toes on the floor. For stiff feet and ankles, you can also kneel with your knees and shins on two or three neatly folded blankets, with the feet hanging over the edge.

To make the pose more comfortable, it is often helpful to roll a wash cloth, sock or soft rope and place it firmly into the crease behind each bent knee, to increase space in the joints.

How to Practice Hero Pose

1. Begin on your hands and knees with the feet and knees about hip-width apart. Point your toes so that the tops of your feet and ankles are on the floor, toes turned slightly inward. Position your block or bolster behind you in such a way that when you lower your bottom you will end up sitting near the front of the prop.

2. From a kneeling position, slowly lower your bottom to the bolster, blankets, block or floor. If your bottom does not make contact with the floor or prop, or if you experience knee strain, add more height. Place enough support under your bottom so that you can sit upright as illustrated, with your hands pressing down into your thighs.

3. To release, come forward to a kneeling position, with your hands on the floor. Follow Hero Pose with Downward-Facing Dog Pose, described in chapter 4.

If Hero Pose feels impossible no matter how many props you use, try practicing this pose kneeling on a

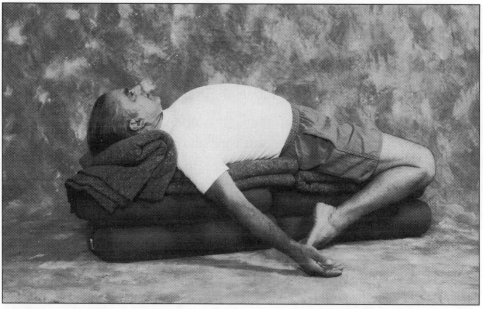

Lying Down Hero Pose is a good way to open the chest and gently stretch the front of the thighs.

Stretch up both sides of the body evenly in this revitalizing pose.

bed. Place a stack of pillows near the edge of the bed in such a way that you can kneel onto the bed with the pillows between your knees and under your bottom. From a kneeling position on the bed, lower your bottom to the stack of pillows. If your feet feel very stiff, position yourself so that your feet hang off the side of the bed.

The above photo illustrates one of the many ways stiff beginners can practice the Lying Down Hero Pose *(Supta Virasana)*. Become comfortable sitting upright in the pose before attempting a lying down variation. The guidance of a teacher is highly recommended.

As you become more comfortable, you can lengthen your spine by stretching your arms upward. Interlock your fingers and note which thumb is in front. Turn the palms out and stretch your arms forward and up. Stretch your trunk up with the arms and draw the arms farther back. Continue stretching upward for several breaths. Release your arms down. Reverse your interlock (opposite thumb in front) and stretch upward again.

Ruth K. Lain at 82:
Yoga Opens Doors to New Friends and Interests

Ruth Lain, 82-year-old yoga teacher, who has practiced yoga since she was 65, demonstrates the Plow Pose.

I feel physically, mentally, emotionally and spiritually better than I ever have in my life. That's an honest fact.

I'm much calmer, and I'm much more stable emotionally and spiritually. Yoga doesn't teach spirituality and meditation, but the support and care of the soul are its by-products.

With yoga, the key word is "aware." Yoga teaches you to be aware of every aspect of your life, including the food you eat. Many people aren't aware of what they do, and yoga helps you with this.

My husband recently died. When anything traumatic happens, yoga is your back-up. When you've been married 52 years, there is such a bond, you cannot just suddenly break that bond. . . . I don't know how I would have survived without yoga. It would have been much harder.

After retiring from public school teaching at the age of 65, I became interested in yoga through Lilias on P.B.S. [Public Broadcasting Service]. At that time my husband and I were living on a farm 40 miles from Corpus Christi, Texas, where there were no yoga instructors—we drove 150 miles to San Antonio. There I had three private lessons before my teacher moved to New York. Totally dedicated to yoga but again without an instructor, I decided to share what I had learned with some of my farm neighbors. Of course, this sharing inspired and challenged me to learn and practice more.

I decided that since I couldn't find a local yoga teacher, I had to become one. Actually, I hauled in unwitting neighbors, relatives and anyone I thought was open enough to try.

My living room class gradually expanded and I started other small classes in abandoned rural schoolhouses and church annexes. After a few years I had six groups in Corpus Christi and was spending a lot of time driving home at night on the highway. My husband and I moved into town and transformed our garage into a studio, called the "Yoga Room," where I now give private lessons and teach nine classes weekly to both men and women of all ages. I also continue with two groups at the Y.W.C.A.

During my lifetime I have had many serious illnesses: 10 operations, a chronic asthmatic condition and a duodenal ulcer. Now because of yoga I feel really good and am never sick. I haven't had asthma since about the first year into doing yoga. I take no medication. Besides physical, mental and emotional health, yoga has opened doors for new friends and interests. I feel truly blessed to be continuously exploring this beautiful way of life.

My personal joy in yoga as well as responsibility to my students encouraged me to study at the Iyengar Institute in San Francisco and to attend workshops with other Iyengar teachers. I've taken week-long and weekend courses with some of the top teachers in the United States. I learned from them, from books and from practicing. There are 8,000 poses in yoga, so there's always something new to learn.

I'm still teaching 10 classes a week and giving five or six private lessons. Recently I had an 80-year-old woman call me and ask, "Am I too old to do yoga?" I asked her, "Can you still breathe? Well then, come for a lesson."

Six

Yoga: A Reliable Companion During Menopause

Actually, aging, after 50, is an exciting period; it is another country. Just as it's exciting and interesting to be an adolescent after having been a child, or a young adult after having been an adolescent. I like it. It's another stage of life after you're finished with this crazy female role.

—Gloria Steinem, interviewed by Cathleen Rountree in *Women Turning Fifty*

Today more women than at any other time in history are approaching or are in the midst of menopause. Within the next 20 years, between 40 and 50 million women in the United States will enter this stage of life.

Menopause itself is changing, in part because the social and economic roles of women have altered so dramatically in the last 50 years, and in part because we understand this transition differently. The work of women like Germaine Greer, Sadja Greenwood, Susan Lark, Christiane Northrup, Gail Sheehy, Susun Weed and many others is redefining this important time in a woman's life. These researchers all recognize the special value of a mind/body discipline such as yoga, especially during menopause. "If you don't already do an hour or more of yoga, t'ai chi, or some other meditative physical activity weekly, begin now," advises Susun Weed in *Menopausal Years, the Wise Woman Way: Alternative Approaches for Women*. And, "It is of the utmost importance for any woman over 45 faced with high-stress professional or personal demands to commit herself to some restorative relaxation measure," echoes Gail Sheehy, in *The Silent Passage: Menopause*.

> Menopause is a time for taking stock, of spiritual as well as physical change, and it would be a pity to be unconscious of it.
>
> —*Germaine Greer*, The Change, Women, Aging and Menopause

What Is Menopause?

Technically speaking, menopause refers to the final menstrual period and the end of a woman's reproductive life. However, most people use the term menopause to mean the year or two of hormonal changes leading up to the cessation of the menses and all of the remaining years of a woman's life.

Menopause, like puberty, is caused by shifts in the balance of hormones within the body. At about midlife, a woman's ovaries, the glands that regulate fertility and the menstrual cycle, reduce their production of estrogen and progesterone. Menstruation becomes irregular and the number of days between periods varies until menstruation finally stops. For most women this occurs in their late 40s or early 50s, although for some this occurs as early as 35.

While these hormonal changes are taking place, some women bleed more frequently than normal or experience dramatically heavier flow. For many women in our culture, changes in their hormonal balance cause uncomfortable "hot flashes" or night sweats. Menopause is associated with a host of other symptoms, including mood swings and depression, urinary problems, vaginal dryness, aching joints, reduced libido and weight gain. Few women experience all of these symptoms, although 75 percent of all menopausal women experience one or two. Some have no major symptoms at all except the end of menstruation.

Menopause also affects women psychologically. Mood swings not unlike those associated with premenstrual tension are reported by some women, while others describe periods of depression, confusion or memory lapses. Many women experience disturbances in their normal sleeping patterns.

However, once the physical changes of menopause are complete, women are free from the powerful hormonal ebb and flow that governed their fertility. They are in a steady state, focused, energized and ready to use the wisdom of their experience. Many women, like Margaret Mead, the famous American anthropologist, feel a new excitement and vitality, which she described as "post-menopausal zest."

While some women sail through menopause virtually symptom-free, others find it distressing. If a woman's identity is tightly bound to being a mother, or if she has not had children, she may grieve the loss of her fertility.

One of the least understood or least tolerated manifestations of the older woman's personality is her withdrawal from the abundantly other-directed behavior patterns of her mothering period, into a more self-directed mode of life, the change more or less coincident with menopause.

—*Barbara Walker,*
The Crone: Woman of Age,
Wisdom and Power

If being sexually attractive is crucial to her feeling valued in her relationships, she may feel the best part of her life is over. Judith Lasater, Ph.D., P.T., in her book, *Relax & Renew: Restful Yoga for Stressful Times*, reminds women that:

> *Perimenopause [just before menopause] comes at a time when other stressful events are challenging women. Sick or dying parents, career demands, a partner retiring or cutting back on work hours, changing financial status, or children leaving home do little to smooth the effects of the roller-coaster ride of changing hormone levels.*

How Culture Affects Our Responses to Menopause

In many cultures menopause is viewed more positively than it is in America, and it is accompanied by fewer negative symptoms. Women may look forward to menstruation's end and the freedom from managing their fertility. In cultures in which old age is honored and women are revered, menopause confers a higher status. In these cultures, menopausal women become the elders who, with their wisdom and experience, guide community life.

As always, our expectations influence outcome. The way we perceive life's stages plays a crucial role in how we experience them. In her excellent book, *Women's Bodies, Women's Wisdom*, Christiane Northrup, M.D., explains how the Celtic cultures view the stages of a woman's life:

> *. . . the young maiden was seen as the flower; the mother, the fruit; the elder woman, the seed. The seed is the part that contains the knowledge and potential of all the other parts within it. The role of*

Though I wrote about living in *The Fountain of Age,* in my personal life I was still being terribly workaholic. So last summer, instead of writing all day I did yoga, I took painting lessons . . . What happens to you after 50, after the reproductive years, more than any other period of your life, depends on what you choose to do. It isn't programmed biologically. To have purpose is essential for vital age.

—Betty Friedan, at age 72

*the postmenopausal woman is to go forth and reseed
the community with her kernel of truth and wisdom.*

In our culture, in spite of many positive changes, aging itself is still hated and feared. In jokes, television shows and greeting cards—all powerful signs of cultural attitudes—we are said to be "finished" or "over the hill" once we reach 50. If we believe that, it's no wonder we perceive menopause as degeneration. However, if we refuse to accept such cultural stereotypes and take charge of the way we think about menopause, it can become the most powerful period in our lives.

Menopause As a Time for Reflection and Re-Direction

Menopause, like other transitional periods in our lives, can be an opportunity to sort out our priorities and celebrate the depth and richness of our experience. If we are at peace with ourselves as mothers or women without children, we may find the freedom from menstruating and managing our fertility liberating. We are at a new stage of our lives. Children, if we have them, are more independent, perhaps even out of the house. Our partners, if we have them, may be experiencing new levels of success or satisfaction. There is time for a different kind of closeness in our relationships. If we live alone, we are more sensitive to our own needs, taking greater pleasure in having things our own way and in the freedom to make our own decisions. We are also more perceptive about ourselves, clearer about our relationships.

Some of us may feel troubled, full of regret for things we've left undone, paths left unexplored. It takes some

When menopause arises as a crisis for us it is sometimes not only the physical symptoms but also the predicament of being a woman undeniably aging in a culture which seems only to value young women.
—Paddy O'Brien, Yoga for Women

courage to face that and lay it to rest, so for a time we may suffer. But in general, our perspective is richer and broader. Years of experience have shaped the way we see the world and made us more confident in ourselves.

What Yoga Offers Women During Menopause

At this juncture, the practice of yoga asanas is extremely beneficial, as it calms the nervous system and brings equipoise.

—*Geeta Iyengar,*
Yoga, A Gem for Women

Yoga can be a powerful tool for helping women experience the passage into menopause as a positive event, both physically and spiritually. For women wanting to break free of lifelong negative habits and addictions, practicing yoga provides strength and support. Because it works by balancing the endocrine system, yoga reduces the effects of menopause's hormonal changes. Not only does regular practice of yoga help ease the physical aspects of menopause, but it inspires a spiritual awakening that helps women open to the power and beauty of this profound change.

How Menopause Can Help Us to Expand Our Practice of Yoga

During menopause, many experienced practitioners discover they are more open to exploring the spiritual aspects of yoga. They feel a deep pleasure in the peace and feeling of unity with the universe that they glimpse during their practice. The menopausal years are a time for women to explore and strengthen their spiritual understandings through their open, flowing and intuitive approach to yoga.

Instead of attempting to mold the body into classic postures, many experienced older women teachers advocate experimenting with yoga in a different way—with less focus on control and more on discovery and exploration. There is richness in finding your body's own way into the poses, exploring your strength, fully experiencing the present while consciously releasing the past and making space for the new.

Yoga Postures to Relieve Hot Flashes, Night Sweats and Other Symptoms

For centuries the classic inverted postures such as Headstand and Shoulderstand, as well as various relaxing forward bends and restorative postures, have been valued for their cooling and calming effects. My students tell me these postures are an effective antidote to hot flashes and other common symptoms of menopause. A long stay in Shoulderstand has a particularly quieting effect on the brain and nervous system. According to 96-year-old yoga teacher Indra Devi, Half and Full Shoulderstands are the most important postures for correcting imbalances or hormonal problems in a woman's reproductive system.

Christiane Northrup, M.D., reports that hot flashes, also known as vasomotor flushes, are experienced by 80 to 90 percent of American women during the menopausal years. During hot flashes the surface temperature of the body rises and sweating occurs, sometimes profusely, often around the head and neck. Some women experience hot flashes for only a few months, others for several years. A small percentage report these disturbances for a decade or more.

Scientists are still not sure what causes hot flashes, perhaps the most infamous symptom of the menopausal years. The most widely accepted theory is that these

When I started yoga in 1958 I considered it a tool, the best and most complete tool available, to live physically and psychologically at optimum level. . . . Now that I am 50 years old yoga has once again become a tool for living. I tend to put more stress on enjoyment both in and outside the field of yoga, using the benefits of a balanced yoga practice to enjoy the miracle of life and one's own potential to the fullest possible.

—*Dona Holleman, yoga teacher, author of* Centering Down

sudden and dramatic increases in surface body temperature occur when your body's thermostat, located in the hypothalamus, is disturbed by hormonal imbalances. Because hot flashes involve the neuroendocrine system, unresolved stress tends to increase them.

Regular yoga practice, as well as other regular exercise such as bicycling or walking, can help alleviate these symptoms. In particular, yoga's inverted poses—such as Handstand, Headstand, Downward-Facing Dog Pose, Shoulderstand, Supported Legs-Up-the-Wall Pose—which position the body completely or halfway upside down, reduce the incidence and intensity of hot flashes and night sweats.

Inverted yoga poses have a measurable effect on what physiologists call hemodynamics—the flow of blood to every organ of the body. They also have a beneficial effect on the glands of the endocrine system that produce the hormones that regulate many of the body's processes. Inverted postures are also cleansing and boost the health of the immune system. According to psychobiologist and yoga teacher Roger Cole, Ph.D., who has conducted extensive scientific studies on the physiological effects of yoga poses:

> Turning the body upside down also tricks the body into believing that blood pressure has risen, because the receptors that measure blood pressure are all in the neck and chest region. The body takes immediate steps to lower blood pressure, including a relaxation of blood vessels and a reduction in the hormones that cause retention of water and salt.

On a more subtle level, according to the ancient yogic texts, the inverted postures affect the flow of *prana*, or life-force energy. They draw the prana inward, toward our vital organs, toward the core of the body and away from the surface—the skin. According to some theories, during hot flashes prana is flowing outward from the center, heating the skin.

For experienced students working under the guidance of a teacher, Lying Back Over a Chair and Supported Bridge Pose, also help reduce hot flashes. (See pages 122-123, 137-139 and 214-215.)

Although most women who experience hot flashes seek relief, some researchers believe the increase in the body's surface temperature might serve a useful function. Vickie Noble, in her book, *Shakti Woman*, reminds us that a high body temperature kills bacteria. She and other researchers feel that hot flashes may actually be healthy. When we attempt to shut down the hot-flash process, we might be interfering with a subtle healing mechanism.

Yoga for Easing Mood Swings and Depression

Shifting levels of estrogen and progesterone, both mood-altering hormones, can result in irritability, anxiety and depression during menopause. "In menopause as in puberty, you are going through hormonal changes," says yoga teacher and grandmother Felicity Green. "When a child is going through puberty, we're patient with her. In menopause you have to be patient with yourself. Women should realize this change is normal and natural and give ourselves some time to be quiet."

Herbalist Susun Weed writes in *Menopausal Years, the Wise Woman Way: Alternative Approaches for Women:*

Joint mobility increases rapidly with the focused attention and gentle stretching of yoga postures. Yoga postures, yoga breathing, and quiet, focused meditation tone and soothe the sympathetic nervous system. Regular practice alleviates anxiety. Yoga— not just the postures, but the breathing and focusing exercises as well—helps create strength in the nerves, adrenals, and heart, making sensitivity and irritability an ally rather than a liability.

Forward Bends

In addition to calming the mind and soothing the nervous system, forward bends (see page 133 for

Forward bends calm the mind and soothe the nervous system.

instruction) gently stretch the spine and lengthen the entire back of the body. They encourage an attitude of surrender and acceptance that relieves stress and alleviates many of the negative symptoms associated with menopause.

A simple Standing Forward Bend inverts the upper body and brings the head below the heart. In this position the pituitary gland is stimulated. This small gland in the center of the brain is involved in the regulation of blood sugar levels and body temperature.

Forward bends also gently compress the abdomen, massaging the uterus and other abdominal organs. When you come out of the pose and release the compression, the organs are bathed in freshly oxygenated blood and you feel refreshed, renewed.

Yoga to Balance Hormonal Changes

Yoga helps modulate mood swings and reduce fear and anxiety by helping to balance a woman's changing hormones. Many of the symptoms commonly associated with menopause, such as aches, pains and irritability, are intensified by stress; therefore, practicing yoga's relaxing, restorative poses helps ease them.

While the ovaries decrease their production of androgenic hormones—the hormones associated with sexual response, libido and a general sense of well-being during menopause—other parts of the body, such as the adrenal and pineal glands, actually increase their hormonal output. Unfortunately, in many older women the adrenals are damaged by habitual stress, poor nutrition, environmental pollution or other problems. The failure of the adrenal glands to perform this compensatory function may produce symptoms such as mood swings, fatigue or depression that are commonly attributed to menopause.

Yoga poses, such as twists and backbends, improve the functioning of the adrenals, helping to increase the amount of estrogen in the body. These poses also stimulate the kidneys, promoting healthy elimination of metabolic by-products.

Inverted Poses to Alleviate Fatigue

For me, these last six months, I just crave going upside down. You just crave it. It takes the place of chocolate addiction.

—Judi Flannery-Lukas, yoga teacher at midlife

Fatigue is a signal that the body needs to rest, repair itself, restore its energy. When you are tired, it is not the time to exert yourself with strenuous exercise. This is especially true during menopause. Inverted poses are famous for reducing fatigue, feelings of heaviness or lethargy. They not only improve circulation and increase blood flow to the brain, they calm and soothe the emotions. These poses literally change our perspective. It never ceases to amaze me how different the world looks after a long stay in Headstand and Shoulderstand, or in easier inversions such as Legs-Up-the-Wall Pose. (See chapter 11, Restful Inversions.)

Menstruating women often feel heavy and tired at certain points in their cycle. This fatigue can be attributed, in part, to congestion in the pelvic organs. Inverted poses are especially beneficial to these women in the 10 days prior to, and the days immediately following, their menstrual period. Inverted poses, however, are not recommended during menstruation itself.

When you feel too tired to do anything, practice supported inversions. If you are not yet familiar with Supported Shoulderstand, then scoot close to a wall and put up your legs. If you fall asleep, it will be a deep, refreshing rest. You can even lie with your legs up against a headboard or wall in bed.

If you have a headache as well as fatigue, practice Supported Half-Plow Pose and Supported-Legs-Up-the-Wall Pose. You can also use bolsters to help you rest in

Ruth Lain demonstrates the One-Legged Shoulderstand. This Shoulderstand variation refreshes the legs and strengthens the back.

various positions. Supported Dog Pose and supported forward bends prior to (or after) supported inversions are also healing and restorative.

Poses to Regulate the Menstrual Cycle

A common problem during the perimenopausal years is premenstrual tension resulting from delayed menstrual periods. Upside-Down Bow Pose and Lying Back Over a Chair, where the pelvis opens and the adrenal glands are activated to produce estrogen, may help regulate the menstrual cycle. Backbends are also known to stimulate the ovaries and fallopian tubes.

On the other hand, some women approaching menopause get their periods too frequently, sometimes

Suza Francina demonstrates Lying Back Over a Chair. Supported backbend poses have a powerful physiological effect. They nourish the nervous system and increase the efficiency of the glandular system.

Women cannot afford to neglect these tremendously beneficial backbends. Note how the whole front of the body is stretching.

More beginning students should practice as illustrated on page 30. Bolsters may be needed under the head to keep the neck and back comfortable.

as often as two or three times a month. Their cycles become erratic. These women often ask whether they should completely avoid inverted poses while they are menstruating. The answer is individual and requires a high degree of body awareness. To help you determine what is best for you, please seek the help of an experienced teacher.

According to yoga master B.K.S. Iyengar:

This "in-between-period bleeding" is not the same as menstrual bleeding. The in-between bleeding may be caused by irritation within the uterus. For example, if after seven days of a regular period the flow has decreased, the yoga practice should be adjusted to further decrease bleeding. Exertion, poses that stimulate the ovaries, may increase the flow and prolong the period.

Yoga to Relieve Excessive Bleeding or Bleeding at Irregular Intervals

If menstrual cycles are normal, inverted poses should not be practiced during menstruation. However, in case of certain menstrual disorders or in the presence of fibroid cysts, inverted poses during the menstrual flow are recommended.

Inverted poses and backbends stimulate the endocrine glands and correct their function. Any endocrine disorder can also be improved by yoga postures and breathing practices (pranayama), which involve the prolonging and restraining of the breath and help regulate hormonal levels in the blood. These practices should be learned under the guidance of a teacher.

Bound-Angle Pose.

Betty Eiler in the Seated Wide-Angle Pose. The centeredness and symmetry of this pose are very soothing.

To relieve bleeding in between normal periods, it is recommended that women practice healing postures such as Seated Wide-Angle Pose, Seated and Lying Down Supported Bound-Angle Pose and inverted poses.

Yoga to Alleviate Pelvic Congestion

Women often experience a feeling of heaviness and congestion in the first months after their periods have stopped. This is an excellent time to make dietary improvements and to consider a body-cleansing program. Increasing fresh fruits and vegetables will help reduce feelings of heaviness in the abdomen.

All inverted poses, forward bends, the Bound-Angle Pose (both sitting upright with the soles of the feet together and Supported Lying Down Bound-Angle Pose, described on page 134) and Wide-Angle Pose, help reduce pelvic congestion. Headstand,

Lying Back Over a Chair, Shoulderstand, Plow Pose and seated forward bend poses have specific influences on the psycho-neurohormonal system. These postures help reduce tension and thus help reduce menopausal symptoms.

Upside-Down Poses After the Menstrual Period

The classic yoga texts, *Light on Yoga* by B.K.S. Iyengar and *Yoga, a Gem for Women,* by his daughter, Geeta Iyengar, recommend that Headstand and Shoulderstand be practiced after every menstrual period to ensure an inner dryness once the bleeding has stopped. Some experienced teachers caution women against switching too quickly to their usual, more active practice after menstruation. Women who follow the recommended gentle, supported forward bend and restorative practice during the menstrual period and then resume a routine of more strenuous standing, backbending and balancing poses immediately after the period stops may run the risk of overexerting themselves. Some teachers believe that this may contribute to health problems as one gets older, especially during the perimenopausal years.

It seems wise to consider the yoga practice immediately following one's period as a healing practice. Inverted poses and their variations heal the reproductive organs, the organs affected by menopause. Iyengar and other respected teachers recommend a long, quiet stay in Headstand or Shoulderstand, without too many variations. Iyengar writes, "Straight asanas keep you cool, calm, quiet, and when that feeling has come, the mind gets subdued."

The long, motionless holding of inversions opens the door to a state of meditation. It is during and after the peaceful practice of inversions that one experiences poise, peace and stillness at a deeper level. A long stay in the poses quiets and strengthens the nervous system, an invaluable gift at all stages of life, but most especially during the menopausal transition.

According to B.K.S. Iyengar:

More than Sirsasana (Headstand), Sarvangasana (Shoulderstand) is helpful because the pituitary gland is not stimulated as much as it is in Sirsasana. That is why the feeling after Sirsasana is very different. . . . That is why I often say that after Sarvangasana the backbends will not be suitable. You do Sirsasana then you do backbending because the pituitary is already active and you can do better backbendings. But after Sarvangasana you cannot do a backbending course.

Standing Poses Also Help Relieve Menopausal Symptoms

Standing poses such as Triangle Pose and Half-Moon Pose, with the support of the wall to open the pelvic region, and revolved variations of these poses are also enormously beneficial in helping women adjust to menopause. Standing poses strengthen the whole body, especially the feet and legs; they are tremendously grounding, helping us to re-establish and deepen our connection to the earth. They build confidence and can help us "stand on our own two feet" again. However, if your legs feel heavy and fatigued, listen to your body and take care not to exhaust yourself with long holds in vigorous standing poses. Standing poses practiced against a wall can be helpful when there is dysmenorrhea—pain in the groin, abdomen or pelvis. A moderate standing pose practice such as Triangle and Half-Moon Pose, standing forward bends and Lying Down Big Toe Pose, may help alleviate pain and fatigue in the legs.

When your legs feel heavy and tired, especially during your period, or if your cycle is erratic, it is refreshing to start your practice with supported lying down postures. Lying Down Hero Pose, Lying Down Bound-Angle Pose or simple Lying Down Cross-Legged Pose are all relaxing and restorative.

Using the chair allows Malchia Olshan to remain in the Shoulderstand much longer.

One-Legged Shoulderstand with a Chair helps beginners keep their backs straight.

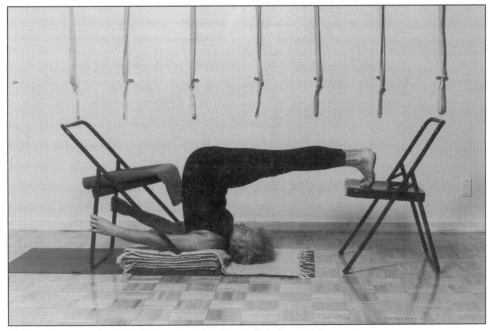

Resting the toes on a chair helps you to straighten and lift your back.

Supported lying down poses followed by Lying Down Big Toe Pose, both straight and with the leg extended to the side, all work to relieve stiffness and tiredness in the legs. When you are not on your period, practice Legs-Up-the-Wall or other inverted poses.

Swollen Legs, Ankles and Varicose Veins

All poses discussed previously, especially inverted poses, are useful for relieving swollen legs and ankles and problems with varicose veins. A long stay (5 to 15 minutes) in simple Legs-Up-the-Wall Pose is highly recommended. Especially if you are standing for long periods of time or involved with sports or athletic activities, as soon as you can conveniently do so, put your feet up on the wall and allow the swelling to decrease. Supported Bridge Pose, Supported Legs-Up-the-Wall Pose, all help decrease swelling of the feet and ankles.

A Word on Stiffness and Starting Yoga Later in Life

People who have not exercised for many years and whose diets are heavy on sugar, caffeine, meat and processed foods commonly experience weight gain and swelling of the joints. A type of stiffness may come into the body that is due not only to lack of movement but also to toxicity. Often these people find it difficult to practice yoga, but they are among those who need it the most.

In these cases, the yoga props described in chapter 3 are especially useful. Without props, people with little energy and many problems often lose heart. Props encourage the body to open and lengthen gradually.

With the help of a wall, people of all ages and conditions can begin practicing the standing poses and simple inversions. Lying back on bolsters and blankets, almost everyone can practice gentle backbends. Forward bends can be practiced sitting on a prop that adds height or with the support of a chair. Again, Supported Lying Down Bound-Angle Pose to open the groin and hips and restore energy is highly recommended.

It is important for women who feel very stiff and rigid prior to and after the time of menopause, to realize that this need not be a permanent condition. If you practice yoga faithfully, the stiffness will leave. With yoga and nutritional support, the body can be cleansed, renewed and rejuvenated. However, without healthy movement, pain and stiffness, especially if masked by medication, can settle deeper, leading to arthritis, osteoporosis and other health problems.

Menopause Practice Guide

The following is a sample sequence of poses to stimulate the ovaries and pituitary gland to produce more hormones. This series has a calming, soothing, quieting effect on the nervous system and, if practiced regularly, helps ease menopausal symptoms. These postures provide a good foundation on which to build a yoga practice that will help maintain a woman's well-being during her menopausal years and beyond.

- Standing Forward Bend with your bottom against a wall (Wall Hang; see page 133)
- Dog Pose (hanging from a strap if available, or with the head supported; see chapter 4, page 73)

- Supported Lying Down Bound-Angle Pose
  (see page 135)
- Supported Legs-Up-the-Wall Pose
  (see chapter 11, page 222)
- Supported Bridge Pose (lying back on bolsters or
  blankets; see pages 138 and 215)
- Supported Child's Pose (see page 140)
- Lying Down Bent-Knee Twist (see page 142)
- Supported Deep Relaxation Pose
  (see chapter 12, page 246)

*Experienced students can practice Headstand (either
with the head on the floor or practiced between two
chairs, as illustrated on page 54) and Shoulderstand,
after the Standing Forward Bend and Downward-
Facing Dog Pose. More flexible students (those who can
bend forward in seated forward bends without strain)
can vary their practice with supported seated forward
bends before Deep Relaxation Pose.*

Standing Supported Forward Bend at the Wall (Wall Hang)

Standing forward bends are halfway upside-down
positions that bring the head and neck below the level
of the heart. In this variation, known as Wall Hang, you
lean your bottom against the wall and bend your torso
forward. The whole back of your body releases and
lengthens, tension flows out of your body, and the mind
quiets down and feels calmer. By remaining with the
head hanging down for a half minute or longer, the
pituitary gland, located in the center of the brain, is
stimulated.

How to Practice
Standing Supported Forward Bend at the Wall

Wall Hang is a relaxing forward bend that brings your head below the level of your heart.

1. Stand tall against a wall with your feet hip-width apart. Keeping your bottom pressing back against the wall, step forward about 12 to 16 inches. The distance from the wall depends on your height and flexibility. Keep your feet hip-width or a little wider apart and breathe out as you release your torso forward.

2. It usually feels relaxing to fold your elbows and allow the weight of your arms and torso to gently stretch your body toward the floor (without bouncing or straining). If tightness in the back of the legs makes this position uncomfortable, practice with your elbows or hands on a chair seat or other support.

3. Keep your legs straight and the quadriceps muscles at the front of the thighs pulled up and active. You will see the definition of the front thigh muscles when you contract them. This action both stabilizes your knee joints and allows the hamstring muscles at the back of the thighs to release.

4. Stay in this Supported Standing Forward Bend for about one-half to one minute. As you become more experienced, you will enjoy remaining in the pose longer. To come up, place your hands on your legs and inhale deeply as you return to standing upright. Step back to the wall and stand tall with your back relaxed against the wall. Feel the calming, soothing effect of this pose.

Supported Lying Down Bound-Angle Pose

Lying Down Bound-Angle Pose can be practiced by simply lying flat on the floor, with the soles of the feet together. However, practicing this position with support from bolsters and blankets opens the chest, abdomen and pelvis and allows the body to relax deeply. Extra support under the forearms, knees and outer thighs makes the pose supremely comfortable. This pose is

This nourishing pose is one of the great gifts of yoga.

beneficial to those with high blood pressure, headaches and breathing problems. It is a key pose to practice during the menstrual period and menopause transition. It relieves tension and constriction in the abdomen, uterus and vagina and may help balance hormonal processes. The centering, balancing effect of this pose helps reduce mood swings and depression.

How to Practice Supported Lying Down Bound-Angle Pose

As in other supported poses, the height of the support under your back depends on your height and flexibility. New students can start with less height or with one or two blankets folded lengthwise.

1. Sit in front of the bolster or blankets. The edge of the support behind you should touch your tailbone. Place a double-folded blanket (or two single-folded blankets) at the far end of the bolster to support your head when you lie back. Sit for a few breaths with the soles of your feet together, your fingers on the floor slightly behind your bottom.

2. Use your arms for support as you lower your back and head onto the bolster or blanket. Tuck the double-folded blanket under your neck and head so that your forehead is slightly higher than your chin. Your head should feel comfortable, neither too high or too low. Be sure the bolster supports you from your sacrum to the top of your head. If you feel discomfort in your back, try repositioning your back on the bolster—you may be too high or too low on the bolster. If you still feel strain, decrease the height under your back.

3. Bring the soles of your feet close to your body. Place a long rolled blanket under your outer thighs. The height of these blankets should adequately support the weight of your legs so that your back and knees relax. Your knees should be level. If one knee is much higher, place extra support under the lower knee. Many people also feel more relaxed with the support of a folded blanket under their forearms. Blankets can be folded lengthwise and carefully positioned to support both the legs and arms.

4. When you feel comfortable, remain still in the pose for as long as you like, 10 to 20 minutes or longer. Observe the quiet flow of your breath. To come out of the pose support your thighs with your hands.

Before sitting up, straighten your legs, allowing them to fall evenly away from the midline. Some people prefer lying flat before sitting up. You can turn to your side with your knees bent, move the bolster out of the way, and lie flat on the floor before sitting up, if you prefer. When you feel ready, bend your knees, turn to your side and slowly sit up.

Supported Bridge Pose

Supported Bridge Pose is a combination gentle supported backbend and mild inverted position. You can clearly see and feel the opening of the chest and heart area as you place your body in this pose. It is restful for the heart and may help balance blood pressure and hormonal secretions. It has a calming effect on the mind and nervous system and helps prevent and relieve headaches. Placing your head lower then the rest of your body with the chest open is soothing, refreshing and removes lethargy and depression. It also helps drain fluid from the legs after long periods of standing.

How to Practice Supported Bridge Pose

1. Study the photo on the next page and note how the support under the back and legs is placed end-to-end to accommodate the length of the body. The height of the support depends on the length of your torso and the flexibility of your back. New students can begin with one or two single-folded blankets and gradually increase the height. Six to 12 inches in height works well for most people. Taller people

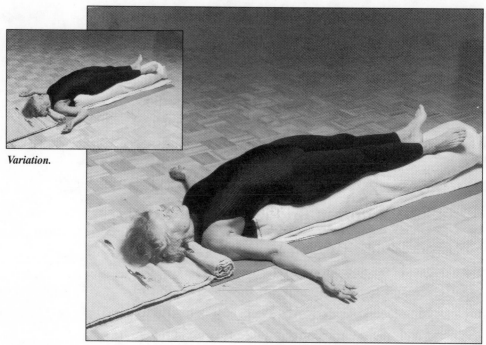

Variation.

Supported Bridge Pose is extremely restful and calming.

can add height to bolsters by placing one or more
blankets on top of the bolsters. Make sure that your
support is level. You can also place a block under
your heels. If your support tends to slope down at
the base, add support under your heels.

2. Sit down on the folded blankets or bolsters, either
with your legs straight out in front of you or strad-
dling the support. Position yourself near the end of
the support so that when you lie down your head is
near the far end. Use the support of your arms as
you lie down. Slowly slide off the end until the
back of your head and shoulders rest flat on the
floor.

3. Notice how you feel. If you had difficulty lowering
your shoulders to the floor, your support may be too

high. If your lower back feels strain, bend your knees and place your feet either on top of the support or on the floor on either side. Relax your throat and chin. Lengthen and release your neck. If your neck feels jammed, experiment with a small rolled towel at the base of your neck, or roll one end of a folded blanket and tuck it under your neck, with the rest of the blanket under your head as illustrated.

4. When you feel comfortable, close your eyes and cover them. Relax your arms out to the side at a comfortable angle, or with your elbows bent, arms relaxing back just above your shoulders ("baby arms").

Stay in Supported Bridge Pose as long as you feel comfortable, up to 10 or 15 minutes. When you feel ready to come out, remove the eye cover and slowly slide in the direction of your head, until your whole back and bottom are on the floor. Relax for a few more breaths with your lower legs supported by the height. Then bend your knees, turn to your side and slowly sit up.

Bolsters are highly recommended both for their height and convenience. Students consistently report that they practice supported poses much more regularly at home after purchasing bolsters. Consider them an investment in your health for years to come! (See chapter 10 for more benefits of supported backbends.)

Child's Pose

Child's Pose is the natural pose babies and young children assume for rest and relaxation. Interestingly, with the arms extended it is also a universal gesture of worship. This pose not only relieves tension in the back

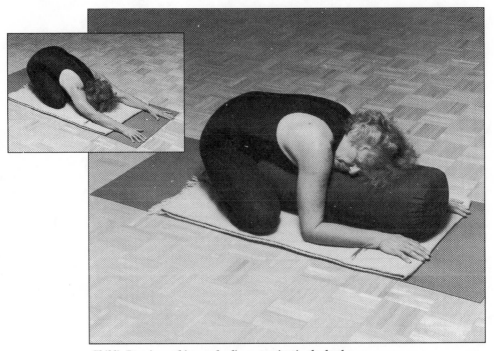

Child's Pose is soothing and relieves tension in the back.

but gives a feeling of comfort and security. Without props it can be practiced between repetitions of active weight-bearing poses like Downward- and Upward-Facing Dog Pose or Right-Angle Handstand. It effectively relieves the lower back ache that comes from standing or sitting still too long. During times when life feels overwhelming and all you want to do is crawl back into bed and hide, try retreating for a few minutes in Child's Pose. With the body supported on a bolster or blankets, it feels like you're giving yourself a warm hug.

How to Practice Child's Pose

1. Kneel on a padded surface such as a carpet, yoga mat or folded blanket with your feet and knees

about hip-width apart, and your bolster or stack of folded blankets placed lengthwise in front of you.

2. Fold more padding under your knees and shins if needed. (See Hero Pose, chapter 5.) Point your toes directly backwards, keeping your feet in line with your calves. Slowly lower your bottom toward your heels. If you experience strain in your knees, place a small towel, folded lengthwise, deep into the crease at the back of your knee to create more space in the knee joints. For discomfort in your ankles or tops of the feet, place a folded towel or sticky mat under the front of the ankles and let the feet hang over the roll, as illustrated in Hero Pose, chapter 5.

3. Separate your knees wide enough to place the bolster or blankets between your thighs as illustrated. To deeply relax and let go in Child's Pose, your torso should be supported by the bolster and your thighs. Your chest should relax on the bolster. Add more blankets if needed. Allow your bottom to move back toward your heels. Many people find Child's Pose easier than Hero Pose, as it can be practiced even if your bottom doesn't touch your heels.

4. If you feel uncomfortable or have trouble breathing, try moving the support in front of you farther away. Some people are more comfortable with their belly off the bolster.

5. Turn your head to the side as illustrated. Relax your chin. If you feel uncomfortable with your head to the side, rest on your forehead with your chin tucked toward your chest. Breathing should feel easy and relaxed. You can hug the bolster or let the arms be passive, as illustrated. Relax in this

position about three minutes. Turn your head to the opposite side when you are halfway through holding the pose. To come out of the pose, place your palms on the floor, under your shoulders. Inhale and sit up slowly on your heels. If your bottom doesn't touch your heels, simply stay in a kneeling position. Then bring one leg forward, placing your foot on the floor. Press your hands on the forward thigh and come back to a standing position. (Refer to chapter 13 on how to use a chair if you have difficulty coming up off the floor.)

People with very stiff knees or ankles may find they need to practice on a thick stack of blankets, either with the toes hanging off or with an extra blanket placed lengthwise under the shins. As in other poses, most problems can be overcome with the help of a teacher or by experimenting with the thickness and placement of the folded blankets and towels.

For people whose feet and knees do not feel comfortable on the floor even with support, or who cannot get up and down from the floor without assistance, practice kneeling and relaxing forward in bed, with pillows or blankets under your bottom and torso, feet hanging off the bed, as described in chapter 5 for Hero Pose.

Lying Down Bent-Knee Twist

I always remind my students, "If you do nothing else, do some gentle twists to relax your neck and back before you go to bed." Gentle floor twists help prevent and relieve lower back pain caused by muscle tension. They also reduce cramps and indigestion and tone the abdominal area.

Gentle twists take the kinks out of your neck and back.

How to Practice Lying Down Bent-Knee Twist

1. Lie down with your knees bent, feet flat on the floor, your upper and lower body in line. If your head tilts back, place a folded blanket under your head. Bring your arms in line with your shoulders, palms up, hands actively stretching away from each other to open your chest. Allow your back to relax toward the floor.

2. Still keeping your feet flat, lift your hips up off the floor to move them gently to the left. Lower your hips back down. Bend your knees toward your chest, and slowly lower your knees to the right.

3. Stretch your left arm away from your knees. Your right arm can stay in line with your shoulders, or you can increase the stretch by placing your right hand

Lying Down Bent-Knee Twist with support feels wonderful after Shoulderstand.

on your knees, gently pressing them down. Stay for a few breaths, allowing your body to release into the pose. When you feel ready, on an exhalation use your abdominal muscles to bring your knees back to the center. Move your hips to the right before slowly lowering your bent knees to the left.

If you feel unusual back or shoulder strain, or if your knees do not touch the floor, place a folded blanket under your knees.

More flexible students can practice this, for example, after Shoulderstand, as illustrated above, with a blanket supporting their pelvis and mid-back.

As you become familiar with the postures in this menopause chapter, please bear in mind that a yoga program for women's health problems may include other poses and variations, which are best learned under the guidance of a yoga teacher. Your instructor may give you specific, individualized instructions to follow. Be aware that there are subtle adjustments and refinements that are beyond the scope of a book. I strongly encourage you to seek and work with a knowledgeable teacher who can help you to make yoga a supportive companion during menopause and for the rest of your life.

Yoga Breathing Practices and Menopause

Pranayama refers to breathing practices that involve the prolongation and restraint of the breath. These practices should be learned under the guidance of a teacher. Pranayama plays a vital role in keeping the body healthy, especially during the later years. Such breathwork eases pressures, breaks up congestion and heals by supplying fresh energy to the pelvic region and the organs of the body, and by stimulating and soothing those areas to bring them back to a normal state.

Pranayama helps relieve menopausal problems by stimulating the autonomic nervous system. For those who wish to explore this subject further, B.K.S. Iyengar in his book, *Light on Pranayama,* explains how these breathing practices directly affect our health by stimulating all the *chakras* where energy is stored. Pranayama releases and distributes the energy of the nervous system. It improves the performance of the respiratory, circulatory, nervous and endocrine systems, alleviating problems associated with menopause. Women at midlife and older are especially encouraged to study and begin the practice of pranayama.

Urinary Problems

At menopause, some women experience urinary problems due to weakness of the pelvic floor which tends to become prolapsed. To counter these difficulties, yoga teachers recommend the practice of a technique called *Aswini Mudra.* Aswini Mudra is similar to the Kegel exercise recommended for pregnant women, in which the sphincters are held and released repeatedly. Regular practice of Aswini Mudra tones the perineal area and pelvic floor and increases blood flow to the pelvic region, encouraging the maintenance of healthy vaginal and urethral tissue.

To practice Aswini Mudra, sit in a comfortable position, preferably on a height to help keep your posture open, your spine lengthening. Gently contract the sphincter and vaginal muscles as if trying to hold back urination. The perineum should be held firm and the pelvic floor should feel as if it is being lifted up. Hold the contraction for a few breaths, then release it for a few breaths. Repeat five to 10 times at first, increasing the repetition and duration as you become more comfortable with the practice. Aswini Mudra can be practiced during daily activities while sitting or standing and during many yoga poses.

A five-step program for bladder control is outlined in the book *Staying Dry: A Practical Guide to Bladder Control* by Kathryn Burgio, Ph.D., Lynette Pearce, R.N., C.R.N.P., and Angelo Lucco, M.D., published by the Johns Hopkins University Press. An estimated one in 10 adults—more than 10 million men and women of all ages—suffer from the annoying problem of urinary incontinence. *Staying Dry* dispels the myths and taboos about incontinence, educates readers about the different types of incontinence and the various treatments available, and teaches a series of exercises and strategies that can significantly improve or cure most types of incontinence in a matter of weeks.

EDITOR'S NOTE: *People who have experienced this medical condition can also obtain a basic packet of information by sending a business size self-addressed stamped envelope and $2.00 to: National Association for Continence (NAFC), P.O. Box 8310, Spartanburg, SC 29305-8310.*

Malchia Olshan:
Hot Flashes Become Cool Breezes
in Forward Bends

At age 65, my student Malchia Olshan is a world whampion swimmer. She says that practicing yoga has helped her achieve better times at swimming meets now than when she was in high school. Currently, she is the world's third-ranked miler in the masters division. She competes regularly in the United States and abroad.

One morning soon after I turned 50, I looked in the mirror and saw that my skin seemed to have dropped overnight. My waist was hip-size and my arms and legs were filled with gravity creases.

"Aha! Get thee to serious yoga classes," my conscience told me.

Handstands, Headstands and Shoulderstands counteract the dropping waist. Stretching keeps the limbs loose and scares away those gravity creases. Hot flashes become cool breezes in Forward Bends.

I am a competitive master swimmer. My secret to swimming is yoga, yoga and more yoga. Restorative yoga calms the body and brain after hard swim workouts. Before my swims, I'm on the deck doing Downward-Facing Dog.

Yoga makes me feel like a kid. The added bonus is that my grandson Martin was so impressed with my Handstand that he invited me to his kindergarten class for Show and Tell so his friends could see Grandma stand on her head!

Felicity Green:
Yoga is Healing and Strengthening
on Every Level

Felicity Green, in her 60s, balances in Headstand with one leg stretching up, the other reaching down.

Felicity Green began practicing yoga after experiencing a traumatic injury to her left shoulder. She had been told by her physicians that her arm and shoulder would never function normally *again. Now, 26 years later, she is able to rotate her arm back and forth behind her head and demonstrate a Handstand, Headstand, backbends and other yoga postures that have helped her to develop strength, flexibility and restore full range of motion to her shoulder.*

Had Felicity waited 10 or 20 years to begin a yoga and thera-peutic exercise program, she would have suffered a functional disability to her left shoulder. The usual consequence of a soft tissue injury is that the body produces pain, inflammation and fibrous tissue in its attempt to heal. However, improvement can still be made even if a person begins yoga many years after an injury, as Felicity confirms:

Hatha yoga is an efficient system that gives you the benefits of cleansing and purifying the body, as well as strength and flex-ibility. You will gain the feeling of greater freedom and aware-ness, but as with anything, it's not what you do but how you do it that's important. Asanas must be practiced with a mature awareness to be constructive.

I started yoga as many people do—for a physical reason. But the more important result has been greater emotional stability. Yoga has been healing and strengthening on every level, and I teach to try to help others to discover this for themselves.

I firmly believe that as human beings, we exist on many lev-els that are intertwined but unfortunately not integrated. The body is the most tangible level and is therefore the most honest feedback system. We carry memories and resistances in our bodies as much as in our minds and psyches. If we pay close attention to the information we gain from the body—for exam-ple, our reaction to challenge—and apply that to other aspects of ourselves, the changes become profound.

My own experience and my observations of students over the years have proved that Mr. Iyengar's approach brings about deep and gradual change, not merely an intellectual "flash in the pan." In this, I think, lies the importance of his work for the West, where intellect is given far too much importance.

My aim and direction through yoga is to learn to cooperate with change in a sensitive, refined manner and thereby create peace and harmony.

Seven

Weight-Bearing Yoga Postures Help Prevent Osteoporosis

Exercise literally can save your life; without it, your body deteriorates. There is no question that if you are inactive your bones will decalcify, leading to osteoporosis, the potentially fatal disease that is so prevalent among older women. Taking time out of your day to attend an exercise class, to take a walk is not frivolous, it's essential.

—LINDA OJEDA, *MENOPAUSE WITHOUT MEDICINE*

The skeleton provides the fundamental structure of the human body. Bones enable us to stand upright and allow us to move when the muscles attached to them are contracted. Bones also protect vital organs. The dense bones of the skull enclose the brain; the slender ribs protect the heart and lungs at the same time as they allow the chest to expand and contract.

How Exercise Strengthens Bones

Our bones are composed chiefly of calcium. They begin developing well before we are born, but are not completed until we are about 25 years old. The reconstruction of bones goes on throughout our lives, in part because bones supply calcium to the rest of the body. Calcium is used by the nerves and in many vital functions. The pattern in which bone rebuilds varies in response to mechanical stress and an individual's activity level.

Bones grow stronger in response to use. In the presence of adequate supplies of calcium in the diet, bone growth is stimulated by any activity in which bones bear weight or muscles pull on them. During these activities, muscles transmit mechanical and bioelectrical signals to bones that cause them to thicken.

If we remain physically active, bones continue a healthy process of regeneration and remain strong. New bone growth, however, is rapidly absorbed if we become immobilized or sedentary. Even athletic and very fit astronauts, living in the weightless environment of space, quickly lose bone mass because their muscles and bones work far less in the absence of gravity.

Osteoporosis is the major cause of bone fractures in older people in general and postmenopausal women in particular. The word "osteoporosis" means "porous bone." As the actual mass of bones decreases, the bones become more susceptible to fractures. In severe osteoporosis, the internal structure of the bones erodes to such an extent that even a slight trauma can cause them to break. Although osteoporosis may affect any bone in the body, the most common sites of osteoporosis-related fractures are the spine (including the neck), wrist and hip.

Fractures of the vertebrae, or bones in the spinal column, occur most often in women within 20 years after menopause and in elderly men. Even tiny fractures in these bones are also the source of the rounded back and loss of height frequently associated with aging. When the bones of the vertebrae are weakened, an ordinary action like bending forward to make a bed or lift a child can be enough to cause a spinal compression fracture. When a fracture occurs, pain may be sudden and severe or chronic and nagging. As additional microfractures develop, the vertebrae collapse into wedge shapes, the upper spine curves forward and height diminishes.

In older osteoporotic men and women, hip fractures frequently occur in the upper portion of the femur or thighbone. Although most of these fractures are surgically repaired, one out of five elderly people with hip fractures dies within a year from complications. Few who recover ever regain complete mobility.

People with osteoporosis are also prone to breaking the bones in the forearm near the wrist. Often, these breaks occur during attempts to stop a fall. Unfortunately, elderly people are prone to falling because of failing eyesight, inner ear disturbances, weak and inflexible muscles, arthritis, slow reflexes, or various neurological conditions.

Osteoporosis and Menopause

Osteoporosis is actually two separate diseases. In one form, osteoporosis develops gradually with age in both men and women. In the other, osteoporosis appears rapidly in women after menopause or after surgical removal of the ovaries. In the second type of osteoporosis, reduced levels of estrogen accelerate bone loss for several years after menstruation ends. Because we are living longer now, we see a great deal more of both types of osteoporosis.

Osteoporosis affects about half the women in the United States over age 45 and 90 percent of those over 75. Although diminishing estrogen levels undoubtedly speed up the loss of bone mass immediately after menstruation ends, recent studies indicate that estrogen deficiency is not the primary cause of hip fractures linked to osteoporosis. The most crucial factor is a sedentary lifestyle. This and other new evidence indicates how important it is for women to remain active and physically fit as they grow older.

Contrary to most people's assumptions about this disease, osteoporosis often appears before menopause. A study of 7,278 hospital discharges following hip fractures shows that the rate for this type of injury for Caucasian women climbs steeply between ages 40 and 44—nearly 10 years earlier than menopause for most women. The hip fracture rate does not increase between the ages 48 to 51 or later, which are usually the menopausal years. Research indicates that bone density actually begins to decrease long before this stage in life sometimes as early as 25 years of age. It is therefore imperative that women take care of themselves well before menopause.

Climacteric refers to the years just before and just after menopause. For most women, the climacteric spans from the early- to mid-40s to late 50s or early 60s, including the pre-menopausal climax years and the post-menopausal years. This entire period constitutes the menopausal years, commonly known as the *change of life.*

During the climacteric, a woman can experience up to 5 percent loss in bone mass, regardless of how much extra calcium she is taking. After that, the rate of loss diminishes and levels out as the body adjusts to lower hormonal levels.

A Healthy Lifestyle Reduces Your Risk of Osteoporosis

Because the risk of osteoporosis is increased by a diet high in salt, protein, caffeine and sugar and by the use of alcohol and tobacco, many holistic practitioners maintain that osteoporosis is simply the result of an unhealthy lifestyle that manifests itself about age 50. The typical American diet promotes calcium loss. The amount of sodium in popular snacks, the sodium and phosphorous in processed foods and the phosphorous in red meat and cola drinks all cause calcium to be leached from our bones and excreted in the urine. High-protein, meat-containing diets may also help to deplete calcium. As little as two drinks per day of alcohol, which is a direct chemical poison to the bone-forming cells, may double a woman's risk for osteoporosis. Smoking may also double the risk.

Stress, too, contributes to osteoporosis. When we are under stress, our blood becomes slightly more acidic, which, over time, removes calcium from bones. When we are more relaxed, our blood becomes more alkaline and the bones retain more calcium. Therefore, regular deep relaxation will enhance any program to prevent osteoporosis.

Because bone is restructured constantly, we need calcium and other minerals every day. To supply adequate nutrition, consider adopting a balanced vegetarian diet, recommended for people at risk of osteoporosis. A nutritionist or holistic health professional can help you determine your individual needs. If you tend not to eat sufficient amounts of bone-healthy foods, you may also want to consider taking a complex mineral supplement. Remember, too, that vitamin D helps the body absorb calcium. Just 15 minutes of sunlight a day on your hands and face will stimulate your body to produce all the vitamin D you need.

Inactivity Leads to Bone Loss

Our bodies are meant to be used. In fact, we now know that prolonged rest is disastrous for the muscular and skeletal systems. A hospital patient confined to a bed for a few weeks will suffer as much wasting in the muscles and bones as someone who has aged a decade.

Deepak Chopra, the world-famous expert on mind/body medicine, recommends walking and yoga as the ideal combination of exercises for strengthening and balancing the body. While walking is an excellent, pleasurable exercise and is practical for most people, it only strengthens the bones in the feet, legs, pelvis and, to a lesser extent, the spine. Exercises that strengthen the muscles of the back, such as weight-lifting, help increase the density of the bones in the vertebra. Swimming, though it stimulates bone growth in the forearms and feet, does not strengthen the vertebral bones. In contrast, yoga strengthens bones throughout the body.

A 30-second stay in Downward-Facing Dog or Full Arm Balance (see chapter 4) is usually long enough to convince even the most skeptical athlete that yoga is, indeed, a legitimate weight-bearing exercise, especially strengthening for the wrists and upper body. Attending even one class will demonstrate yoga's superior ability

Aerobics for your heart or weight-training for your muscles may achieve some limited good, but these are not comprehensive enough. . . . The ideal is to balance the whole system, mind and body. It is also vital that exercise give more energy than it takes, a consideration many people tend to ignore. What is needed is a new approach to exercise in which the aim is not to sweat and strain or to pound muscles into shape. . . . The body is not just a shell or a walking life-support system. It is your self intimately clothed in matter. Getting back in touch with this intimacy is very reassuring and delightful, particularly for people who have given up on exercise and become virtual strangers to their bodies.

—Deepak Chopra, M.D.,
Perfect Health: The
Complete Mind/Body Guide

over other form of exercise to remove years of stiffness from the spine, hips and rest of the body.

Look for Exercise Opportunities in Daily Life

To complement your yoga practice, remember to walk in the fresh air in natural light as often possible. Look for exercise opportunities in your daily routines. Get in the habit of walking and riding your bicycle for transportation as well as recreation. Studies show that in spite of all the weight-loss programs, a greater percentage of the American population is overweight than ever before. I firmly believe that this would naturally change if we left our cars at home more and used our legs for daily transportation. Whenever practical, do errands on foot instead of in your car. This can increase the joy of everyday life and help both you and your community to be healthier.

While most of us must make a conscious effort to get enough exercise, people who exercise too much also increase their risk of osteoporosis. For example, female athletes who train excessively or who suffer eating disorders often stop menstruating and become osteoporotic. Premature loss of menstruation is a sure sign that young bones are at risk.

Yoga Helps Prevent Falls by Improving Balance

Yoga not only strengthens bones, it actually helps to prevent falls in the first place by improving balance and coordination, posture and body mechanics. If an older

For me, a long five- or six-mile walk helps. It is at these times that I seem to get recharged. If I do not walk one day, I seem to have on the next day what Van Gogh calls "the meagerness or what is called depression." For a long time I thought dullness was just due to the asphyxiation of an indoor sedentary life. But I have come to learn otherwise. For when I walk grimly and calisthenically, just to get exercise and get it over with, I find I have not been recharged. But if I walk in a carefree way, without straining to get to my destination, my neck and jaw are loose and I look at the sky or the lake or the trees or wherever I want to look, then I am living in the present. And it is only then that the creative power flourishes.

—Brenda Ueland, a writer who set an international swimming record for over-80-year-olds and lived an active and vital life until her death at the age of 93.

person should accidentally stumble and fall, yoga reduces the degree of trauma and likelihood of fractures by strengthening the muscles and making the body more flexible. Strong muscles are better able to control and absorb the impact of falls. Moreover, older people who practice yoga feel stronger, more agile and confident and can get up from falls far more easily than those who are frail and afraid.

Over the years, I have known several students who came to yoga with severe balance problems and needed to use canes. Naturally, these people did not feel secure in the middle of the room. However, they could begin to practice standing poses against a wall, with or without wall ropes. Or they could practice at home, holding onto a built-in counter or sturdy table, as described in chapter 3. It is especially important that students with balance problems not be relegated to practicing yoga in chairs because without weight-bearing exercise, these are the very people who are most often headed for wheelchairs.

Yoga Helps Prevent Height Loss

According to Dr. Christiane Northrup in *Women's Bodies, Women's Wisdom,* decreased height is not always the result of bone loss. Years of poor posture, lack of stretching, or feeling weighed down by life's burdens can also make elderly people shorter than they once were. Some height loss results from the shrinking of spaces between vertebral discs, even when bone density is good. Dr. Northrup observed the least height loss in her patients who regularly practiced yoga. I believe this is because yoga helps to keep the discs between the vertebrae plump and supple and the spaces between them open. Many of my older students report that after practicing yoga for a while, they regain their youthful height.

One-Legged Downward-Facing Dog Pose. The weight-bearing benefits of Downward-Facing Dog Pose are intensified by shifting the weight from four limbs to three.

Yoga for Preventing and Counteracting Osteoporosis

The best yoga program for preventing osteoporosis is a balanced practice of asanas (postures), including standing, inverted, seated and lying-down postures. Such a practice stimulates bones to remain strong and healthy as well as improves self-image, coordination and posture. Yoga effectively reeducates the body, correcting posture by elongating the spine and stretching and opening the chest.

If you already have osteoporosis, you can practice yoga safely by using props and taking special care to maintain proper alignment. Use extra caution practicing movements such as forward bends that put pressure on already weakened vertebrae. If you are new to yoga, avoid weight-bearing on the spine in poses such as Shoulderstand or Headstand until vertebral fractures heal and your bones have regained more normal strength. As I do for other physical conditions, I highly recommend you seek and work with a qualified yoga instructor.

Betsy Goodspeed:
Yoga Is As Important to Me As Music

I've played the harp professionally for over 30 years. Around the age of 57, I stopped because it just wasn't worth the pain. I had a chronic stiff neck, my right leg was longer than my left, the left hip joint was shot to pieces, the left shoulder was practically frozen, the upper right arm was swollen, and I was so stiff and sore that I had to pull myself out of bed every morning by sheer will power.

Besides playing the harp, I also play the piano, sing, write music and spend long hours at the computer writing. These life-giving creative endeavors took a heavy toll, resulting in punishing pain, endangering my health, perhaps even shortening my life. I paid this stiff price because I told myself that it was worth it.

I've met hundreds of musicians who suffered aches and pains because of the hours spent perfecting their art. Most had terrible posture—the first clue that a problem exists. How often do we see a dedicated musician with excellent posture? The two just don't seem to go together.

Knowing that I was used to discipline, my chiropractor advised me to take up yoga. My first class was a disaster. Accustomed to starring, I was humiliated by not being able to perform the simplest stretch. I asked for a private lesson to help me design an at-home practice.

Six weeks after practicing on my own at home, I joined another beginners' class. In the months that followed I developed a whole new attitude toward exercise. My whole life changed. Yoga gradually became an essential part of my focus—physically, intellectually, emotionally and therefore spiritually. That sounds intense, really overboard, like a new convert to some way-out religion, but I don't know how else to put it.

In school we learned how to play volleyball and other sports. We also did bouncing calisthenics, which can compact our bodies rather than lengthening them. In my earlier adult years I was busy raising three kids and performing on television. I figured I was getting enough exercise keeping up with daily life. What a common excuse that is! The fact is, even cars that are constantly driven and not maintained end up in a junkyard. That's where my neglected body was headed.

Now, five years later, yoga is as important to me as music. At 65, I'm more healthy and energetic by far than I was at 40, and to my delighted surprise I have been found worthy of sharing my yoga knowledge and experience with a beginning senior citizen class in my mobile home community.

By nature I'm hyper and highly energetic. Yoga provides a peaceful, tranquil place for me to just be. Stretching slows me down and makes patience possible. Yoga reminds me that there's no hurry to perfect a piece of music or write the final scene in a story. I will still be here tomorrow, and the music or story will benefit from me taking a breather.

My grown children, who are not as limber as I, can see me stretching farther and becoming stronger with every passing month. I can't help asking if I'm getting younger as they're growing older.

Jean Getchel:
Yoga Is Simply a Joy to Do

For all the reasons one can give for doing yoga, I would say first of all that it is simply a joy to do. One works with nature to create a calmness of spirit in a way that nothing else can duplicate. Yoga is also completely non-competitive and without negative impact on joints. Joints, instead, are made more flexible. Yoga is a gentle form of movement that greatly reduces injury, relieves stress as it develops concentration, builds strength, tones muscle, and helps develop balance.

I have had scoliosis, a lateral curvature of the spine, since my teenage years. Since discovering yoga, I have found that its practice has stopped the progression of the curve and has strengthened the muscles that help keep my spine as straight as possible.

I also had a stroke three years ago that briefly paralyzed my right side, leaving the muscles on that side weak. Yoga has helped to strengthen those muscles and to make me aware of the imbalances of the body so that I can work to correct them.

Eight

Yoga Techniques to Prevent or Overcome Arthritis

Just as a pebble produces ever widening concentric ripples on the surface of a still lake, the positive effects of a yoga practice spread into all aspects of one's life. As muscle strength, joint flexibility and neuromuscular coordination improve, one develops the ability to move with ease and confidence. This vitally important enhancement of mobility brings with it independence, self-determination and renewed interest in life. As exercise tolerance improves, yoga's salutary effects cascade into a positive, self-perpetuating cycle, replacing the pernicious negative cycle of inactivity, deterioration and depression.

—MARY PULLIG SCHATZ, M.D.
YOGA JOURNAL, MAY/JUNE 1990

Over 40 million Americans are struggling with arthritis. The term "arthritis" means "inflammation of the joints" and refers to a number of diseases that produce deterioration in various joint structures, which causes pain and immobilization. Severe arthritis often results in loss of function and deformity. In extreme cases, people who suffer with these diseases are no longer able to live independently. I have had personal experience assisting people with arthritis who were wheelchair-bound. I understand the extreme suffering involved in this disease.

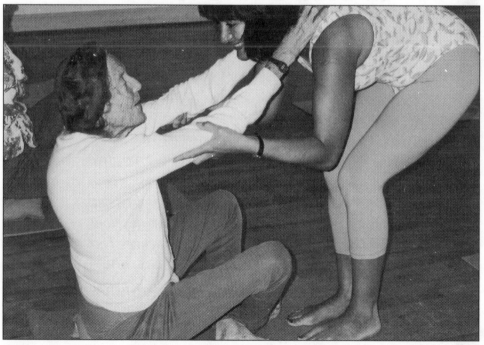

Students with painful arthritis often need special encouragement.

There are two main types of arthritis: osteoarthritis and rheumatoid arthritis. Rheumatoid arthritis is a chronic inflammatory disorder, resulting in stiffness in the joints and muscles, joint erosion and pain.

Osteoarthritis is a degenerative disorder that erodes the cartilage that cushions areas where bones rub against one another. The roughening of this normally smooth tissue makes moving difficult and painful. Osteoarthritis frequently occurs in people who are overweight or whose joints are painful from extreme overuse.

The cycle of arthritis begins with joint pain and swelling. As you probably know from even minor injuries, most of us respond to pain that occurs during movement by keeping still. However, we now know that one of the worst things for someone with arthritis is inactivity. Instead, regular gentle movement helps reduce pain and maintain mobility. Physicians are discovering that an appropriate yoga-based exercise program expands range of motion without stressing or straining joints.

To remain healthy, joints must move and bear weight. Without weight-bearing, bones become fragile and prone to fracture and collapse—the condition we call osteoporosis, discussed in chapter 7.

Physical movement promotes health in many systems of the body. Movement increases circulation which, in turn, reduces swelling and increases delivery of oxygen and nutrients to the joint tissues to facilitate their healing. With immobilization, a cycle of deterioration begins: muscles shorten from lack of stretching, which, in turn, creates deforming contractures. Unused muscles lose strength. This weakness, coupled with joint swelling, makes the joints unstable. Joints in this condition are vulnerable to dislocation, increased injury and pain. As arthritis patients feel weaker and more vulnerable, they become increasingly fearful and dependent; they feel tired and depressed.

Unlike other forms of physical exercise, yoga has something for everyone. No one is excluded. People with chronic disease and disabilities face "can't" at every turn in their lives: they can't play golf, can't play tennis, can't run, can't over-exert themselves, can't walk without canes, some can't walk at all. But everyone can do yoga. In yoga, there are no can'ts. Yoga can be modified and adapted to suit the needs of everyone.

—*Lorna Bell, R.N.*, Gentle Yoga, For People with Arthritis, Stroke Damage, Multiple Sclerosis and in Wheelchairs

Because movement is crucial to so many physiological processes, the arthritic person's health tends to deteriorate. The normal functioning of the immune system declines, infections and illnesses occur, and the person becomes frustrated and depressed. This cycle is self-perpetuating.

Generally, physicians advise regular gentle exercise for people with osteoarthritis because it tones muscles and reduces stiffness in joints. Yoga is an ideal form of exercise because its movements are fluid and adaptable. Moreover, students progress gradually, beginning with simple stretches and strengthening poses, and advancing to more difficult postures only as they become stronger and more flexible.

If necessary, you can begin yoga with gentle movements sitting in a chair or lying on the floor. Simple "one-joint" movements will gently loosen your joints and relax your muscles. These therapeutic yoga-based movements also improve your breathing and help relieve physical and emotional tension.

You can gradually add weight-bearing standing postures, with the help of a wall or table, wall ropes, chairs, blocks and other props described in chapter 3, to improve your strength, flexibility and balance. You can do standing postures holding on to a sturdy chair, wall or table, so that you'll feel safer and steadier. A regular exercise program is not only physically beneficial; it is extremely valuable psychologically because it increases your confidence that even though you have arthritis, you're not going to become a prisoner in a wheelchair.

Nellie Eder, an octogenarian yoga teacher who overcame arthritis with the help of yoga and proper nutrition, recommends self-massage and acupressure to help bring warmth and improved circulation to the joints. It is advisable to seek the services of a qualified massage therapist or other health professional familiar with holistic therapies to help you on your journey back to health.

I encourage you to seek the help of a teacher and to make yoga postures, breathing, relaxation and meditation techniques an important part of your daily routine. These techniques improve your respiration throughout the day as well, helping to relieve anxiety and pain. You can use yoga's breathing and relaxation techniques any time to relieve stress and tension. Remember that calm, slow, rhythmic breathing helps to release both physical and emotional tension by flooding the body and brain with oxygen. Remember, too, that the regular daily practice of deep relaxation—quieting the mind and body for even a few minutes at a time—is restorative to every cell of your body and encourages healing.

Arthritis need not mean a prognosis of inevitable disability. By practicing consistently, every day, you can reduce pain, build your strength and live with health and renewed energy.

Guidelines for a Yoga Program for Arthritis

Yoga's approach to arthritis recognizes the interaction between the mind, body and spirit. Combining medical evaluation, nutrition, yoga, massage and other holistic therapies can break the debilitating cycle of arthritis. Yoga helps people with arthritis to be aware of their physical limitations without being paralyzed by them.

An intelligent, non-mechanical, individualized stretching and strengthening program is one of the keys to restoring health to arthritic joints. Properly aligned

Moving hurts, but not moving destroys. Incorrect movement harms, but intelligent movement heals.
—*Mary Pullig Schatz, M.D.*

movements designed to strengthen weak muscles and stretch those that have shortened are crucial to restoring stability and range of motion.

Keep the following tips in mind as you exercise:

- **Respect pain**. People with arthritis must learn the difference between the beneficial feeling of muscles stretching and the pain that signals harm. They must distinguish between the normal discomfort of moving stiff joints and the pain caused by a destructive movement or an excessive demand on a joint. Never bounce. Bouncing causes reflex tightening of muscles and can result in torn muscles and tendons. Sudden or severe pain is a warning. Continuing an activity after such a warning may cause joint damage. In general, if pain lasts more than two hours after an activity, ask someone who understands good alignment to check how you are practicing. If your alignment is good, consider easing up on your effort and experiment with holding those positions you suspect are causing pain for less time. Try moving more slowly and practicing more regularly. Most problems can be overcome by paying close attention to your body's innate feedback system.
- **Balance work and rest**. Conserve energy. Balance activity and rest. This principle applies to exercise as well as to daily activities. Overwhelming fatigue is counterproductive and may even be harmful. Weakened, fatigued muscles set the stage for joint instability and injury. Balance your active yoga session with yoga's deeply relaxing restorative poses. Restorative poses will help your internal healing processes to work.

- **Breathe properly**. Without fully expanding your lungs, the muscles you are exercising cannot be adequately supplied with oxygen. Holding your breath while stretching inhibits relaxation. Smooth, peaceful, rhythmic breathing through the nose reduces pain and tension and increases the feeling of deep relaxation that follows a stretching session.

- **Maintain muscle strength and range of motion**. Think about the way your body works, then use each joint in its most stable and functional anatomical plane. Avoid extending your limbs abruptly or in unnatural directions. Also, be careful not to hold a single position for too long. There is no set answer to the perennial question, "How long should I stay in the pose?" Long enough so that a healthy change has been made. Not so long that your body feels unhealthy strain or stiffens up from leaving muscles in a static position too long. Avoid mechanical repetitions and counting while exercising. Watch the flow of your breath and your body's response to a particular pose or exercise. Learn to tune into what your body is telling you.

- **To help maintain muscle strength and range of motion, learn to use yoga props**. Review the benefits of props in chapter 3. Remember, the more problems you are experiencing, the more useful yoga props are. Props allow you to hold poses longer so you can experience their healing effects. By supporting the body in a yoga posture, muscles can lengthen in a passive, non-strenuous way.

  The use of props helps improve blood circulation and breathing capacity. Props help you stretch, strengthen, relax or improve your body alignment.

Standing poses with the help of a table or other prop can be a life-saver for people with arthritis.

By providing more height, weight or support, props help you extend beyond habitual limitations and teach you that your body is capable of doing much more than you think it can. This is especially important for people who are coping with arthritis.

Yoga props make postures safer and more accessible. People with arthritis may already be quite stiff by the time they start yoga. Props allow them to practice poses they would ordinarily have great difficulty in doing. Props help conserve energy and

Helping a new student with stiff legs and shoulders to stretch.

allow people to practice more strenuous poses
without overexerting themselves. Props are invaluable for people with arthritis, who must learn to
exercise without strain and fatigue.

- **Warm up**. Although affected joints should be
moved through a full range of motion at least once
a day, it is unrealistic and possibly harmful to expect
to attain full range of motion on the first try. Work
into a pose gradually. It is often helpful to relax in a
hot bath or shower before you practice.

Extended Side Angle Pose.

Standing poses remove stiffness from the hips and knees. People with arthritis can practice against a wall and make the pose easier by using a block or chair for support.

- **Walk**. Walking is the ideal companion to an intelligent, therapeutic stretch-and-strengthen program. The tranquilizing effect of its moderate rhythmic exercise decreases pain. The movement and weight-bearing aspect of walking improves joint health. Equally important, walking can take you outdoors, in touch with nature, the greatest of all healers—uplifting to the mind and spirit. Pace yourself, walk where there are places to rest and stop there when you feel fatigued. Be in the moment and walk with awareness of the beauty around you.

Yoga for Arthritic Hips, Knees and Hands

The areas most commonly affected by arthritis are the hips, knees and hands. Like the knees, the hips often develop flexion contractures or are immobilized because tendons and muscles have shortened. These limit full straightening of the thigh at the pelvis. As a result, the hip joint frequently remains somewhat bent. With decreased movement, the muscles and soft tissues around the hip shorten, causing the joint space to decrease, and putting additional wear and tear on the gliding surfaces. If a person becomes more sedentary in an effort to minimize pain, bones and cartilage receive less weight-bearing stimulation. Bone spurs may even develop to further limit movement.

Prolonged pain and the natural impulse to immobilize the knee can cause flexion contractures, shortened muscles and tendons that immobilize the knee in a bent position. When muscles around the knee, including the front thigh muscles (quadriceps), shorten, this decreases the space in the joint.

Lack of exercise also weakens the thigh and calf muscles. Their strength provides stability and support for the knee. When the soft tissues of the joint swell, this causes compression and reduces space in the joint even further.

Standing poses such as the Right-Angle Pose are crucial for stretching out contractures and building supportive strength in the hip, buttocks and thigh muscles. Moving the head of the femur in the hip socket in the standing poses and in the range of motion exercise helps distribute synovial fluid, lubricating the joint's surfaces.

The same standing poses recommended for hips are also critical for knee rehabilitation. They provide stretching to relieve contractures, create more space in the knee joint for synovial fluid circulation, and develop the strength of the thigh and calf muscles to more adequately

In the Right-Angle Pose, you feel supremely alert and confident.

support the knee. The psychological benefits of experiencing your own strength increase and stamina improve are also important. The ability to "stand on your own two feet" cannot be underestimated.

Modified Hero's Pose

Modifying positions that require bent knees such as the Hero Pose, or *Virasana,* are important for gradually restoring flexibility in the front thigh muscles or quadriceps. Postures such as Virasana also initiate a squeezing action in the soft tissue of the knee, which helps reduce swelling. (Refer to chapter 5 for instructions.)

Yoga Stretching Positions for the Hands

When arthritis develops in the hands, their normal movements are altered and the fingers begin to slant outward toward the little finger side of the hand. As swelling over-stretches the joint-stabilizing structures, inflammation often causes dislocation in the joints. Frequently, the fingers and wrists become discolored and deteriorate. Sometimes muscular shortening makes it impossible to open the hand fully or to separate the fingers. Swelling in the wrist can cause pain and numbness in the hand (carpal tunnel syndrome).

In the following exercises, hands and wrists should not be placed in any position that accentuates or encourages deformity. Every movement should be designed to move the hand back toward normal.

Namaste (Prayer) Position

Sit or stand and press your palms together in prayer position. This position helps to stretch the muscles in the hand and straighten the fingers. If you have arthritic wrists or carpal tunnel syndrome, practice *Namaste* with forearms touching.

1. Gently press the palms and fingers of both hands together. As you breathe smoothly and evenly, encourage the fingers to move toward the thumb side of the hand. Hold for several breaths. Release the pressure but keep the hands together for a few more breaths. Then repeat the effort three or four times.
2. Gently, firmly and evenly press the palms together. Smoothly open the fingers and spread

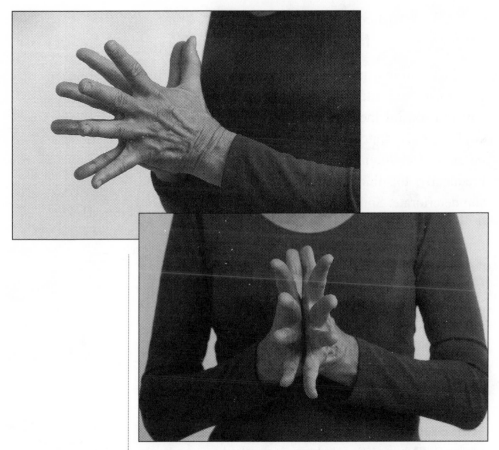

Stretching the hands in Prayer Position helps straighten the fingers.

them as wide as possible. Try to spread them evenly, moving them more and more toward the thumb side of the hand. Hold and stretch for a few breaths. Release and repeat.

3. Firmly and evenly press the palms together, especially the parts of the palm at the base of each finger. Stretch the fingers backwards, away from each other, gradually increasing the V-shaped space between them. Again encourage the fingers to move toward the thumb side of the hand. Encourage your fingers to stretch for three or four

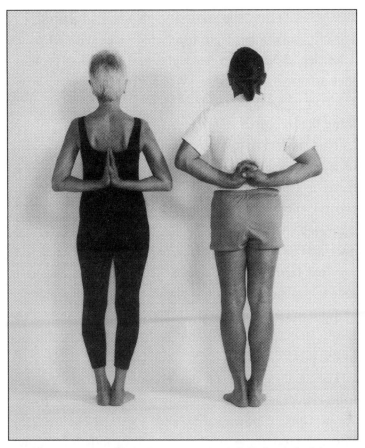

Practicing Namaste or simply clasping your hands behind your back improves your posture. These positions prevent stiffness in the hands and wrist.

more breaths. Release and repeat three or four more times.

One Student's Program for Arthritis in Knees and Hips

"I might be in a wheelchair or much worse, if not for yoga," said one 78-year-old student, who has been attending class two or three times a week for the last four years. Diagnosed with osteoarthritis in her knees two years before she started yoga, she had pain in her left hip, pins in her right hip (which was without pain), a left leg shorter than her right, and a hearing problem that affected her balance. When she first started yoga, she had to sit down or hold onto a support to do anything that required standing on one foot.

This student first started standing poses with the help of a wall, chair and wall ropes. She practiced at home using a kitchen counter or table. As her balance improved, she began to practice the poses in the middle of the room. At first, because she could not hold her balance and her legs and hips were stiff, she practiced with her feet fairly close together. Three years later, she is able to practice with her feet wide apart, and her balance and alignment are beautiful. She enjoys staying in the poses. In class she is usually the last person to come out of the pose.

The healing core of the program this student has followed consists of standing poses including Right-Angle Pose, Downward- and Upward-Facing Dog Poses, Legs-Up-the-Wall Pose, and various lying down and seated poses. During her first and second year of yoga, standing poses were most often practiced with her back on the wall, leaning into the wall and holding wall ropes. As her balance improved, she practiced near a wall for safety, but without depending on the wall. She now practices in the center of the room. She generally starts class by relaxing with her legs up the wall in various positions—legs straight up, legs wide apart, soles of the feet together and ankles loosely crossed. This legs-on-the-wall routine is followed by Child's Pose and Downward- and Upward-Facing Dog. Her most challenging position, kneeling in Hero Pose with a big bolster under her bottom and padding under feet, is generally practiced at the end of class. This student also enjoys practicing seated forward bends, Bridge Pose, floor twists, Shoulderstand-at-the-Wall and, most recently, Right-Angle Handstand.

Nellie Eder:
One Octogenarian Yoga Teacher's Experience with Yoga and Arthritis

Nellie Eder (right), with one of her many students. At 89, Nellie is an inspiration to people of all ages.

Over the years I have met many people who started yoga as therapy for arthritis and who credit gentle stretching combined with chiropractic care, improved nutrition and an inner-body cleansing program, including juice fasting and colon hydrotherapy, for the recovery of their health. Nellie Eder, an 89-year-old yoga teacher in Palm Springs, California, turned to yoga when she was 50 to combat an advanced case of crippling rheumatoid arthritis that began in her early 30s.

"The pain was so bad," she told me, "I used to step outside and scream." Both her physical and social activities were severely curtailed by constant pain. Her doctors offered pain pills. By the time she reached 50, Nellie was in the advanced stages of arthritis and headed for a wheelchair.

At this time she became friends with a woman who introduced her to natural and holistic health concepts, which were not as widely recognized 35 years ago as they are now. Nellie said it gradually dawned on her that all the pain pills and other medications were killing her.

"This friend told me about yoga and a vegetarian diet," she continued. "I took all the pills and dumped them. I changed my diet and stopped eating meat and sugar. I also took colon hydrotherapy treatments and ate a cleansing diet of mainly fresh fruits and vegetables. I found that the yoga stretching benefited my body, and the accompanying meditation allayed my pain and enabled me to get complete control of my body."

Nellie has spent the past years teaching yoga at various senior centers. At one nursing home she met a 94-year-old woman who couldn't get out of a wheelchair. Nellie worked with her patiently and gently, showing her how to improve her breathing and stretching and strengthen her legs. According to Nellie, within three months the woman was walking without a cane.

This spirited octogenarian also teaches students how to do foot and hand reflexology (pressure-point massage therapy), which she feels is increasingly useful as we grow older. She believes elderly people need to explore natural, holistic, nutrition-oriented therapies instead of traditional medications.

After falling, fracturing her pelvis and severely spraining her sacroiliac joint, Nellie recently recovered her ability to walk. "At the hospital, the orthopedist told me, 'I'm sorry but you're not going to walk again.' 'Well,' I said, 'that's the second time in my life the doctors told me that I would never walk!' " To me, Nellie looks as if she could win a dance competition.

Nellie tells her students, "Don't come to me unless you practice every day. Nobody can help you unless you make up your mind to follow through."

Lolly Font:
Aging People Respond Beautifully to Yoga

Lolly Font in Full Lotus Twist Pose.

Lolly Font is a teacher with an M.A. in Education. At 43, after raising five children, she became paralyzed with arthritis in her spine, neck, shoulders and hands. Yoga became her saving grace.

Obvious to me now but not then, I was in the process of a deep transformation. Miraculously, I discovered yoga, and I broke through the paralysis and stiffness in my body as the asana practice washed its magic through my being. Never an athletic person, I was overwhelmed by the beauty of the postures and the feelings of accomplishment and satisfaction I derived from attaining mastery over each pose. The expanded breathing, stretching, postural alignment and balance challenged my arthritic joints and created a new life for me.

Within four years my marriage broke up. This was not without pain. The pain in my joints and the pain in my heart were healed over a long tumultuous period of deepening practice.

Lolly went on to study at the Iyengar Yoga Institute in San Francisco, work with senior Iyengar teachers in the United States and Greece, and travel to India, where she studied with B.K.S. Iyengar. Afterward she earned a master's degree in transpersonal psychology and now does dreamwork. She and three partners began a yoga center. She is now a 70-year-old grandmother of eight and leads yoga retreats in Hawaii.

When I first trained as a yoga teacher, I asked for guidance in my daily meditation. I received a vision of a stooped, aging woman with a round back. At that moment I knew I had to teach seniors. I started a program for seniors at the Jewish community center. My students were my greatest teachers. They taught me about the effects of aging on older bodies and how yoga could reverse common problems. Learning by doing appeals to me and that is what I did.

Aging people are not touched very much, and they responded beautifully to the adjustments in the postures. Because they have lived in their bodies for a long time, they become unaware of subtle changes. When these subtleties multiply, they may expand into larger problems. I was able to point these physical problems out to them and help to avert or minimize some adverse outcomes. Maturing adults have physical problems such as difficulties with hands, feet, walking, eyesight, diet and sleep. In the yoga class they can discuss these problems and gain support from the teacher and the group. Moving together in a group also creates a pleasant social environment and encourages friendships among like-minded people outside of class.

Nine

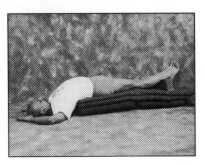

Opening the Heart
with Yoga

Your Heart: Where Body, Mind
and Spirit Converge

A lot of people swear by yoga—they love that more than anything else we do. It's another form of mindfulness, and it has the added benefit of reversing disuse atrophy and really toning the body. It's a full-body musculoskeletal strengthening and conditioning exercise. Yoga postures redirect the energy flow in your body and your mind. When you use yoga as a door into awareness of the body, it can teach you all sorts of things. You get to know your body on a totally intimate level. . . . When you're practicing yoga, you're releasing tension, and the tension isn't always in the body, it can also be in the heart, in the mind, or in feeling states. Release of that tension can put you back in touch with yourself on a very deep level. It's an inner experience of coming home.

—Jon Kabat-Zinn, Ph.D., interview with Bill Moyers in *Healing and the Mind*

Heart disease is the single most common cause of death in affluent countries like ours. In most Americans, heart disease is caused by a narrowing or blocking of the coronary arteries that deliver blood to the heart. The physical factors that contribute to heart disease—inherited risk factors, diabetes, high blood pressure, elevated serum cholesterol, deposits of plaque in the coronary arteries, smoking, poor diet, lack of exercise, job pressures, stress and other lifestyle-related considerations—explain only in part the cause of heart disease.

Western techniques, such as drugs and surgery, can be very helpful in a crisis, but they are limited. . . . The stress management techniques derived from yoga address the more fundamental issues that predispose us to illness. And while yoga is a very powerful system of stress management, these techniques were designed for something much greater—as tools for transformation.
—*Dean Ornish, M.D., Dr. Dean Ornish's Program for Reversing Heart Disease*

If the arteries are obstructed, blood flow is reduced, which reduces the amount of oxygen the heart receives and impairs its normal functioning. Yet heart disease is highly individual. Someone with relatively little obstruction in the coronary arteries can be incapacitated by angina, or squeezing chest pains, while another person with far more severely obstructed arteries may not be aware of any problem. Some people have run marathons with 85 percent of their coronary arteries blocked, while others with no sign of arteriosclerosis have dropped dead of a heart attack. It is clearly impossible to explain heart disease exclusively from physical causes.

Over 300 years ago, William Harvey, the founder of modern heart physiology, understood that the mind and emotions affect the health of the heart. "Every affection of the mind that is attendant with either pain or pleasure, hope or fear," he said, "is the cause of an agitation whose influence extends to the heart."

In our era, Norman Cousins, after surviving a massive heart attack, wrote in *The Healing Heart*:

> *The human heart is not sealed off from countless processes that take place within the human body. It is a point of culmination, a collection center for all the malfunctions or deficiencies that exist in the body as a whole. It is a zone of infinite vulnerability to all the anguishes and insults and provocations of mind, soul and body.*

Dr. Dean Ornish's Prescription for Reversing Heart Disease

"Your heart is the place where your body, psyche and spirit all converge," writes Dean Ornish, M.D., in his book *Dr. Dean Ornish's Program for Reversing Heart Disease*. This famous and popular physician teaches that the health of the cardiovascular system involves all aspects of a human being—physical, emotional and spiritual.

Ornish's healthy lifestyle "prescription," which includes yoga postures and meditation, won the backing of mainstream medicine in the early 1990s, when one of the nation's largest providers of health insurance, Mutual of Omaha, announced that it would cover the $5,000-per-year cost of his program.

Ornish's program prescribes a low-fat vegetarian diet; yoga-based stress management techniques, including stretching, meditation, breathing and deep-relaxation techniques; a program to stop smoking; support-group

With tears and time you can heal any pain. Not doing so causes ill health and drains energy. Rage, hate and revenge are the biggest killers.
—Elisabeth Kübler-Ross

sessions; the development of communication skills; and directions for safe and moderate exercise.

His landmark research validates this holistic approach. After just one year on the program, his patients show a measurable reduction of blockages in their coronary arteries and sufficient relief from symptoms to make angioplasty or bypass surgery unnecessary.

In a *New Age Journal* interview, Dr. Ornish explains that:

> *Bypass surgery is much like clipping the wires to a fire alarm and going back to sleep while your house burns down, or like mopping up the floor around a sink that's overflowing without also turning off the faucet. It is now known that after five years, on the average, half of all the bypassed arteries have clogged up again, and within seven years, 80 percent have. And angioplasty, a newer technique in which a small balloon is inserted in the artery and then inflated to open blockage, doesn't fare any better. Within four to six months an estimated one-third have closed up again.*

In America, more money is spent on treating heart disease than any other illness—estimates range from $78 billion to $94 billion annually. More than $7 billion a year is spent on bypass surgery alone. When a doctor performs bypass surgery on a patient, the insurance company pays at least $30,000. If he or she performs a balloon angioplasty, the insurance company pays at least $7,500. If a doctor spends the same amount of time teaching that patient about nutrition and stress management, the insurance company pays considerably less.

Data/studies show that exercise and meditation can reverse the biological markers of aging—bone density, strength of muscles, blood pressure, aerobic capacity and others. It is now well-established that you can take a 97-year-old man or woman and put (him or) her on an exercise program and show reversals in the aging process. If people meditate regularly, their levels of DHEA—a hormone that usually declines with age—rise. Blood pressure drops, cholesterol levels balance, hearing and vision improve.
—*Deepak Chopra M.D.,*
Ageless Body,
Timeless Mind

The primary concept of Ornish's approach is that the farther back in the causal chain of events we can begin treating a health problem, the more powerful and lasting the healing will be. He describes how all the traditional risk factors—cholesterol, blood pressure, age, gender, genetics, smoking, diabetes, obesity, sedentary lifestyle, etc.—explain only half the reason some people get heart disease and others don't.

His nonsurgical, non-drug therapy, healthy-lifestyle approach to solving the great problem of heart disease has been endorsed by the American Heart Association and the President's Task Force on Health Reform. The "heart" of the program includes spending an hour a day practicing stretching and breathing exercises, deep relaxation and meditation. Ornish's program recognizes the importance of daily relaxation periods for preventing future heart deterioration. He recommends the practice of restorative yoga postures to ease stress and live a more balanced life.

Daily stretching, relaxation and meditation are crucial for people already suffering from high blood pressure and heart disease. Dr. Ornish and other health professionals who advocate the use of yoga techniques recognize that the course of heart disease is highly individual, and that our ability to overcome any health problem is greatly enhanced when we learn to consciously quiet and relax the mind and body.

> Yoga may be a centuries-old Eastern philosophy and art practiced by a variety of cultures, but it is also the finest, most adaptable form of combined physical and mental refreshment available today.
> —Dawn Groves, Yoga for Busy People

> If you think you can't spare 20 minutes a day for stretching and relaxing, I can almost guarantee that you are the very person that needs it the most!
> —Judith Lasater, Ph.D., yoga teacher and author, Relax & Renew: Restful Yoga for Stressful Times

Chronic Tension and Heart Disease: The Fascinating Fight-or-Flight Response

Herbert Benson, M.D., coined the phrase "relaxation response" to describe the profound physical and mental responses that occur when we consciously relax. According to Benson, the relaxation response is "a physiological state characterized by a slower heart rate, metabolism and rate of breathing, and lower blood pressure and slower brain-wave patterns." The opposite physiological response, characterized by a rapid heart rate, increase in blood pressure and quick, shallow breathing, is known as the "stress reaction" or "fight-or-flight response."

The fight-or-flight response is instinctual and deeply imprinted in our nervous system. It evolved to help us cope with immediate danger or situations of acute stress. Can you remember the last time you felt this reaction? All of us have experiences that trigger anger and anxiety.

By observing your own reactions to events in your life, you can become aware of your physical responses to emotions such as fear, anger and anxiety. When you are angry or anxious, your whole being becomes involved. Your breathing becomes rapid and shallow, to provide you with more oxygen either to do battle or run away. Observe how your muscles begin to contract. This is nature's way of preparing you to attack or protect yourself. Your metabolism and heart rate speed up, boosting your strength and energy. The amount of blood pumped with each heartbeat increases.

Next time you feel angry, see if you can feel your digestive system begin to shut down, so that more blood and energy can be diverted to the large muscles needed for fighting or running. If you are really angry, the arteries in your arms and legs begin to constrict and blood chemistry changes, so that clots form more quickly to conserve your blood should you be wounded.

According to heart and stress specialists, our bodies are designed to cope with acute stress, not the chronic stress of modern everyday life. When our survival mechanisms are chronically activated, they begin to exhaust our internal organs and nervous system. Arteries constrict not only in the arms and legs but also in the heart and brain. Blood clots are more likely to form inside the coronary and cerebral arteries, increasing blood pressure, sometimes leading to coronary thrombosis or stroke.

During emotional stress, our skeletal muscles constrict, causing pain and tension in the neck, back and shoulders. In addition, the internal smooth muscle lining the coronary arteries contracts, resulting in spasm in the heart muscle itself. Dr. Ornish describes the effect of chronic stress on the heart this way:

> *Under conditions of intense chronic stress, even the muscle fibers inside the heart itself can begin to contract so vigorously that the normal architecture of these fibers is disrupted, damaging the heart muscle. To me this is an amazing metaphor: The inability of the heart to relax causes the heart's muscle fibers to constrict to the point that it damages itself—like clenching your fist so hard and for so long that the bones and knuckles in your hand begin to break.*

Obviously, stress reduction is not a luxury but a health-promoting and potentially life-extending technique. Begin now to become aware of your breath and take time to practice slow, calm, even breathing. It's the first step to feeling more relaxed.

Enlightenment: The Ultimate Safeguard Against Stress

Yoga reverses the stress syndrome by bringing your body into the "relaxation response." However, ancient yogis and sages, wise men and women through the ages, have taught that the ultimate solution to stress is not simply learning how to manage it, but learning to see events in such a way that we don't create stress in the first place. Very often, the stress we blame on outer events is the product of our own emotions.

Coping with tension means understanding that stress comes not only from events in our daily lives but from how we react to them. We have all seen people who react entirely differently to the same situation: One will be very stressed while the other will just smile and say, "That's not important. I won't let it bother me." If you feel you are a person who increases the stress in your life emotionally, you may want to experiment with the way you see things. Altering your vision of events is sometimes a matter of simply reviewing priorities, examining what truly matters to you, or evaluating whether you will honestly care about whatever is bothering you a week, a month, or a year from now. Visual thinkers might draw a simple picture of the situation as they see it now, being sure to include a symbol of themselves, then moving the self step-by-step away from the center of the situation. How do they feel about it then? Positive habits of mind can be learned and practiced until we no longer feel at the center of the storm of daily life, but can observe, actively participate and remain at peace.

Practicing yoga is another way of finding your own

All religions encourage self-sacrifice, but when we are ill we pray to God to heal us. How inconsistent we are! To be simple, to appreciate what has been given to us, and to take care of our body, is an act of humility. . . . Do not fight your body. Do not carry the world on your shoulders like Atlas. Drop that heavy load of unnecessary baggage and you will feel better.
—*Vanda Scaravelli,*
Awakening the Spine

calm center. Gradually, as you discover the pleasure of sitting quietly or practicing the standing poses, you may realize that something has changed, that you feel a quiet joy and see a subtle shift in your perspective. As one of my students said, "It was like discovering an entirely new banquet of foods I had never experienced before." Because the practice of yoga unfolds gradually and is different each time, we can explore these subtle feelings and understandings throughout our lives. Until enlightenment comes, however, it helps to learn breathing, stretching and relaxation techniques for releasing mental and physical anxiety.

To understand the value and effect of relaxation techniques, keep in mind that tense, contracted muscles and shallow rapid breathing are major stress signals to the brain. If you can feel for yourself how during times of stress your entire sympathetic nervous system becomes stimulated (i.e., heart rate, blood pressure and muscular tension increase), you can more easily grasp how slowing down your breathing and stretching tense muscles sends out the opposite signal of relaxation, which not only calms and quiets the mind but relaxes the heart, balances your blood pressure, lowers cholesterol levels and helps to return all of the systems of the body to a healthier state.

Monday Morning Heart Attacks

"To be healthy, acknowledge the fact that if you try to take care of your health without taking care of what gives your life meaning, it's fighting the battle with one hand tied behind your back," says Larry Dossey, M.D., a specialist in mind/body medicine.

I believe the holistic healing systems of the future will combine the tremendous body of the "analyzed" knowledge of the traditional medical profession with the "synthesized" knowledge of the higher body/energy systems. The future holistic healing systems will diagnose and prescribe healing for all the energy bodies and physical body simultaneously as needed by the patient and incorporate both the inner and the outer healing processes. Medical doctors, chiropractors, homeopaths, healers, therapists, acupuncturists, etc. will all work together to aid the healing process.

—*Barbara Ann Brennan,*
Hands of Light:
A Guide to Healing
Through the Human
Energy Field

You can treat blood pressure, stop smoking, get your cholesterol level down—and do those things perfectly—but that is not covering the bases. We know that most Americans under the age of 50 who have their first heart attack have none of these major risk factors present. They check out fine! That's a way of saying that the meaningless, purely physical approach can't account for even a simple majority of first-time heart attacks.

Dr. Dossey's books eloquently challenge the most ingrained assumptions of modern medicine, such as the notion that the body is a machine; that it is fully automatic; that thoughts, attitudes and feelings are of little consequence; that consciousness is simply the product of chemicals in the brain; that there is nothing higher than the brain mechanism; that when the body dies, that is the end of it.

According to leaders in the field of mind/body healing, medicine rests on a set of assumptions that, when examined from a philosophical or scientific point of view, become extremely shaky. As Dossey says, most of us plod along without ever examining the fundamental questions about our life and health.

One fascinating signal that told him we ought to be looking at health from many angles is the Monday morning heart attack phenomenon. In the United States, more people have heart attacks on Monday morning between 8 and 9 A.M. than any other time. In a study done in the early 1970s, Dossey determined that the best predictor of heart attack was none of the major risk factors—not smoking, high blood pressure, diabetes or cholesterol—it was level of job satisfaction. In his book,

Meaning and Medicine, he asks, "What is the meaning of a person's job to them? What does a job represent or symbolize in their mind? Without bringing in meaning, we can't understand the origin of the most common killer in our society: heart disease."

Dossey claims that one of the first steps toward health is to simply acknowledge and honor the fact that we need to perceive our lives as meaningful, feel that our efforts are worthwhile and understand that even our illnesses have meaning. "Then you can go from there. It's impossible to spell out in any one person's life, what might follow. How should one person sort out their meanings?"

Can you imagine going for a free cholesterol check-up and having a medical technician ask you whether or not you're happy with your life? It's much easier to focus on cholesterol and blood pressure or to treat heart attacks when they occur. A doctor has a tough enough time getting a patient to exercise, eat right and cover all the physical bases. Nonetheless, as challenging as it is to examine ourselves mentally, emotionally and spiritually, the whole field of psychoneuro-immunology is demonstrating that thoughts are not simply ethereal notions that pass through the brain. Apparently our thoughts penetrate the body and affect every major system in the body—particularly the cardiovascular and immune systems—as much as, or even more than, food and exercise.

Scientific research continues to affirm the value of disciplines such as yoga. One of the most effective ways to lower cholesterol levels is simply to sit down twice a day for about 20 minutes and clear and quiet the mind. Cholesterol levels drop by about one-third. Dr. Dean Ornish says, "Even a few minutes of meditation a day brings real benefits."

How Posture Affects the Health of Your Heart

Notice how you are sitting as you read this. Don't change anything, simply observe yourself. Now deliberately slump forward as much as you can until you feel your internal organs pressing in on each other. How are you breathing now and how do you feel? These observations tell you how your everyday posture—the way you sit, stand and walk—affects your respiration, circulation and the health of your heart. Chronic slouching decreases circulation to all your vital organs.

When the chest is collapsed, your diaphragm barely moves down as you inhale. This keeps you from completely filling your lungs. It prevents the "chest pump" from helping to return blood to the heart. If the abdomen and chest are compressed most of the time, the lymphatic vessels, arteries and veins serving the vital abdominal organs may collapse, even partially. This reduces the circulatory cleansing of all these tissues, and minimizes the transportation of vital nourishment to the cells and the circulation of important hormones that regulate the body's processes. This compromises overall health, including the health of the heart.

One of yoga's most immediate effects is a dramatic improvement in our posture. People of all ages sigh with relief as the chest opens and they can breathe freely. For those with rounded shoulders and chronically stooped posture, basic standing poses begin to open the chest and expand the breathing process. More vigorous poses such as Upward and Downward Dog, both from the floor and stretching back and forth with wall ropes, further stretch the muscles of the front of the body, expand the chest and increase breathing capacity, as well as strengthen the back, chest and shoulder muscles.

Gentle supported backbends and various restorative postures expand the chest, lungs and rib cage without effort. These passive poses are useful for everyone, but are especially recommended after healing from heart surgery, and should be practiced with the guidance of a qualified instructor.

Breathing Freely, Relaxing Deeply Reduces Stress

Note in the photo on page 246 how the folded blankets positioned evenly under the back of the head and length of the spinal column, open and expand the chest, allowing the body to breathe freely and deeply relax.

All body/mind disciplines teach that the breath is a bridge between the body and the mind. Most present-day stress reduction techniques are based on this recognition. Conscious, calm breathing is one of the simplest, yet most elegant and effective ways to combine awareness of our physical body with consciousness of our mental and emotional state. Breathing is also a powerful means of relaxing your sympathetic nervous system, which governs the stress response, and accessing the healing effects of your parasympathetic nervous system, which governs the relaxation response.

Being aware of your breath and softly, calmly, regulating it in the way that I describe next sends a direct message of relaxation to the mind and nervous system.

The following technique comes from the teachings of B.K.S. Iyengar. It can be practiced either sitting upright or lying down. Even though the four steps look very

Breathing is the essence of yoga. Breathe naturally, without forcing. No pressure, no disturbance, nothing should interfere with the simple, tide-like movement of our lungs, as we breathe in and out. . . . Breathing is the most important part of our yoga practice.

—*Vanda Scaravelli, Awakening the Spine*

We've mentioned breathing during childbirth but very few people recognize the importance of breathing during the dying process. When I am in the presence of a person who is very ill and dying, I rarely talk; instead I go into three-part breathing on my own. I don't know exactly what happens, but there is an effect on the other people who are present. Perhaps as we reduce our stress, those around us can relax and let go
—*Mary Dale Scheller,*
Growing Older, Feeling
Better in Body, Mind
and Spirit

simple, I recommend that you read the instructions all the way through at least twice before you begin. Note that what makes this way of breathing different from our usual, habitual way of breathing is the pause at the end of each exhalation.

Breathing to Calm the Mind and Reduce Stress

1. Inhale slowly, gently, without strain, through your nose.
2. Exhale slowly, gently, without strain, through your nose.
3. Pause briefly without straining or attempting to hold your breath.
4. Repeat steps 1, 2 and 3. Continue breathing in this calm, simple way for several minutes.

To review: at the end of every exhalation, pause without straining for a second or two before inhaling again. By pausing briefly after each exhalation, you may notice a spontaneous, natural, unforced continuation of the exhalation. This additional release of the breath completes a true, normal exhalation. In our habitual way of breathing, especially if we are tense, the exhalation is incomplete. We start each exhalation without allowing the previous exhalation to come to its natural conclusion. If you practice this new way of breathing regularly, your breathing will deepen naturally, effortlessly, and you will find yourself feeling calmer and more relaxed, even under what once seemed like stressful circumstances.

In daily life, practice watching your breath. As you observe its flow, inhale and exhale calmly through your nose. Do not try to impose a deeper breath on your natural breathing rhythm. By being aware of your breath and pausing without strain after each exhalation, you will feel a natural urge to inhale more deeply and freely. Follow this urge, without straining or forcing, and then return to normal breathing. Remember to practice this relaxed way of breathing whenever you want to calm down and relax—when stuck in traffic, while waiting in line or at the doctor's office, whenever you feel anxious or upset.

When lying down to relax, close your eyes and look down toward your heart. Allow your eyes to relax. Looking downward toward the chest helps your brain to relax. To help you relax further, cover your eyes, using an eyebag, if available (see the Resources section) or a damp folded facecloth. The gentle pressure of a light weight over the eyes reduces their involuntary movements and releases tension in the eyes and forehead.

When using folded blankets or pranayama (breathing) pillows to support your body, as described in the prop chapter, note how the shape of the blankets expands and opens the chest, and supports and encourages the natural curves of the spine. The body generally responds with delight and a sigh of relief upon being positioned on folded blankets or long, firm pillows. I consider their use absolutely essential for learning to breathe freely and relax deeply.

Some Facts About Breathing

• Changes in your body affect your breathing. When you walk or run uphill or otherwise exert yourself, your oxygen requirements go up and you breathe faster and deeper.

- Changes in your mind affect your breathing. If you are anxiously anticipating the dentist's drill, or about to deliver a speech, your breathing is apt to be rapid and shallow. Conversely, when you are sitting on a hill watching a sunset, getting a massage or petting your dog, you are more likely to breathe slowly and deeply.
- Changes in your breathing affect your body. Athletes, martial artists, singers, dancers and gymnasts all use breathing techniques to improve their performances.
- Changes in your breathing affect your mind. When you are worried, a few deep breaths can help calm you down. Deep breathing wakes up a tired mind.
- Changes in your posture dramatically affect your breathing. It is impossible to breathe deeply and freely if you slump.

There are many books and schools of thought on breathing techniques for reducing stress. The first step is to have the presence of mind to notice the way you breathe in both relaxed and tense situations. I believe that any changes made in breathing should be gentle and very gradual. When experimenting with deepening your breath, the rules are: Stop if you feel any strain, and inhale and exhale naturally for several cycles before you begin again. Remember to close your lips softly and breathe through your nose.

Stress and the Immune System

Stress depletes the body's reservoirs of vital energy. Stress also tends to speed up our metabolism to the point that even when we do have a chance to slow down and recuperate, we find we are unable to unwind and

My regimen is targeted specifically to stress. Stress ages you. Yoga and meditation help me connect with my inner-self rather than my tensions. This is not to downplay calisthenics, weights and aerobics. But I know for a fact that they don't help my state of mind, or help my skin glow in quite the same way yoga and meditation do.

—*Raquel Welch*

relax. Many people, as they are introduced to yoga's deep relaxation practices, discover that their bodies and minds need to re-learn what it feels like to experience a deep level of relaxation.

A potentially life-saving principle to remember is: The way you react to stress influences immunity more than the severity of the actual stressful event. Of the personality traits associated with low resistance to illness, feelings of helplessness are the most destructive. Just knowing that you can do something constructive when life seems overwhelming boosts your immune system. Even if the actual situation cannot be changed, reducing the body's reaction to stress by practicing relaxing yoga positions and peacefully observing the breath allows the mind and body to break the stress cycle, recuperate, and restore equilibrium.

EDITOR'S NOTE: *For further information on Dr. Dean Ornish's program for reversing heart disease, other resources and educational opportunities please write: The Preventive Medicine Research Institute, 900 Bridgeway, Suite 1, Sausalito, CA 94965, 415-332-2525.*

Alice Stevens:
Maturing at a More Satisfying Pace

Alice Stevens is a yoga instructor for the American Heart Association. She also teaches cardiopulmonary resuscitation (CPR) as part of a cardiac rehabilitation program. Unfortunately, she says, these days more and more young men and women are becoming afflicted with heart disease.

Yoga has enhanced my strong qualities, made me more aware of my not-so-hot ones, and helped me mature at a more satisfying pace. It takes a lot of energy and good spirits to thrive in a busy and happy life.

Older students sometimes need more encouragement until their confidence perks up and they start enjoying themselves in class. So many women and men who are older come into class for a whole new physical adventure, and I want it to be a good experience for them. At first they are easily intimidated, but once they begin to feel at home and the poses become familiar, they get stronger. It's wonderful to see them bloom.

I have found that the Downward-Facing Dog is the best pose in yoga for older students to regain or build posture, upper-body strength, flexibility, and to stretch those concrete hamstrings. I just love it when an elegant matron proudly tells me that she's teaching her husband the Dog Pose so he'll stand up straight.

Ina Marx:
If Your Life Has Purpose You
Can Never Grow Old

Ina Marx enjoys an exhilarating stretch in the Camel Pose.

Ina Marx, 76, is the author of Yoga and Common Sense *and* Fitness for the Unfit. *After an accident that left her severely disabled, she made a personal commitment to yoga.*

My own uphill climb began at age 40. I was totally devoid of self-confidence. Self-love and self-realization were not in my dictionary. I had struggled along with limited capacities for most of my life, plodding along as best I could. My life seemed a never-ending series of personal tragedies.

When I was 30, I was working in a luxury resort on weekends while my husband stayed home, taking care of our young daughter. I stayed overnight in a ramshackle firetrap reserved for the hotel's help. I was pregnant again as well. One night, as the staff members were sleeping, a fire broke out. I was in a room on the third floor, where many of us were trapped as flames spread and consumed the wooden staircase.

What saved me were my instincts. The only way to escape was through the window. There were two in the dormitory. One was always stuck and the other was blocked by a crowd of other panic-stricken occupants. I smashed the window that would not open. When the flames came unbearably close, I jumped, plummeting onto a concrete drive. By the time the fire truck arrived, the building had burned to the ground and 10 people were dead.

I had fractured my back and pelvis, cracked ribs and sustained internal injuries. My psyche was broken as well. I lost my unborn baby. First I was placed in a body cast. Later I graduated to a steel corset, which I was told I would wear for the rest of my life. Because I was in constant pain, I became addicted to tranquilizers, barbiturates, painkillers and three packs of cigarettes a day. I suffered from fatigue, insomnia and obesity. I spent all my time seeing a succession of physicians and physical therapists. Life held no meaning for me and I often contemplated suicide.

After two suicide attempts, I sought psychiatric treatment. I began to function better but broke down again. I found a new therapist and my mental attitude improved. Physically I was still a wreck, but decided to hang in there and try every healing method I heard of. One day someone said, "Try yoga!" I latched onto it as a last resort. At first, it seemed hopeless. I could not bend in any direction. However, I was determined to stick to it because I had caught a glimpse of a more hopeful future.

Discipline and hard work were the keys, and consistency was the vital factor. I didn't turn my whole life over to yoga, for I still had my family to take care of, but I practiced conscientiously for at least one hour each day. My back grew stronger. After three months I threw away my corset. My figure improved, I was healthier and more energetic. For the first time in years, I fell asleep easily. I gradually eliminated the tranquilizers and barbiturates that had been a staple in my diet. Quite naturally and painlessly, I gave up smoking.

As my body changed, so did my mental outlook. My sense of values changed as well as my attitude toward people. I developed more patience, kindness and tolerance. At age 40, I realized that I am responsible for my state of being and that I could attain any goal I desired.

I refused to be defeated by the laws of probability and defied the medical advice, "You must learn to live with your pain." I disproved the orthopedist's prescription, which would have confined me to a wheelchair for life. Although my back injuries have knitted, subsequent examinations have shown deterioration of several discs and a congenital curvature of my spine. Those who view my X rays find it incomprehensible that I can even walk. Yet my back is strong, supple and free of pain.

According to medical and societal standards, I am old. If I applied for a life insurance policy or a job, I would probably be rejected. But I maintain that age is irrelevant. I feel younger, healthier, am stronger, have more energy, vitality and optimism than I did in my earlier years. While I do not follow the popular expectations for aging, neither do I yearn for youth. There is no way to escape forever the exterior signs of aging. Old age, however, is a state of mind that can be changed. If your life has a purpose, you can never grow old.

Robert Whiteside:
Agile at 80, Nimble at 90

Robert Whiteside is author of Agile at 80, *a short, motivational book, encourage older adults to exercise, eat right and keep a fresh outlook on life.*

I think that one of the things people most dread and fear about old age is aches and pains, or being wheelchair-bound, a burden to themselves and others.

I myself and many others have demonstrated that such a fate is not necessary, if one follows nature's laws through proper nutrition, exercise and a positive mental attitude.

I'm an octogenarian now and still jog three miles before breakfast and play tennis a couple times a week. I also carry out an active professional career—traveling a third of the time.

If I were to write my book over again (I am shooting for *Nimble at 90*), I would emphasize yoga even more, and also more use of fresh vegetable juices.

Traces of old injuries pop up if I don't behave myself and each day remember to practice my yoga exercises to keep everything supple and in its proper place. It's marvelous what the human body will do, given half a chance.

The wisdom of the body—knowing how to sound alarms, receive supplies, do repairs, strengthen muscles, heal cuts, join a broken bone together—makes the best computer archaic.

Once you get started it's an upward spiral. Don't be ashamed of beginning modestly, but *do* begin. And do not get discouraged if some emergency keeps you from your whole program. Just do as my yoga instructor advises: start the next day where you left off.

Any time you are tempted to quit, consider the alternative. As my son-in-law quips, "It's better to be over the hill than under the hill!"

Ten

Backbends Open Posture, Lift Spirits and Expand Perspective

*The intention, the long term goal, is to become com-
pletely fluid, completely liquid and sinuous. As I get
older I'd like to be that. I'd like to have explored the
entire range of my body's abilities. It's not that I'm
afraid of getting old. I just want to get old in a
certain way. I want to get old gracefully. I want to
have good posture, I want to be healthy, and I want
to be an example to my children.*

—STING, INTERVIEWED BY GANGA WHITE,
IN *YOGA JOURNAL*, NOV/DEC 1995

People who just talk about, read books on, and go to lectures on spiritual matters are, to use my favorite analogy for this, much like the person who wants to learn to swim before getting into the water. Obviously, you must get into the water in order to discover that it can be trusted to hold you up; only then is learning to swim possible. . . . Practicing is the getting into the water—the acting on, rather than only listening to or talking and reading about spiritual matters. . . . You must act—practice.

—GEORGE JAIDAR, THE SOUL: AN OWNER'S MANUAL, DISCOVERING THE LIFE OF FULLNESS

Health is freedom. . . . As the sun opens the flowers delicately, unfolding them little by little, so the yoga exercises and breathing open the body during a slow and careful training. When the body is open, the heart is open. There is a transformation in the body's cells. They work in a different way and a new growth is possible.

—*Vanda Scaravelli,* Awakening the Spine

Backbends counteract many of the changes often considered to be a normal part of the aging process. As we grow older, the spine degenerates and we become shorter. Roundness of the upper back, known as kyphosis, is so common in older people in this culture that it is almost accepted as an inevitable part of the process of aging. As the back curves forward, the chest sinks, the lungs are at least partially compressed, and respiration is compromised. This has a negative impact on every system of the body, especially the cardiovascular system.

Bending backwards counteracts all of these tendencies. To breathe freely, we must release the tension in our chest, the chest cavity must be able to expand fully, and the muscles surrounding the rib cage must be free to stretch. Backbends free the chest, open the rib cage and encourage healthy breathing. After a series of backbends the whole body becomes charged with oxygen, making us feel exceptionally energetic and alive.

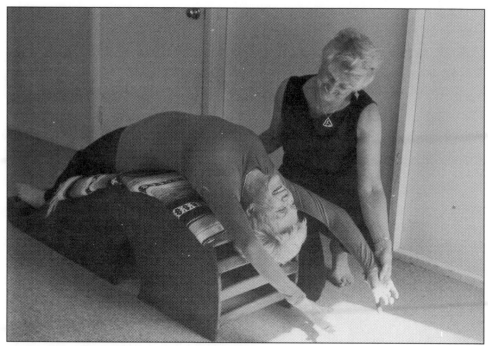

Backbends bring greater blood supply to the discs and nerves of the spinal column.

Backbends strengthen and refresh the entire body. Gently and gradually arching the body backwards lengthens and tones the entire spinal column, opens the shoulder joints, stretches the surrounding muscles, tendons and ligaments, and stimulates production of synovial fluid, the lubricant in the joints. Calcium deposits, bursitis and the general tension many of us carry in the upper back, neck, shoulders and arms can be relieved with backbends. Weight-bearing, active backbends also strengthen the arms, wrists and legs and increase circulation to the organs in the pelvis. In addition, backbends massage the kidneys and adrenals and stretch and stimulate the liver.

During a backbend, the front of the body is stretched to the maximum, especially the chest area. The opening of the chest opens the heart center. Backbends relieve

Whatever is flexible and flowing will tend to grow, whatever is rigid and blocked will wither and die.
—*Tao Te Ching*

depression by dispelling the heavy-hearted feeling we have when we are sad. Stretching backwards lifts your spirits, changes your perspective and makes both the mind and body feel light and free.

Begin practicing backbends gradually. If you are stiff, never force a backbend; instead, prepare the body by working on standing poses and practicing gentle supported backbends. As always, use common sense. You don't start a yoga stretching program with backbends any more than you would hike uphill 15 miles if you hadn't taken a long walk in many years. If you have back problems such as degenerative disk disease, or if you have high blood pressure, a history of stroke or heart trouble or other serious illness, do not attempt the more strenuous backbends without the guidance of a qualified teacher. Also, do not practice backbends during pregnancy.

Bridge Pose to Strengthen the Back (*Setu Bandha Sarvangasana*)

This pose is frequently prescribed by chiropractors and other doctors as part of a back-strengthening program. It is most effective done slowly and thoroughly, with yoga awareness of the breath, rather than repeated mechanically. Like Downward-Facing Dog and standing poses, it should be part of a core program for those who begin yoga after 50.

This pose can be done after standing poses or as part of a lying-down-on-the-floor series. For experienced students, it is safe preparation for more active, advanced backbends. Almost everyone of any age can enjoy doing the Bridge Pose from the first days of their practice.

Bridge Pose.

This beginning backbend will strengthen the back and open the chest.

1. Lie on the floor or sticky mat, your knees bent, feet placed hip-width apart, fairly close to your bottom.
2. Relax in this position, watching your breath, allowing your belly to soften and relax. Wait until you feel your abdomen drop (a pleasant, relaxed, hollow feeling), and the back of your waist drop to the floor as your lower back releases and lengthens. Anchor the soles of your feet to the floor, turning your toes in slightly.
3. When you feel relaxed, tuck your tailbone under, toward your heels. Continue drawing your tailbone under, until your bottom naturally begins to lift off the floor.
4. Keeping your feet in line with your hips, press down into your feet even more and continue to slowly lift as much of your spine off the floor as you comfortably can.

5. Continue pressing down into your feet and stay in the pose long enough so that you feel your back, buttock and thigh muscles strengthening. Exhaling, release your spine slowly back to the floor.

6. Relax and repeat, lifting up higher to the tops of your shoulders.

Placing a strap around your feet and knees, as illustrated on the previous page, will remind you to keep your feet active, pressing firmly into the ground, and your knees parallel, in line with your hips. A block can be placed as shown by bringing your toes close to your bottom, lifting your pelvis high up and carefully positioning the block under your sacrum. While using props for Bridge Pose is optional, they can be especially helpful for beginners over 50.

Gentle Supported Backbends Create Breathing Space

Healthy circulation and respiration are fundamental to good health. Backbends, like many other yoga postures, stimulate and strengthen these two vital systems, which nourish, cleanse and invigorate the entire body.

The gentler, passive, supported backbend positions, practiced by lying over bolsters, stacks of folded blankets or yoga props such as the backbending bench in the photo on page 209, can generally be safely practiced by people with high blood pressure, heart disease, breathing difficulties and other stress-related problems. In fact, for such people these postures are especially beneficial.

Backbends make breathing easier by opening the rib cage and passively stretching and lengthening the chest muscles.

The young, the old, the extremely aged, even the sick and the infirm obtain perfection in Yoga by constant practice. Success will follow him who practices, not him who practices not. Success in Yoga is not obtained by the mere theoretical reading of sacred texts. Success is not obtained by wearing the dress of a yogi or a recluse, nor by talking about it. Constant practice alone is the secret of success. Verily, there is no doubt of this.
—Hatha Yoga Pradipika, 1:64-66

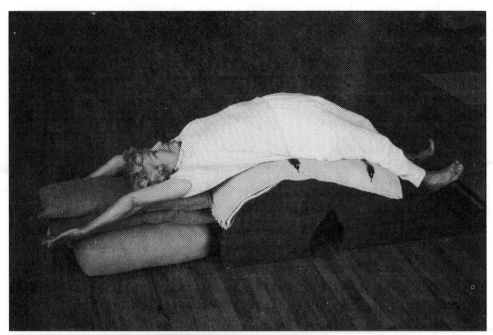

An 84-year-old student in a supported backbend.

Both the more strenuous active backbends and the gentler, passive backbend positions stretch the abdominal organs and expand the space in which they function, as well as improve blood supply to the area. When the body is placed in a passive supported backbend, muscles stretch, the chest and rib cage open, and breathing becomes easier, eliciting the soothing, restorative relaxation response.

To further understand the value of supported backbends, consider the changes that occur in our respiratory system after years of poor posture aggravated by emphysema, asthma or other breathing difficulties. When the chest is collapsed in the presence of lung disease, the lungs lose their normal elasticity, so it takes more muscular effort to move air in and out. If the chest is closed, tense and tight, and breathing is difficult, the muscles of the neck and shoulders (not meant for this purpose) are recruited to keep us breathing at least enough to keep the body alive.

When posture is chronically poor and the head is held forward, the upper back rounded, the neck and shoulder muscles are always partially contracted. In this chronic state of tension, the neck, shoulder and upper-back muscles require more oxygen and energy at the same time as their blood supply diminishes. Lactic acid builds up in the tissues, causing the muscles to become sore and stiff and respiration to be further restricted.

The body constantly monitors blood pressure, blood oxygen levels and muscle tension. Shallow breathing reduces the amount of oxygen available to tissues and cells. When the muscles are also chronically tense, the body interprets these signals to mean it is in a state of emergency and responds with the "stress response" discussed in chapter 9. Supported backbends counteract and reverse all of these postural and respiratory problems by deeply relaxing the muscles and opening the chest and rib cage. As muscular tension is released, circulation improves. The deep relaxation that takes place when one remains in a supported backbend, even for a short period, is healing and rejuvenating for the whole body.

Study the photo of Supported Bridge Pose on the following page and note how the support under the back and legs is placed end-to-end to accommodate the length of the body. The height of the support under your body depends on the length of your torso and the flexibility of your back. New students can begin with one or two single-fold blankets and gradually increase the height. Six to 12 inches in height works well for most people. Taller people can add height to the bolsters with one or

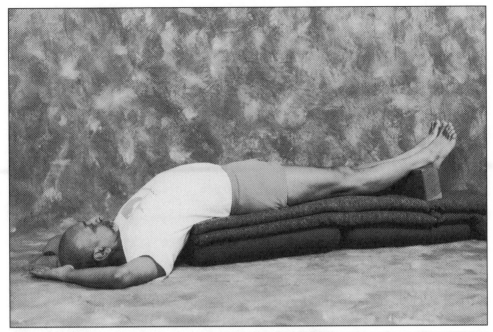

Yoga teacher Ramanand Patel demonstrates Supported Bridge Pose.

more blankets on top of the bolsters. You can also place a block under your heels, as illustrated. For details on practicing Supported Bridge Pose, review chapter 6, page 137.

B.K.S. Iyengar:
It Is Never Too Late to
Practice Yoga

You are as young as your spine is flexible.

B.K.S. Iyengar, born December 12, 1918, is one of this century's foremost exponents of yoga. The author of Light on Yoga, Light on Pranayama, The Tree of Yoga and many other books, Iyengar still performs a rigorous daily yoga practice that would challenge anyone at any age.

It is never too late in life to practice yoga. If it were, then I should have stopped my practice long ago. Why should I do so now? Many Indian yogis reach a certain point in their lives and say they have reached *samadhi* [the highest state of spiritual evolution], so they don't need to practice any more. But I have not said that up to now. Why not? Learning is a delight, and there are many delights to be obtained through the practice of yoga. But I am not doing it now for delight! In the early days delight was the aim, but now it is a by-product. The sensitivity of intelligence which has been developed should not be lost. That is why the practice has to continue.

If you have a knife which you do not use, what happens to it? It gets rusted, does it not? If you want to go on using it, you have to sharpen it regularly. With regular sharpening you can keep it sharp forever. Similarly, having experienced samadhi once, how do you know that you are going to remain alert and aware forever? How can you say that you can maintain it without practice? You may forget and go back to enjoying your life in the same way as you did before. Can a dancer or concert performer give a fine performance if they have not practiced for a year? It is the same for a yogi. Though one may have reached the highest level, the moment one thinks one has reached the goal and that no practice is required, one becomes unstable. In order to maintain stability, practice has to continue. Sensitivity requires stability. It has to be maintained by regular practice.

You may be 50 years old, or 60 years old, and ask yourself whether it is too late in life to take up yoga practice. One part of the mind says, "I want to go ahead," and another part of the mind is hesitating. What is that part of the mind which is hesitating? Perhaps it is fear. What produces that fear? The mind is playing three tricks. One part wants to go ahead, one wants to hesitate and one creates fear. The same mind is causing all three states. The trunk is the same, but the tree has many branches. The mind is the same, but the contents of the mind are contradictory. And your memory also plays tricks, strongly reacting without giving a chance to your intelligence to think.

Why is an old man fond of sex? Why does his age not come to his mind at all? If he sees a young girl, his mind will be wandering, even though he may have no physical capacity. But ask him to do a little yoga, or something to maintain his health, "Oh, I am very old," he says. So the mind is the maker and the mind is the destroyer. On one side the mind is making you and on the other side the mind is destroying you. You must tell the destructive side of the mind to keep quiet—then you will learn.

Fear says that as you get older, diseases and suffering increase. Your mind says you should have done yoga earlier, or you should have continued and not stopped in your youth. Now you say you are very old and perhaps it is too late, so you hesitate. It is better just to start, and when you have started, maintain a regular rhythm of practice.

At a certain age the body does decay, and if you do not do anything, you are not even supplying blood to those areas where it was being supplied before. By performing asanas we allow the blood to nourish the extremities and the depths of the body, so that the cells remain healthy. But if you say, "No, I am too old," naturally the blood circulation recedes. If the rains don't come, there is drought and famine, and if you don't do yoga—if you don't irrigate the body—then when you get drought or famine in the body as incurable diseases, you just accept them and prepare to die.

Why should you allow the drought to come when you can irrigate the body? If you could not irrigate it at all, it would be a different matter. But when it is possible to irrigate, you should surely do so. Not to do so allows the offensive forces to increase and the defensive forces to decrease. Disease is an offensive force; inner-energy is a defensive force. As we grow older the defensive strength gets less and the offensive strength increases. That is how diseases enter into our system. A body which carries out yogic practice is like a fort which keeps up its defensive strength so that the offensive strength in the form of diseases will not enter into it through the skin. Which do you prefer? Yoga helps to maintain the defensive strength at an optimum level, and that is what is known as health.

I have been doing yoga for over 50 years and have taught many thousands of students in the five continents of this globe. Sadly, there are teachers of yoga who know very little and claim to teach. The problem comes not from the art of yoga, but from the inexperience of the teachers and also from the impatience of the pupils. If a person who cannot stand tries to walk, he will break his legs, and so it is with yoga. In Western countries particularly, people want above all to do *Padmasana*, the Lotus Pose. They say, "I think I can do it!" Unfortunately, the thinking is in the head, but the doing is in the knee! If you do not understand the intelligence of the knee and you force it to follow your brain, then the knee will break. But if you understand the stiffness as well as the mobility of the knee, and go step by step to remove the stiffness and increase the range of mobility, then there is no danger at all. If there are accidents in yoga, it is not the fault of yoga, but of the aggressiveness of the pupil who does it.

So you can all do yoga. The Queen of Belgium started doing Head Balance at the age of 86. Nothing happened to her. I hope there will be no confusion about what I am saying. You can do it, but do it judiciously, knowing your capacity. If you try to imitate me, naturally you will suffer, because I have been doing it for a century. You have to wait to reach that level. Yoga cannot be rushed.

Excerpted from *The Tree of Yoga*, by B.K.S. Iyengar, pp. 31-34, 1988. B.K.S. Iyengar. Reprinted by arrangement with Shambhala Publications, Inc.

Eleven

Restful Inversions:
The Elixir of Life

I started yoga during a sad period, because I had lost my husband and I was very run down. With yoga I could survive. . . . It is such a shock when someone near you dies. Yoga helped me. I didn't know it would help, because I did it like I did tennis or a game—it was fun for me. But it went much deeper than I could understand at that point. I saw this later on. . . . A new life came into my body. In nature the flowers blossom in the spring and then again in the autumn. I felt this.

—Vanda Scaravelli, Adapted from an Interview by
Esther Myers and Kim Echlin, *Yoga Journal*, May/June 1996

Legs-Up-the-Wall Pose

I recommend this restorative pose to everyone. It is sheer bliss!

The benefits of Legs-Up-the-Wall Pose confirm the wisdom in the old-fashioned phrase, "Put your legs up." Few things are easier and more refreshing after standing upright all day than lying on your back and elevating your legs on a wall.

Relaxing with your legs up on a wall is a safe and soothing way to get used to inverting your body. I have taught this pose to many people who had difficulty getting down to the floor and scooting their body close

enough to the wall. If lying close to a wall is truly not fea-
sible for you, a friend or companion can bring the "wall"
to you in the form of a tall, straight-backed, sturdy chair,
and you can elevate your legs against its back. If even getting down on the floor is not practical for you, consider
placing your bed beside a bare wall. Then you can relax
with your legs up the wall while you're lying in bed.

How to Practice Legs-Up-the-Wall Pose

1. Sit on the floor beside a wall, knees bent, with one
   shoulder and hip touching the wall.
2. Now roll your weight onto your opposite elbow and
   shoulder and swing around to bring your legs up the
   wall so that you end up lying on your back. Many
   beginners inadvertently move too far away from the
   wall when they first try this, so you may need to
   wriggle your bottom closer to the wall. Most people
   feel more comfortable with their bottom a few
   inches away from the wall, especially if their legs
   are quite stiff. Position yourself so that your back
   and your legs feel relaxed. At my yoga center, students learn this pose within reach of the wall ropes
   and pull on the ropes to get close to the wall.
3. Be sure your pelvis is relaxing on the floor and that
   your back feels comfortable. If you are uncomfortable lying with your bottom close to the wall, scoot
   back a few inches until your body feels more
   relaxed, or try bending your knees a bit to ease the
   backs of your legs.

As in other lying down positions, it is important that
your head and neck feel comfortable. Notice if your

Yoga is the best thing for regularity. As long as I do my yoga, I can throw away my Metamucil. If I don't do yoga I have to get back on it. . . .

—*Octogenarian yoga student commenting on the benefits of inverting the body*

head tilts back, and your chin is higher than your forehead. If so, place a firm, folded blanket under your head and neck so that your forehead is slightly higher than your chin. An eyebag or other cover over your eyes will deepen the feeling of relaxation.

Stay in the pose about five minutes at first, then gradually increase the time. Let your mind follow the soft rise and fall of your breath. Allow your breathing to slow down and relax, as described in the section on breathing and the heart in chapter 9. When you are ready to come out of the pose, remove the cover from your eyes and lie still for a few more breaths with your eyes open. Then bend your knees toward your chest, turn to your side and with the help of your arms uncurl slowly back to a sitting position or to Child's Pose.

You can also conveniently follow Legs-Up-the-Wall Pose with a few minutes of active stretching before you sit up. Stretch your inner thighs by slowly widening the legs, or do a lying-down variation of the Bound-Angle Pose by bending your knees and bringing the soles of your feet together. Or, you can cross your ankles, allowing the wall to support your feet. Some people feel more comfortable with a folded blanket under their bottom.

If you practiced Legs-Up-the-Wall at the beginning of your yoga practice, you can follow it with the invigorating Downward-Facing Dog Pose, since you are already kneeling on the floor. If done at the end of your practice, follow with Deep Relaxation Pose (chapter 12) or other relaxing positions.

In the 20 years that I have taught people of all ages and conditions, I have not come across anyone who did not feel the benefits from this simple inverted pose. People who were bedridden, sometimes incontinent, found the pose a great relief, especially when their legs and feet were swollen from retaining water. For people who spend long periods in bed or in a wheelchair, this pose is even more important. Lying in bed with the soles of the feet together, the feet elevated on a pile of extra large firm pillows or a stack of firm, neatly folded blankets, provides a relief similar to Legs-Up-the Wall Pose.

Cautions: People with heart problems, neck problems, eye pressure, retinal problems or hiatal hernia should use caution with all inverted poses. However, Legs-Up-the-Wall Pose is recommended for people with mild hypertension because it can help normalize blood pressure. Such people should gradually increase the length of time spent in the pose until they are spending at least 15 minutes a day.

Beginning yoga students with high blood pressure, heart problems or other medical considerations are advised to help the body become accustomed to inverting by regularly lying down with the legs resting on the seat of a chair, or practicing in bed with a stack of firm pillows or folded blankets. For some, this position will be appropriate for several weeks, then they may graduate to relaxing with the legs up on the wall.

These simple inverted positions are especially beneficial for people who stand for long periods of time, those whose legs and feet swell easily, or anyone with varicose veins.

Supported Legs-Up-the-Wall Pose
(Viparita Karani)

In Sanskrit, *Viparita* means "inverted." *Karani* refers to a particular type of practice. This pose is also known as a variation of Supported Shoulderstand.

Mental clarity comes
from a clean colon
and a straight spine.
—*Hindu proverb*

Viparita Karani is a gentle, supported Legs-Up-the-Wall inverted pose that can be practiced by almost everyone. It is a safe, non-threatening position that most people can hold long enough for gravity to return the blood from the extremities to the vital organs. This pose can generally be practiced by people who are too weak or debilitated for other inversions. Viparita Karani is especially valuable during times of fatigue, low energy, illness and stress.

The way this restful, restorative yoga posture gently inverts the body without effort, it is almost physiologically impossible not to relax deeply. The "work" lies in learning how best to arrange your body—but once the body is in the right position, your job is to let go of all effort and just enjoy the feeling of relaxing.

Viparita Karani is considered the most healing of the yoga restorative poses. When we turn upside down, gravity helps the venous blood—which otherwise tends to pool in the legs—to return easily to the heart. In people whose heart rates are elevated because they have not been receiving an ample supply of blood, Viparita Karani reduces the heart rate by improving the blood flow into the chest. In this gentle supported inversion, as in other more active inverted postures, the weight of the blood in the feet, legs and abdomen stimulates blood pressure receptors in the neck and chest to reduce the constriction of the arteries throughout the body. This reduces blood pressure.

Part of the soothing effect derived from Viparita Karani is due to the angle of the torso. Note in the photo how the stack of blankets positioned under the pelvis brings the torso into a gentle supported backbend, while the legs are supported by the wall. As we lie in the pose, we can imagine that its shape creates an internal water-fall, as the fluid in the legs cascades down to the abdomen and spills over into the chest, toward the heart. The blood seems to cascade like a gentle waterfall toward the heart in a smooth, controlled flow. This waterfall effect creates a peaceful, soothing sensation.

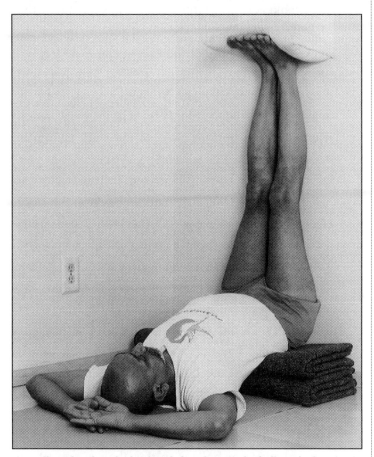

A sandbag placed on the feet by a helper deepens the feeling of relaxation in the Viparita Karani pose.

How to Practice Viparita Karani

1. Have available three firm blankets, or yoga bolsters or other firm cushions. If you use blankets or sofa

cushions, they must provide a firm, level support. The width of the blanket depends somewhat on your height and flexibility. For most people, the edge of the blanket can be placed at the waist. This placement allows the back to curve in such a way that the lower lumbar spine is protected. Note that the blankets support the buttocks in such a way that the ribcage is spread. The chest should appear and feel very open, so that the breath can flow freely.

2. It is generally helpful to first be familiar with Legs-Up-the-Wall Pose. Sit with your side to the wall, your folded blankets within easy reach. Keeping your bottom close to the wall, swing the legs up the wall, supporting yourself on your elbow and forearms. When your bottom is close to the wall, relax your arms and lie back on the floor. As described in Legs-Up-the-Wall Pose, most people feel more comfortable with their bottom a few inches away from the wall, especially if their legs are quite stiff. Position yourself so that your back and legs feel relaxed.

3. Beginners should wait to proceed until this feels comfortable. The next step is to place one or two blankets under your bottom, with your lower back supported. When this feels comfortable, a third blanket or fourth blanket can be added to open the chest even more. Experiment with the height of your blankets or bolster so that their support feels just right for your body—not too high or too low. If your neck or shoulders are uncomfortable, experiment with a small folded towel under the head or shoulders. The neck must feel comfort-able—without any tightness or pinching at the

nape. If blood flow to the head is obstructed, the brain cannot relax.

4. Close your eyes and cover them with a soft folded cloth or eyebag. Observe the rise and fall of your breath. Allow your heart and chest area to relax and open. Stay in the position a minimum of 5 minutes —10 to 15 minutes is preferable. You may find yourself falling asleep.

5. When you are ready to come out, bend your knees and slide your body away from the wall. Lie on the floor for a few more breaths before you turn to your side and slowly sit up.

Viparita Karani refreshes the heart and lungs, works to restore depleted energy and rebuild energy reserves. It is deeply relaxing during times of stress and tension and has a beneficial effect on the immune system. As we stay in the pose, the agitation and fatigue that accompany stress fade away.

Supported Shoulderstand
(Salamba Sarvangasana)

The Shoulderstand, known as the queen or mother of the yoga postures, is one of the greatest gifts the ancient yogis have given us. (See photos in the menopause section, chapter 6, pages 128-129, and on pages 234-235 in this chapter.) This posture brings health, happiness and harmony to the mind and body. It restores our energy reserves and brings a feeling of quiet and lightness. It is practically a panacea for most ills of the body, considering the circulatory, respiratory, eliminative and endocrine benefits. According to B.K.S. Iyengar, the Shoulderstand helps to regulate metabolism by stimulating the thyroid and parathyroid glands at the base of the throat, a region that is gently compressed during Shoulderstand. Because this pose allows fluid to drain naturally from the lungs and it effectively expands the chest, Shoulderstand is excellent for people suffering from respiratory problems.

The inversion of the body in Shoulderstand profoundly affects the circulatory system. When we are in a standing position, blood must be pumped against the force of gravity back to the heart from the lower extremities. In Shoulderstand, blood flows to the heart effortlessly, without taxing the heart or constricting the arteries. It is this boost to the circulatory system that gives the Shoulderstand its fatigue-relieving properties.

Please read the instructions for the Shoulderstand all the way through several times before proceeding. While you can learn how to practice this pose on your own from a book or tape, these tools are even more valuable as a way of reinforcing what you learn in class. Especially when it comes to the safe practice of inverted poses, I cannot emphasize too strongly that the guidance of an instructor is invaluable for learning these healing, rejuvenating poses. (See also Cautions at the end of this chapter.)

How to Practice Supported Shoulderstand at the Wall

The support of a wall or a chair placed at the sacrum helps you to straighten your back, open your chest, relax and stay in the pose longer. It is advisable to have a teacher assist you with placing the chair the first few times you practice this way. I suggest that you be familiar with Shoulderstand at the Wall before practicing with a chair.

In Shoulderstand, the back of your head is on the floor, with the shoulders evenly placed about an inch from the edge of the blanket. The purpose of elevating the shoulders on a stack of blankets is to protect the neck. Keep your neck soft and relaxed. In Shoulderstand the cervical vertebrae, the smallest and weakest of the spinal bones, are in a most vulnerable position. The raised platform of folded blankets is used to lift the neck away from the floor. The tops of your shoulders come close to the edge of the blanket, the back of your head touches the floor and the neck retains its normal curve.

Putting your feet on a wall during Shoulderstand can help you straighten your back and gradually learn to balance in this pose.

The bones of the neck should not press into the floor. To avoid injury, do not turn your head in Shoulderstand. Keep your chin in line with your chest.

Prepare a set of three to five folded blankets, with the folded edges neatly together. In general, the height of the blanket should be two to three inches, and wide enough to support your shoulders and elbows. Shoulderstand may seem easier without having to fold all those blankets, but there is a tendency for the neck to feel strain and pressure from the unaccustomed weight of the body. The height of the blankets varies according to the length and flexibility of your neck. Your teacher can help you to determine the right number of blankets to use.

1. Place the firm, folded blankets near a wall, with the folded edges away from the wall. The exact distance away from the wall depends on the length of your torso.
2. Sit on the blankets beside a wall, knees bent, with one shoulder and hip touching the wall, as in Legs-Up-the-Wall Pose.
3. Now shift your weight onto your opposite elbow and shoulder and swivel your trunk around to bring your legs up the wall, so that you end up lying on your back, your shoulders well supported by the blankets, the back of your neck and head off the blankets.
4. Many beginners inadvertently slip off the blankets when they first try this pose. As in Legs-Up-the-Wall Pose, you may need to wriggle your bottom closer to the wall, so that your shoulders remain on the blankets while your legs are up the wall. If your shoulders fall off the blankets when your bottom is

close to the wall, try placing the blankets a little farther away from the wall, to accommodate the length of your torso.

5. Bend your knees, press the soles of the feet against the wall, raise your hips and chest off the floor until your back feels straight. Firmly support your back with both hands, bringing your elbows closer together. Make sure your shoulders have remained on the blankets and the back of your head on the floor. The neck should feel comfortable without pressure or pain. Do not turn your neck once you are in this position. If you feel comfortable, stay for several minutes, gradually increasing your stay in the pose. When you are ready to come down, place your hands back on the floor, bend your knees, and gently lower your back and bottom to the floor.

You can relax with your legs up on the wall a few more minutes by sliding back off the blankets till your shoulders touch the floor. When you feel ready to sit up, bend your knees, turn to your side and slowly sit up. Or you can remain lying down and follow this pose with some gentle floor twists.

Shoulderstand and Plow Pose with Chairs

Study the photos on the following page carefully and note the positioning of the props. If you are new in class, it is important to have an opportunity to see your teacher or another student demonstrate the pose before you attempt it. Again, your teacher can help you determine the number of blankets that works best for you. In the beginning, you may need the help of a teacher to position

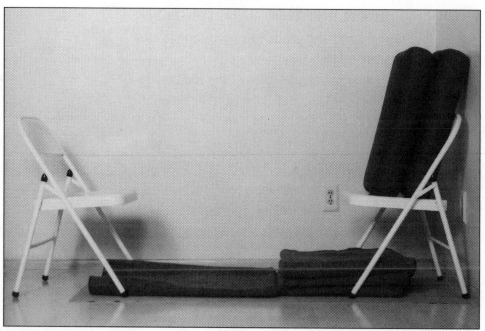

A set-up of two chairs, blankets and bolsters as shown can help you safely practice Shoulderstand and Plow Pose.

the chairs. Later, you can use the chairs independently. For added security, place the back of the chair against a wall for Shoulderstand. Two bolsters can be placed on the seat of the chair, as illustrated, to further increase your stability, openness and the length of time you remain in the pose.

1. Position the neatly folded blankets so that when you lie down, your shoulders are still on the blankets. Your bottom should be near the front edge of the chair and your lower legs should rest on the seat. You should be close enough to the chair to grasp its front legs.

2. Place a second chair far enough behind you so that when you lower your legs, your feet will rest on the seat of the chair.

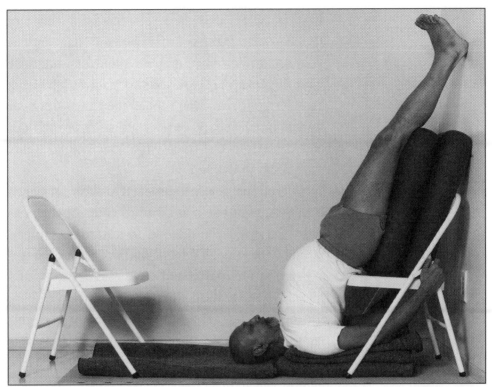

Reversing gravity reverses the aging process. Supporting the hips and back of legs makes this pose secure and relaxing.

Notice the vertical position of the back and 90-degree angle formed by the legs.

3. Bend your legs and place your feet on the edge of the chair seat. Press your feet into the chair, and lift your torso till you are on your shoulders. Interlace your fingers and stretch your arms toward the front chair. If your arms don't touch the floor with your fingers interlaced, place your arms flat on the floor and press into the floor with your palms. Relax your neck, throat and face.

4. When you feel ready to proceed, place your hands firmly on your back, as close to your shoulders as possible. Bring your elbows in toward each other, in line with your shoulders, and open your chest.

5. When you feel stable and secure here, exhale and lift one leg up to the ceiling. On the next exhalation, lift the other leg. Hold for a few breaths, straightening your back as much as possible.

6. Exhale and lower one leg at a time onto the chair behind you. When your feet are secure on the chair, release your hands off your back and reach for the legs of the chair. Move the chair toward you until it touches your back.

7. With your arms inside the legs of the chair, hold the back legs of the chair, near the seat as illustrated. When you feel ready to proceed, lift one leg at a time to the ceiling, pulling the chair toward you and firmly grasping the back legs. Stretch up through your legs and feet. With practice, you will come higher up on your shoulders and your body will be straighter. As you feel more comfortable, increase the length of time in this position.

8. Lower your feet one at a time to the chair, until the tips of your toes touch the chair seat. Stretch your bottom up toward the ceiling, lengthening your

spine. Hold the position about half a minute, gradually increasing your time in the pose.

9. There are several ways you can come out of the pose. With your feet on the chair, push the chair at your back away from you (or to the side if it is braced against a wall). Place your hands on the floor behind your back and lower your back slowly down, keeping your head on the floor. Bend your knees and place your feet on the floor. Slide back off the blanket until your shoulders are completely on the floor. Relax in this lying down position for a few more minutes.

Cautions: As discussed in chapter 2, inverted positions, whether done with special equipment or on your own, are only as safe as people learn to practice them. While simple inversions such as Legs-Up-the-Wall Pose and Viparita Karani can generally be learned on your own, other inverted poses should ideally be learned under the guidance of an experienced teacher.

People with the following conditions should not practice inversions without the supervision of a physician or an experienced yoga teacher: high blood pressure; glaucoma or detached retina; hiatal hernia; heart problems or stroke; epilepsy, seizures, or other brain disorders; temporary conditions such as acute infections of the ear, throat or sinuses; obesity; conditions requiring aspirin therapy; chronic neck problems, whiplash or osteoporosis. People with neck injuries or conditions such as cervical spondylosis must take care to avoid yoga poses that involve weight-bearing directly on the spine, as in Headstand and Shoulderstand.

Indra Devi:
Yoga Defies Aging

Indra Devi at age 94.

After a year of fruitless phone calls, I finally managed to catch up with Indra Devi, one of this century's most renowned and beloved yoga teachers. Born Eugenie Peterson on May 12, 1899, in Riga, Russia, she explains in the introduction to her book, Forever Young, Forever Healthy, *that even as a child she was irresistibly drawn to India. She left Moscow in 1920 to become an actress and dancer, and traveled throughout Europe, where she met J. Krishnamurti. In 1927 she arrived in India. During the first four months of her stay there, she traveled extensively. She met Mahatma Gandhi and other leading spiritual and political figures. After many mishaps and adventures, she unexpectedly became an Indian movie star, and renamed herself Indra Devi.*

She married a Czechoslovakian diplomat and her spiritual aspirations fell by the wayside as her life became one of parties and worldly pleasures. During this period she developed severe health problems, which lasted several years and almost killed her. A remarkable series of events led her to study yoga with T. Krishnamacharya, a highly regarded yogi who also taught T.K.V. Desikachar and B.K.S. Iyengar. In 1939 Indra Devi followed her husband to Shanghai, China, where she opened her first yoga school.

Returning to India in the mid-1940s, she retreated to the Himalayas, where she wrote the first of a series of bestselling yoga books that were eventually translated into 10 languages. From then on she has lectured on yoga all over the world. In the late 1940s, when people were still confusing yoga with yogurt, she established a yoga school in Hollywood, California, where her students included Gloria Swanson and Olivia de Haviland.

On the day of my interview with Indra Devi, I arrived half an hour early. She was still in her nightgown and slippers, her wild white hair uncombed.

"Oh," she laughed, eyes twinkling and looking every bit like some mischievous, impish child clad in a wise old woman's body, "usually people are late!" The fact that she did not dash into the bathroom to dress made me feel like a close friend or member of the family. "Well," I thought, "the First Lady of Yoga has nothing to hide and doesn't give a hoot about first impressions." I also noted with great relief that she still painted her nails pink—laying to rest my fears that one must renounce frivolous feminine rituals in order to tread the spiritual path. I expected to have to wait until Indra Devi dressed and finished her breakfast, but she motioned for me to sit at the table and get started. Here is an excerpt of what Indra Devi had to say:

I know you're writing a book of yoga for people who are aging. Just a few days ago I was interviewed for the newspaper, the radio and the television, all at the same time. The radio man asked, "How does it feel—aging?" I said, I don't know how it feels. Ask someone who's aging. I don't know anything about aging.

People don't have to wait to do yoga until they're aging. But what can you do? Usually that's how people are. Not until something happens to their body—then they start thinking about it. People allow their body to become old, stiff, fat or whatever by doing everything wrong. From exercising, to eating, to thinking.

If you think yourself old, you will become old. Youth comes from inside. Of course, for people over 80 who are already aging, yoga can do a lot of things for reversing the aging process. They can start with yoga exercises that everybody can do. But just doing only the physical part of yoga is not enough. You have to do things that affect your mind. You must make your mind like an arrow when you want to do something, and not be distracted.

Whatever is in the physical world can be translated into the spiritual world. In the physical world, if you want to be a musician, like a pianist, what do you have to do? Hours and hours of practicing. The same thing in the spiritual world. You want to achieve something, you have to work, but a different kind of work. The practice of yoga is invaluable. Yoga affects all of our faculties, physical, mental and spiritual.

I start with the physical body, deep relaxation and deep breathing. Deep breathing is an exercise you don't do all the time. You start with five or six deep breaths, filling the entire lungs, starting from the lower part and on upward. But very few people take the time to practice this. Most of our life we use only one-third of our lung capacity, the upper part. Many people don't even know they have a lower part. The lower part of your lungs is not for decoration. You first start with five, six or seven deep breaths. You can practice this in the fresh air two or three times a day. And then gradually more. Sixty deep breaths a day guarantees you to be in good health and good spirits.

I practice every day. Some people say, "Oh, I don't feel the need." What do they mean? You breathe every day. You eat every day. That's why you should continue to practice the yoga exercises. You might not feel the need today, but three years later you will feel the need! You should not practice yoga only when you are in pain or discomfort, then give it up because the pain is gone.

Twelve

Savasana, Pose of Deep Relaxation

When a member of my extended family died, I did not feel inspired to practice my usual yoga routine. In response to my grief, I decided to practice only restorative poses, and did so almost every day for a year. It was practicing these poses, I believe, that helped me through this extremely painful period. They opened me in a way that allowed me to accept my pain and to recover from the emotional drain and fatigue.

—JUDITH LASATER, PH.D., P.T. *RELAX & RENEW: RESTFUL YOGA FOR STRESSFUL TIMES*

In Sanskrit, *Sava,* or *Shava,* means "corpse." In this pose, the body lies on the floor face up and completely relaxed, while the mind is at ease and alert. The eyes are closed, the arms relaxed at the sides, the palms up. The body remains as still and motionless as a corpse.

The influence of this asana on the body and the mind—from relaxation, to surrender, to death, and even after death—is incredible. If you do not want to be a living corpse, then the purpose of life has to be established. If you want to be an active participant in your life and not a parasite, then the dynamic interdependence between life and death has to be recognized, and the two have to meet in directed and concentrated interaction.

In Shavasana, relaxation is the first attempt to surrender, to let go. As the mind follows the flow of the breath, the ripples of the mental lake slowly subside. With continued practice, the senses are gradually withdrawn and become still. Passion, egocentricity, self-importance are, for the moment, put to rest. Rest becomes an important word whose meaning expands with experience. Shavasana, the corpse posture, gives a new understanding of death, of the need for surrender. The body at rest can do its repair work. Sufficient rest allows the body to recuperate from the driving forces of the emotions and the ambitions of the mind. The benefits—physically, mentally, emotionally—are profound. In that state of peace and quiet and inner harmony, one can perceive a vision of the Light that is present in both life and death.

—Swami Sivananda Radha
Hatha Yoga: The Hidden Language

Although *Savasana,* the classic Pose of Deep Relaxation, might seem like the easiest posture to practice, those who delve deeply into the art of yoga realize that it is among the most difficult poses to master.

In Savasana, the body lies still and silent—empty of agitation, mimicking a corpse. The mind is in the present, alert and aware, serene, detached, observing while the body releases deeply and lets go. In this state of conscious relaxation, the body and mind are recharged, refreshed and rejuvenated. This condition of deep rest allows the body and mind to be profoundly quiet and at peace.

Setting aside at least five minutes at the end of your yoga session to consciously relax is healthful for your whole being. To consciously relax removes fatigue from the body, soothes the nerves and brings a feeling of pleasant calmness and happiness. Deep relaxation—which is very different from flopping in front of the television and falling asleep—is healing. Physiologically, such relaxation is characterized by a slower heart rate, metabolism and rate of breathing, lower blood pressure, and slower brain wave patterns.

The yogic art of relaxation known as Savasana puts down in precise steps how relaxation and recuperation take place. Savasana is a posture that simulates a dead body, and evokes the experience of remaining in a state as in death and of ending the heartaches and shocks that the flesh is heir to. It means relaxation and therefore recuperation. It is not simply lying on one's back with a vacant mind and gazing, nor does it end in snoring. It is the most difficult of yogic asanas to perfect, but it is also the most refreshing and rewarding.
—B.K.S.Iyengar
Light on Pranayama

The Art of Conscious Relaxation

Learning to consciously relax takes practice. Choose a warm, quiet place where you will not be disturbed. Have an extra blanket available in cooler weather to cover yourself if needed. Remove contact lenses or glasses.

In this pose the body is placed evenly on the floor. If your head tilts back or feels uncomfortable, place a folded blanket under your head and neck, thick enough to bring your forehead slightly higher than your chin. If your back feels uncomfortable when you straighten your legs, then bend your knees and place a pillow or folded blanket under your knees to help you relax or support your calves on the seat of a chair.

1. Sit on the floor with your legs straight out in front of you. Lean back on your elbows and check that your upper body and legs are in line. If not, adjust the position of your torso or your legs so that your chin, center of your chest, navel and pubic bone extend directly toward a center point between your heels.

2. Lower your back slowly to the floor. Place a blanket under your head as needed. Allow your back to sink toward the floor.

3. Turn the upper arms out, slightly away from the trunk, palms up. Extend the arms out of the shoulders, then allow them to relax. Soften your hands, allowing your fingers to curl naturally.

4. Extend the legs and feet. Allow them to drop toward the floor and fall evenly away from the midline. Soften your toes.

5. Close your eyes. Relax your lips and tongue. Let your lower jaw be loose, your lips barely touching, your tongue passive. Allow your mouth to feel as if you are about to smile. Feel your face muscles soften and relax. Let your eyes sink deep in their sockets. Allow your body to release and sink toward the floor.

6. As you lie still, let your mind follow your breath. Quietly observe the peaceful in-and-out flow of your inhalation and exhalation. Listen to the beat of your own heart. Each time your mind starts to wander into the past or future, bring your awareness back to the breath.

Stay lying still for about five or 10 minutes, or longer.

Let go of the past—become "dead" to the old and prepare the way for something new.

Deep relaxation is also considered a preparation for conscious death. Letting go of the fears and morbid misunderstandings associated with the death of the physical body may help prepare us for the mysterious transition from life to death.

Regular practice of deep relaxation will leave you feeling refreshed, rejuvenated and optimistic.

Note again that it is of utmost importance that your head and back position are comfortable. Check that the height under your head is thick enough so that your forehead is slightly higher than your chin. Your head should not drop back and pinch your neck. Experiment with the thickness of the blanket under your head until it feels right. Keep your knees bent if your back feels more comfortable in this position.

As described in chapter 3, an eyebag over your eyes will help you relax. An eye covering helps quiet the mind by creating darkness and removing visual stimuli, by relaxing the muscles around your eyes and by calming the involuntary movements of your eyes.

Savasana Variation

Relax deeply by lying evenly on two or three blankets folded lengthwise under the lower back, chest and head, with an extra blanket placed under the head so that the forehead does not drop back. Supporting the length of your spine and back of the head with firmly folded blankets opens the chest, allowing the breath to flow freely—especially beneficial to people with heart and respiratory problems. For those who suffer from depression and low energy, a relaxing position that opens the chest can be a lifesaver.

Refer back to chapters 3, 6 and 10 to review the use of eyebags, bolsters and folded blankets for deep relaxation.

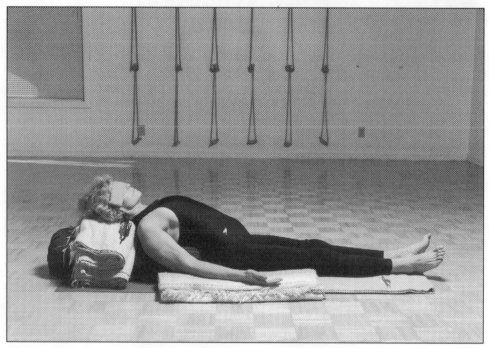

Find the right amount of support for your body to make this pose truly relaxing.

Diana Clifton:
Letting Go of Pain, Learning to Relax

Diana Clifton at age 73 makes difficult poses look easy.

To be in a class with Diana is to experience the soul of yoga. Her extraordinary knowledge of poses and gentle, deeply penetrating manner open fresh lines of perception, depth and self-inquiry. She reminds one of the river in Siddhartha—*sharing its secrets endlessly with those who hear, and yet in the ordinary passage of its flow, simply carrying many across from one side to the other.*

—Anonymous Student

Diana Clifton, an inspiring teacher who lives in England, did not begin practicing yoga or become health- and nutrition-conscious until she was in her 40s, at which time she says she was someone who was in poor health with anemia, constipation, nervousness and fatigue.

Three decades later, she makes difficult postures look effortless, reminding students half her age that yoga postures are natural—young children can do most of them with ease! Like many older people, Diana experienced the loss of a spouse. We discussed many aspects of yoga and life—including aging, death and depression. Diana described the year she spent nursing her sick husband and the overwhelming grief and exhaustion she felt when he died.

We all have to face death. I think you have to die while you're still alive. That means renouncing everything before your death—goods, wealth, friends, relations, loved ones. Mind you, you can't take anything with you when you die. Older students know that. This realization might bring them to a point of silence. You see, I believe that death is such a natural thing. It's only the death of the body. We are not the body. So, strictly speaking, we don't die anyway. It's only the physical that goes—that includes our possessions, of course. But after all, you close your eyes and they're gone. So, it's our senses we hold on to most.

There is great duality in life. One is joy—one is pain. We all embrace joy, but we do not embrace pain. And why is that? Maybe we could look at our pain for a little while and become detached from it. It's hard to do. A lot of the pain is stress. Yoga can teach you how to let go, how to relax.

Older people starting yoga can lie flat on their backs, for a start. A support under their neck and head, and a yoga bolster or folded blankets under the back to open the chest, will help them to breathe and relax. Ask students to watch their breath and look at their own thoughts. If someone comes to yoga, something inside them has prompted them. It may not be merely because of their suffering, physically. It may be more than that. It may be because they haven't found any comfort in their religion. As it is now, a lot of people don't find comfort in religion. And if they can discover what's inside themselves, so that they don't need a religion, so that they don't need love from another, because the love is within them; if they can discover the love within themselves, and if you, as a teacher, have got love within yourself—at a certain point, maybe, once the ego has dropped—there is serenity. Once the ego has dropped, they can relax. Teach them how to relax lying on their back in Savasana. A short session of gentle stretching first will make it easier for them to relax.

For depression, you have to open the chest. Because if the chest is concave, they won't get over their depression. So they must open the chest by lying on the floor, placing bolsters or folded blankets under the chest. And then have them watch their breathing.

One thing you can tell them while they're lying in Savasana is to keep the inhalation and exhalation the same length. So, while they're working on their breathing they can think about this. It not only helps the student to concentrate, but it's also better, from the breathing point of view, if they exhale completely. Say they count to 10 on the exhalation—then they should count to 10 on the inhalation. If one is longer than the other, they have to shorten the long one, not to try to overdo it. And also—if they can, pause at the end of the exhalation, because often there's a lot of air still there and, if they pause, that's the deepest point in relaxation. You tell them about this way of breathing, but also explain not to do it if it's a strain. You help them understand that the breathing is cleansing.

Thirteen

Tips on Teaching and Hints for Home Practice

As your body becomes more flexible, you become more flexible in how you interact in your life and more possibilities open up. People become more aware that they can do more things as a result of establishing strength and then becoming more flexible. When you start to do things that you never did as a child, as I have found in my 50s and 60s, it's just incredible. You feel that life can just keep opening up for you, instead of closing down.

—TONI MONTEZ, YOGA TEACHER

Older students come to yoga with various levels of abilities and medical histories. Some athletic 60-year-olds are soon practicing

Handstands at the wall. Others gain confidence by beginning with simpler poses done lying on the floor. Both teacher and student need to remember that, "Expectations determine outcome." With rare exceptions, I expect students of all ages to practice the vital weight-bearing standing postures that safely stretch and strengthen the whole body. For older beginners, it helps to introduce the poses at a slower pace and encourage students with balance or other problems to make intelligent use of walls, wall ropes, chairs and other props.

Don't treat me like an old lady!
—78-year-old student to her teacher

As I stated in the introduction, older people start yoga for most of the same reasons as other people. They want to feel better, have more energy, improve their posture, strength and flexibility. However, many older students also come to yoga when everyday activities become more and more difficult for them. They hope the practice of yoga will help them remain mobile and return them to their normal ability to function, instead of becoming increasingly incapacitated, and in some cases, facing the very real possibility of spending the last period of their lives having to depend on other people.

The Three Gifts of a Yoga Program for Seniors

1. Yoga improves posture and breathing, thereby improving the health of all the systems of the body.
2. Yoga restores and maintains normal mobility and a healthy range of motion in order to function well in daily life.
3. Yoga helps us to grow and expand psychologically and spiritually. Yoga offers a foundation for conscious living, conscious aging and conscious dying.

The yoga practices presented here are specifically geared to people who need to start exercising gradually. Older beginners who leap enthusiastically into Full Arm Balance during their first class can refer to the more general yoga books listed in the Resources section as their guides.

Guidelines for Teachers of Older Beginners

Many older people have parts of their bodies that have become virtually immobilized. Over time their longstanding physical problems have become magnified. Often these students accept these limitations as an inevitable part of aging and have withdrawn their awareness from their trouble spots, such as a shoulder, knee, foot or hand. It's almost as if that part of the body no longer belongs to them. As they practice the yoga postures, their awareness and sensitivity return. At the same time as this can bring an awakening and new sense of aliveness, it can also increase these students' perception of their other physical problems. Some students then blame them on doing yoga.

The main adjustment for teachers of older beginners is to work at a slower pace. Older people need to conserve their energy. A slower pace allows them to get deeply in touch with what is happening to their bodies during the postures.

Some of the most common physical problems of older students are high blood pressure, osteoporosis and arthritis. A teacher must be aware of how to adapt a student's yoga program to each condition.

- **For beginners with high blood pressure, practicing inverted poses such as Headstand or Shoulderstand**

are not recommended. However, as these students progress, gentle inversions should become a regular part of their practice. Of course, students with hypertension should be encouraged to work in a more relaxed way, to minimize stress and tension.

- **Students with osteoporosis should be extra aware in yoga poses.** Any pose that might place stress on a brittle bone should be practiced very carefully.

- **Arthritic students also need a lot of encouragement**. Although it is essential for them to continue moving the arthritic region, the pain such motion causes induces many students to immobilize the area. They have to be coaxed, bit by bit, to bring mobility back. Often the role of the yoga teacher is to help gently awaken the areas of the body that the older student has deemed frozen or hopeless.

Look at what your
people need.
*—B.K.S. Iyengar to
teachers*

A teacher must also be sensitive to the energy level and emotional condition of older students. As I look around at my group of seniors, I am aware that while many of them are having the time of their lives, others are going through difficult periods in their lives. Several are primary caretakers for a dependent spouse, other family member or friend. For them, yoga class may be one of the rare moments that they do something for themselves. For others, yoga class is the social highlight of their week. Often someone in the class is grieving and has little energy. This student may also be on medication for depression. The class atmosphere should be both physically and spiritually uplifting so it doesn't add any additional stress to a student's psyche. A well-rounded class includes invigorating poses to stretch, strengthen and revitalize, and restorative poses to relax and let go.

Ten Important Tips for Teaching Older Students

1. Remember to speak loudly and clearly. Many older people are hard of hearing. Make a point to stand near students who have difficulty hearing. If a student does not respond to directions, it may well be because he or she cannot hear you or, in some cases, clearly see you. You may need to stand directly in front of a student, demonstrating while repeating directions louder and clearer. Instructions must be easy to follow. Most of the time they are better understood when they are visual.

2. Take time to demonstrate what happens when a pose is done with healthy alignment; but, on the other hand, be open to allowing students to learn and explore the poses in their own way as long as they don't hurt themselves. Older students may resist and question more than younger ones, especially, perhaps, if you, the teacher, are much younger.

3. Emphasize the benefits of each posture repeatedly, to motivate students to practice and to help them remember the poses. Share the following points:
   • Breathing reflects and affects our selves and our lives. Learning to breathe more deeply and evenly through the nose increases ventilation and reduces stress on the heart.
   • Asanas release energy in the nervous system and stimulate glands. Most of us are energized by yoga and feel increased vitality.

4. Be playful and creative and make sure your students spend time laughing, relaxing, letting go and having fun.

Old people don't need a younger person telling them what to do.
—*Advice from a seasoned teacher*

5. Be ever on the alert that you do not impose your own myths about aging on your students. It continues to be a revelation for me as a teacher to witness the extent of the progress made by my older students who practice. There may be long periods of time when students, afraid of losing their balance, do standing poses in ways that do not look right. Outwardly, you see little progress. But if you keep demonstrating, directing and encouraging, one day you will be amazed, as I have been, to see almost everyone in the class suddenly grasp what you've been saying all along.

6. A group dynamic is at work in all classes, but it may be even more noticeable in older groups. The octogenarian who comes to class for the first time and sees people practicing yoga in chairs will assume that this is what is expected and act accordingly. Conversely, the more that older people see their peers practicing more challenging poses, the more they will expect of themselves.

7. Create an atmosphere where students feel comfortable to explore the poses at their own pace. Keep in mind that as people grow older, they very often feel unstable and lose their balance. Remember that if your ability to stand and walk becomes increasingly unsure, the rest of life feels less and less secure. Yoga can help seniors regain confidence. The standing poses re-establish the body's connection to the earth. Students develop a sense of rootedness, stability and balance. Standing on one's own two feet is not generally emphasized in other forms of exercise, yet is invaluable to older people.

8. Always do some type of inversion poses, and gradually encourage students to hold them for longer periods. Inversions counteract the aging process by reversing the gravitational pull on the internal organs and improving the venous return to the heart.

9. Relaxation is often difficult for older people. If you have a class of older adults with many physical problems, which as a group is not very strong or limber, it is generally advisable to begin with simpler lying-down poses and work up to standing postures. Some students may experience discomfort lying down and complain about the hardness of the floor. Check their head and neck position to see that the arch at the back of the neck is not extreme—that the head is not tilted back with the forehead lower than the chin. Hyperextension of the back of the neck can restrict the blood supply to the brain. Place pillows, books (or whatever height-increasing prop is available) under the head, and pillows or folded blankets under the thighs if the lower back does not relax into the floor.

10. While as a society our view of aging is changing by leaps and bounds, the myth that older people are still expected to go downhill will not disappear overnight. Bear in mind that many older people feel increasingly helpless, that they are losing control over what happens to them. As they develop and progress through yoga, their confidence returns when they find the downward spiral can be reversed.

Hints for Getting Down to and Up from the Floor

Few experiences are as demoralizing for older people than falling and being unable to get back up off the floor. If this happens while they are home alone, it can mean waiting many hours until someone comes to help them.

The main concern of many of my older students during their first yoga class is how to get up from and down to the floor. New students often tell me, "I'll just sit in a chair instead of going on the floor." For these people, regaining confidence in moving to and from the floor is the most important part of their first lesson. I assure them that it will get easier with practice. Experienced students are generally sympathetic and give newcomers suggestions and encouragement.

Students unsure about getting up from the floor will find it helpful to lower themselves to the floor, and raise themselves back up, holding onto a sturdy chair braced securely against a wall. Be sure to position the chair so that it will not slip or collapse. A folded blanket on top of a sticky mat on the floor near the seat makes kneeling less painful.

Each student is unique, and safety—avoiding injury—is always my first concern when assessing the strength, flexibility and coordination of someone new. I may stand nearby to make sure the student does not fall, but generally I offer minimal assistance. It has been my experience that the majority of people who come to class unassisted can relearn to get up and down from the floor with ease and confidence.

If I see a student struggling to get up, I suggest, "Try turning to your other side," or "See which side, which

One day I found myself looking around for something to get hold of to help me get up. Here's what I found to avoid that: From a kneeling position I raised my right knee, put my right hand on it, and with the help of a push from the floor with the left hand, I was up without trouble. Maybe I've invented an exercise. Anyhow, I do it two or three times a day.
—Zilpha Main,
Reaching Ninety My Way

arm, which leg, is stronger." It usually does the trick. If the problem is simply that the student is very weak or stiff and just isn't in the habit of getting up and down off the floor, I often just leave that person alone. Aside from providing a chair or other sturdy object, I let these students figure it out for themselves—which is often what they prefer, rather than drawing attention to their dilemma. I might assure struggling students that it doesn't matter if it takes five or 10 minutes to get up and that, as with everything else, this ability improves with practice. My students who first had difficulty hardly think twice about it now.

Remember this Golden Rule about almost any physical activity: Use it or lose it. Suggest that students avoid sitting in chairs whenever possible; sit on the floor the way children do. Urge them to rise from and descend to the floor at least once a day. When the neuromuscular pathways for getting up and down are used regularly, it is easier to recover if, by chance, you should slip and fall.

Practical Hints for Moving to and from the Floor

1. Stand next to a sturdy chair or other piece of furniture near a wall. If balance is very unstable or leg muscles are very weak, try standing in between two chairs or other stable supports on which you can brace or hold. From standing, holding on to the support as needed, step forward and begin to lower your knees to the floor.

2. Kneeling on both knees, lower your bottom toward your heels. In our chair-oriented culture, where

hips and leg muscles are often very stiff, many older people have difficulty lowering their bottom to their heels. Holding the chair as needed, try sitting to the side of your feet. Use your arms to lower yourself to the floor. To stand back up, reverse this process. Again, be assured that with daily practice your muscles will strengthen and the movement will become easier. I cannot overemphasize the importance of maintaining the ability to get up and down from the floor. Practicing will also make it easier to get up from chairs and other places, such as the toilet, even if the seat is low.

Wall ropes hanging in many yoga centers are extremely useful for people with balance problems and difficulty getting up and down. They can be ordered for home use from the Resources section.

Barbara Uniker:
I Do Not Accept the Stereotype of Old Age

Barbara Uniker's **joie de vivre** *lights up morning classes at the Ojai Yoga Center.*

About 20 years ago, a friend insisted that I try yoga. It sounded far out, but I agreed, and became rather fascinated. I was surprised to find out that I could even do a Headstand, and other advanced poses. After a year or so, my teacher, Suza Francina, left the area and I discontinued yoga.

As I aged, I became aware that even though I ate nutritiously, did aerobics and was active in community affairs, there was another dimension to be considered. I found myself facing the inevitable diminution of vision, hearing, memory, etc., all of which leads to a loss of confidence and pride. I felt that I was on a downward spiral. It was oh-so-easy to slip into not making an effort, letting things slide, instead of taking steps to do something about situations.

While I was musing about what to do, Suza returned, and I started regular classes with her, knowing that with her both mind and body are stretched. Spiritual values are also part of the picture. Now, the discipline of yoga spurs me on to make more of an effort to learn new poses and hold them longer. The satisfaction derived from success turns the spiral upward and flows into other aspects of living.

Occasionally younger people join our class, and often they are unable to do poses that we practice every class. I enjoy a certain wry satisfaction about that. It encourages me even though I can no longer do a Headstand without help. I know how much yoga has done for me.

It is a sad thought that gravity literally pulls me downward every day of my life. The worst of it settles in my tummy. It cheers me up to know that yoga reverses this pull with its constant stretching upward and inverted poses. While it creates no miracles, it at least gives me an even break.

Another joy from yoga comes from the feeling of relaxation and peace at the end of class. Often I come in tired after early-morning aerobics and various chores and errands. Since the mind as well as the body is cleared in yoga and the spirit is refreshed, it is like starting a new day.

Since concentration is essential to yoga, yoga is an especially good remedy for one of the more difficult problems of aging. For me, yoga classes are an invaluable time for quiet concentration, stretching the mind as well as the body, being spiritually aware and experiencing deep relaxation. The last several years since I have returned to yoga have been especially rewarding ones, telling me that I do not have to accept the stereotype of old age— for I still have much to enjoy in life.

Beatrice Wood: Age Is Only a Number

Beatrice Wood at 100—exuberance, daring and joy.

My lifelong friend, Beatrice Wood, a world-renowned potter who took her first ceramics class around age 40, was asked by an interviewer how she felt about aging. Without hesitating she replied: "I have nothing to say. I'm not concerned about it. I may be 100 to you, but to me, I'm 30, so I have no problem."

Over the years, whenever I've eaten with her, I've always noticed that she picks away, like a bird, at her nutritious vegetarian meals, saving room for the most important part—dessert. She reminded me why she does when she turned 103.

"My friends who didn't eat chocolate died a long time ago," she explained. "I don't overindulge. Just a small amount. It makes me feel human."

When I asked her how she felt about dying, she replied: "Isn't death amazing? We all have to die and it's something we can't accept. I have absolutely no concern; when I'm tired I often think I'm practically there. But then, on the other hand, I still want to do quite a few things and I shall be enraged, in heaven, if I can't get them done. I'd go to the great teachers and say, 'Why didn't you let me stay longer?'"

Personally, I think one of her secrets of longevity is that she laughs all the time. "What can one do when one is up against the absurdity of life but laugh," she once told me, adding, "It is curious, but if one smiles, darkness fades."

When I asked about her deep interest in so-called spiritual matters she answered, "I'm very interested in what could be called spirituality, but it doesn't stop me from being very naughty in conversation and even in action. These last few years I've been more or less a nun, but if I fell in love I wouldn't hesitate not to be a nun."

Beatrice meditates every day. She says, "It is in silence that new thoughts come. If we divert the mind with too much distraction, it becomes scrambled like eggs."

Appendix: Understanding the Anatomy of the Spine

The average human spine is about 27 inches in length and has four normal curves. The lower back and neck are concave curves; they dip into the body. The tailbone area and rib cage are convex curves—they move out. For optimum health of the spine while stretching, all four curves should be maintained, with no excess (shortening) or rounding in any segment of the spinal column. Any alteration in one curve will affect the curves above or below.

A long, gentle S curve of the spine helps maintain proper spacing between the bones of the spine, the vertebrae. Lengthening the spine to create space between the vertebrae is vital to our health because nerves connected to the organs and structures of the body branch out from the spinal cord between the vertebrae.

The curves of the spine are formed by the different shapes and thicknesses of the individual vertebrae, which are separated and cushioned by disks of cartilage and water. The weight and stress of the body pass through each vertebrae. Between the bones of the spine, disks act like hydraulic "shock absorbers" and allow for movement in all directions. If the curves of the spine become distorted, the spaces between the vertebrae are compressed. This may cause various problems to the disks which are vulnerable to the changing shape of the spine under pressure. The organs and other parts of the body which are stimulated by the corresponding nerves may decline.

By age 30, the blood supply to the disks gradually lessens. In the adult spine, all nourishment to the spine comes from movement. Fluids are drawn in and flushed out of the disks by stretching, lengthening and moving the spine in all directions—forward, backwards, sideways and twisting. If the disks are not nourished, they start to shrink and lose their elasticity, becoming more prone to injury such as herniation and pressure on the sciatic nerve root. When a disk is damaged or ruptured, a gelatinous matter oozes from it and severe pain results from the pressure of the bones of the vertebrae on the spinal nerve roots.

As the human body passes maturity, the disks gradually shrink, causing the body to lose height. This is a complex process about which new discoveries are being made.

However, for the health of the back, careful stretching to lengthen and create space between the vertebrae is essential to allow the disks to return to a more youthful condition.

Resources

For Yoga Books and Videos

Rodmell Press

2550 Shattuck Ave., #18
Berkeley, CA 94704
(800) 841-3123
Fax: (510) 841-3191
e-mail: rodmellprs@aol.com

Yoga Journal's Book and Tape Source

2054 University Ave.
Berkeley, CA 94704-1082
(800) 359-YOGA (M-F, 9 A.M.-5 P.M., P.S.T.)
Fax: (510) 644-3101

For Props

Bheka Yoga Supplies

P.O. Box 147
Carlsberg, WA 98324
(800) 366-4541

Fish Crane

P.O. Box 791029
New Orleans, LA 70179
(800) 959-6116

Half Moon Yoga Props

2137 W. First Ave., Suite 2
Vancouver, B.C. V6K 1E7, Canada
(604) 731-7099

Hugger-Mugger Yoga Products

31 W. Gregson Ave.
Salt Lake City, UT 84115
(800) 473-4888

Living Arts

2434 Main St., 2nd Floor
Santa Monica, CA 90405
(800) 2-LIVING

Tools for Yoga

P.O. Box 99
Chatham, NJ 07928
(201) 966-5311

Yoga Mats

P.O. Box 885044N
San Francisco, CA 94188
(800) 720-YOGA

Yoga Pro Products

Box 7612
Ann Arbor, MI 48107
(800) 488-8414

Yoga Props

3055 23rd St.
San Francisco, CA 94110
(888) 856-YOGA

Finding a Teacher in Your Area

B.K.S., Iyengar Yoga National Association of the United States

554 Orme Circle N.E.
Atlanta, GA
(800) 889-YOGA

Provides a complete listing of certified Iyengar Yoga instructors. Visit their Web page at http://www.iyoga.com/iynaus/

International Association of Yoga Therapists

20 Sunnyside Ave., Suite A-243
Mill Valley, CA 94941
(415) 332-2478

Provides an international network and resource guide for yoga and yoga therapy.

Yoga International's Guide to Yoga Teachers and Classes

RR 1, Box 407
Honesdale, PA 18431
(800) 821-YOGA

Regional, national and international listings of yoga teachers, as well as certification programs and yoga associations. Updated annually as a supplement to the January/February issue of *Yoga International*.

Yoga Journal's Yoga Teachers Directory

2054 University Ave., Suite 600
Berkeley, CA 94704
(800) 359-YOGA

Directory of who's who in yoga today in the United States, Canada and throughout the world. Updated annually in the July/August issue of *Yoga Journal;* also available as a special supplement.

Resources for Conscious Aging

Omega Institute for Holistic Studies
260 Lake Drive
Rhinebeck, NY 12572-3212
(800) 944-1001

Information on Workshops with the Author

Ojai Yoga Center
P.O. Box 1258
Ojai, CA 93024
(805) 646-4673
Fax: (805) 640-8232
e-mail: sfrancina@aol.com

About Yoga in the Ojai Valley

Surrounded by majestic mountains between Los Angeles and Santa Barbara, the Ojai Valley in California is one of the most beautiful and sacred places on earth—a true Shangri-la only 30 minutes from the ocean. Your yoga vacation can include daily classes at the Ojai Yoga Center, hiking, visits to nearby mineral hotsprings and many spiritually oriented organizations, art galleries and cultural events. The Ojai Yoga Center is a fully equipped center located within easy walking distance of hotels, motels and shops. Write or call for a brochure.

Bibliography

Bringing home a book to read is like having the author over for a visit. I've always been an avid reader, and it is impossible to list all the books that have influenced my perspective on yoga and life. The following publications were consulted during the writing of *The New Yoga for People Over 50,* and many are quoted or cited in the text. Many more excellent books are available on all aspects of yoga, health, the stages of life, aging, conscious dying and death. I encourage you to browse through your local bookstore and read those books that speak to you.

Yoga, Health and Aging

Alberg, Maria. *The Yoga Workbook for Seniors,* Sandpoint, Idaho: Moon in the Pearl, 1993.

Beauvoir, Simone de. *The Coming of Age.* New York: G.P. Putnam's Sons, 1972.

Beeken, Jenny. *Yoga of the Heart.* Hampshire, England: White Eagle Publishing Trust, 1990.

Bell, Lorna, R.N. and Eudora Seyfer. *Gentle Yoga for People with Arthritis, Stroke Damage, Multiple Sclerosis or People in Wheelchairs.* Berkeley, Calif.: Celestial Arts, 1987.

Bender, Ruth. *Be Young and Flexible After 30, 40, 50, 60.* Avon, Conn.: Ruben Publishing, 1976.

_____. *Yoga Exercises for Every Body.* Avon, Conn.: Ruben Publishing, 1976.

Bianchi, Eugene C. *Aging as a Spiritual Journey.* New York: Crossroad Publishing Co., 1989.

Brennan, Barbara. *Hands of Light: A Guide to Healing Through the Human Energy Field.* New York: Bantam Books, 1988.

Breslow, Rachelle. *Who Said So? A Woman's Fascinating Journey of Self-Discovery and Full Recovery from Multiple Sclerosis.* Berkeley, Calif.: Celestial Arts, 1991.

Burgio, Kathryn, Ph.D. *Staying Dry: A Practical Guide to Bladder Control.* Baltimore: Johns Hopkins University Press, 1989.

Chopra, Deepak. *Ageless Body, Timeless Mind: The Quantum Alternative to Growing Old.* New York: Harmony Books, 1993.

_____. *Perfect Health: The Complete Mind/Body Guide.* New York: Harmony Books, 1991.

_____. *Quantum Healing, Exploring the Frontiers of Mind/Body Medicine.* New York: Bantam Books, 1989.

_____. *Unconditional Life.* New York: Bantam Books, 1992.

Christensen, Alice, and David Rankin. *Easy Does It Yoga for People Over 60.* Cleveland: Saraswati Studio, 1975.

Clow, Barbara H. *Liquid Light of Sex: Understanding Your Key Life Passages.* Santa Fe: Bear & Company, 1991.

Couch, Jean. *The Runner's Yoga Book.* Berkeley, Calif.: Rodmell Press, 1990.

Cousins, Norman. *The Healing Heart.* New York: Avon, 1983.

Criswell, Eleanor. *How Yoga Works: An Introduction to Somatic Yoga.* Novato, Calif.: Freeperson Press, 1987.

Delany, Sarah, and Elizabeth Delany. *Having Our Say: The Delany Sister's First 100 Years.* New York: Kodansha America, 1993.

Desikachar, T.K.V. *Patanjali's Yoga Sutras.* Madras, India: Affiliated East–West Press, 1987.

Devi, Indra, *Forever Young, Forever Healthy.* Englewood Cliffs, N.J.: Prentice Hall, 1953.

Dossey, Larry. *Beyond Illness: Discovering the Experience of Health.* Boulder, Colo.: Shambala, 1984.

_____. *Meaning and Medicine: A Doctor's Stories of Breakthrough and Healing.* New York: Bantam Books, 1992.

_____. *Recovering the Soul: A Scientific and Spiritual Search.* New York: Bantam Books, 1989.

Douillard, John D.C. *Invincible Athletics: Awakening the Athlete in Everyone.* Lancaster, Mass.: Maharishi Ayur-Veda, 1991.

Dworkis, Sam. *ExTension: A Twenty-Minute-A-Day, Yoga-Based Program to Relax, Release and Rejuvenate the Average Stressed-Out Over 35-Year-Old Body.* New York: Poseidon Press, 1994.

Dychtwald, Ken, and Joe Flower. *Age Wave: The Challenges and Opportunities of an Aging America.* New York: Bantam Books, 1990.

Erdman, Mardi. *Undercover Exercise.* Englewood Cliffs, N.J.: Prentice Hall, 1984.

Estés, Clarissa Pinkola, Ph.D. *Women Who Run With the Wolves.* New York: Ballantine Books, 1992.

Evans, William, Ph.D., and Irwin Rosenberg, M.D. *Biomarkers: The 10 Keys to Prolonging Vitality.* New York: Simon & Schuster, 1991.

Feuerstein, Georg, and Stephen Bodian. *Living Yoga: A Comprehensive Guide for Daily Life.* New York: Putnam, 1993.

Folan, Lilias. *Yoga and Your Life.* New York: Macmillan, 1981.

Friedan, Betty. *The Fountain of Age.* New York: Simon & Schuster, 1993.

Gottlieb, Bill, ed. *New Choices in Natural Healing.* Emmaus, Pa.: Rodale Press, 1995.

Greer, Germaine. *The Change: Women, Aging and Menopause.* New York: Ballantine Books, 1991.

Groves, Dawn. *Yoga for Busy People.* San Rafael, Calif.: A New World Library, 1995.

Holloman, Dona. *Centering Down.* Italy: 1981.

Iyengar, Geeta S. *Yoga, a Gem for Women.* Spokane, Wash.: Timeless Books, 1990.

Jaidar, George. *The Soul: An Owner's Manual.* New York: Paragon House, 1995.

Johns Hopkins Medical Letter, eds. *The Johns Hopkins Medical Handbook: 100 Major Medical Disorders of People over 50.* New York: Rebus, Inc., 1992.

Kabat-Zinn, Jon. *Wherever You Go, There You Are: Mindfulness Meditation in Everyday Life.* New York: Hyperion, 1994.

Kelder, Peter. *Ancient Secret of the Fountain of Youth.* Gig Harbor, Wash.: Harbor Press, 1985.

Laird, Joan. *Ageless Exercise: A Gentle Approach for the Inactive or Physically Limited.* Williamsburg, Mich.: Angelwood Press, 1994.

Lasater, Judith, Ph.D., P.T. *Relax & Renew: Restful Yoga for Stressful Times.* Berkeley, Calif.: Rodmell Press, 1995.

Lieberman, Jacob. *Light, Medicine of the Future.* Santa Fe: 1991.

Luby, Sue, and Richard Onge. *Bodysense: Hazard Free Fitness Program for Men and Women.* Winchester, Mass.: Faber & Faber, 1986.

Main, Zilpha Pallister. *Reaching Ninety My Way.* Los Angeles: Zilpha Pallister Main, 1984.

Mehta, Mira. *How to Use Yoga: A Step-by-Step Guide to the Iyengar Method of Yoga, for Relaxation, Health and Well-Being.* New York: Smithmark, 1994.

Mehta, Silva, and Mira and Shyam Mehta. *Yoga the Iyengar Way.* New York: Alfred Knopf, 1990.

Montagu, Ashley. *Growing Young.* New York: Greenwood Press, 1989.

Moyers, Bill. *Healing and the Mind.* New York: Doubleday, 1993.

Murphy, Michael. *The Future of the Body.* Los Angeles: Jeremy P. Tarcher, Inc., 1992.

Myers, Esther. *Yoga & You: Energizing and Relaxing Yoga for New and Experienced Students.* Toronto, Canada: Random House of Canada, 1996.

Nelson, John E., M.D., and Andrea Nelson, Psy.D. *Sacred Sorrows: Embracing and Transforming Depression.* New York: Tarcher/Putnam, 1996.

Noble, Vicki. *Shakti Woman.* San Francisco: Harper, 1991.

Northrup, Christiane, M.D. *Women's Bodies, Women's Wisdom.* New York: Bantam Books, 1994.

O'Brien, Paddy. *Yoga for Women.* San Francisco: HarperCollins, 1994.

Ojeda, Linda. *Menopause Without Medicine.* Claremont, Calif.: Hunter House, 1989.

Ornish, Dean, M.D. *Dr. Dean Ornish's Program for Reversing Heart Disease.* New York: Random House, 1990.

Padus, Emrika, ed. *The Women's Encyclopedia of Health & Natural Healing.* Emmaus, Pa.: Rodale Press, 1981.

Pelletier, Kenneth R. *Longevity: Fulfilling Our Biological Potential.* New York: Dell Publishing Co., 1981.

Perez-Christiaens, Noëlle. *Sparks of Divinity.* Paris: Institut de Yoga B.K.S. Iyengar, 1976.

Pilgrim, Peace. Peace Pilgrim. Santa Fe: An Ocean Tree Book, 1982. Available from Friends of Peace Pilgrim, 43480 Cedar Ave., Hemet, CA 92344.

Radha, Swami Sivananda. *Hatha Yoga: The Hidden Language.* Boston: Shambhala, 1987.

Rountree, Cathleen. *On Women Turning 50.* San Francisco: Harper, 1993.

Scaravelli, Vanda. *Awakening the Spine.* New York: HarperCollins, 1991.

Schatz, Mary Pullig, M.D. *Back Care Basics: A Doctor's Gentle Yoga Program for Back and Neck Pain Relief.* Berkeley, Calif.: Rodmell Press, 1992.

Scheller, Mary Dale. *Growing Older, Feeling Better In Body Mind & Spirit.* Palo Alto, Calif.: Bull Publishing, 1993.

Sheehy, Gail. *New Passages: Mapping Your Life Across Time.* New York: Ballantine, 1995.

_____. *Pathfinders.* New York: Bantam Books, 1981.

_____. *The Silent Passage: Menopause.* New York: Random House, 1991.

Smith, Bob. *Yoga for a New Age: A Modern Approach to Hatha Yoga.* Englewood Cliffs, N.J.: Prentice Hall, 1982.

Steinem, Gloria. *Revolution From Within: A Book of Self-Esteem.* Boston: Little, Brown & Co., 1992.

Stewart, Mary. *Yoga Over 50: The Way to Vitality, Health and Energy in the Prime of Life.* New York: Simon & Schuster, 1994.

Tobias, Maxine, and John Patrick Sullivan. *Complete Stretching.* New York: Alfred A. Knopf, 1992.

Tobias, Maxine, and Mary Stewart. *Stretch & Relax.* Tucson, Ariz.: The Body Press (HPBooks), 1985.

Ueland, Brenda. *If You Want To Write.* Saint Paul, Minn.: Graywolf Press, 1987.

Walker, Barbara G. *The Crone: Women of Age, Wisdom, and Power.* San Francisco: Harper, 1988.

_____. *The Women's Encyclopedia of Myths and Secrets.* San Francisco: Harper, 1983.

Ward, Susan Winter. *Yoga for the Young at Heart: Gentle Stretching Exercises for Seniors.* Santa Barbara, Calif.: Capra Press, 1994.

Weed, Susun S. *Menopausal Years, the Wise Woman Way: Alternative Approaches for Women 30–90.* New York: Ash Tree Publishing, 1992.

Weininger, Ben, and Eva L. Menkin. *Aging Is a Lifelong Affair.* Los Angeles: Guild of Tutors Press, 1978.

White, Timothy. *The Wellness Guide to Lifelong Fitness.* Berkeley, Calif.: University of California at Berkeley Wellness Letter, 1993.

Whiteside, Robert L. *Agile at 80.* Pukalani, Hawaii: Robert L. Whiteside, 1989.

Wood, Beatrice. *Playing Chess with the Heart: Beatrice Wood at 100*. San Francisco: Chronicle Books, 1994.

Yogananda, Paramahansa. *Autobiography of a Yogi*. Los Angeles: Self-Realization Fellowship, 1946.

Books and Publications by and About B.K.S. Iyengar

The Art of Yoga. London: Unwin Paperbacks, 1985.

Body the Shrine, Yoga Thy Light. Bombay: published by B.I. Taraporewala for Iyengar's 60th birthday, 1978 (chapter on Yoga for Women by Geeta Iyengar).

Iyengar: His Life and Work. Palo Alto, CA: Timeless Books, 1987.

Light on Pranayama. New York: Crossroad, 1981.

Light on Yoga. New York: Schocken, 1979.

Light on the Yoga Sutras of Patanjali. London: HarperCollins, 1993.

70 Glorious Years of Yogachrya B.K.S. Iyengar. Bombay: Light on Yoga Research Trust, 1990.

Tree of Yoga. Boston: Shambhala, 1989.

Use of Props (Iyengar describes the benefits of props in old age); *Yoga and Medical Science: Yoga for Overall Health* (Iyengar); *Effect of Yogasanas on Metabolism of a Cell* (Dr. Karandikar); *Asana, Pranayama and the Circulatory System; Asana, Pranayama and Coronary Tuning; Asana, Pranayama and the Nervous System* (Dr. Krishna Raman); *Symposium on Hypertension,* and other related articles cited in text.

Effect of Asanas and Pranayama on the Endocrine System, (Dr. Karandikar); *Symposium: Women's Problems, with Geeta Iyengar; Yoga and Medical Science: Yoga for Overall Health* (Iyengar).

Publications on Conscious Dying and Death

Ansley, Helen Green. *Life's Finishing School—What Now—What Next? A Ninety Year Old's View of Death and Dying a Good Death*. Sausalito, Calif.: Institute of Noetic Sciences, 1990.

Graber, Anya Foos. *Deathing: An Intelligent Alternative for the Final Moments of Life*, rev. ed. With a preface by Ramamurti S. Mishra, M.D. York Beach, Maine: Nicolas–Hays Inc., 1992.

Krishnamurti, Jiddu. *On Living and Dying.* San Francisco: Harper, 1992.

Kübler-Ross, Elisabeth. *Death: The Final Stage of Growth.* Englewood Cliffs, N.J.: Prentice Hall, 1979.

Levine, Stephen. *Healing into Life and Death.* New York: Doubleday, 1987.

Nearing, Helen. *Loving and Leaving the Good Life.* Post Mills, Vt.: Chelsea Green Publishing Co., 1992.

Ring, Kenneth. *Life at Death.* New York: Coward, McCann & Geoghegan, 1988.

Rinpoche, Sogyal. *The Tibetan Book of Living and Dying.* San Francisco: Harper, 1992.

Wilber, Ken. *Grace and Grit: Spirituality and Healing in the Life and Death of Treya Killam Wilber.* Boston: Shambhala, 1991.

Periodicals and Additional Publications

Yoga Journal:

Burke, David. "Sri Chinmoy: Athlete of the Spirit" (September/October 1983).

Carrico, Mara. "Yoga with a Chair" (May/June 1986).

Cavanaugh, Carol. "Staying Young with Yoga" (September/October 1983).

_____. "Viparita Karani: Supported Inverted Pose" (November/December 1983).

Cogozzo, Linda. "Hatha After Hip Surgery" (May/June 1985).

Farhi, Donna. "Adho Mukha Svanasana: Downward-Facing Dog" (January/February 1994).

Francina, Suza. "Nutritional Aspects of Arthritis" (November/December 1977).

Hall, Rosemary. "Seated Sun Salutation" (March/April 1986).

Iyengar, B.K.S. "The Art of Relaxation: Savasana" (September/October 1982).

Kilmuray, Arthur. "The Safe Practice of Inversions" (November/December 1983); "Yoga: A Doctor's Prescription for Asthma" (May/June 1983); "Understanding Twists" (September/October 1984); "Sarvangasana: Shoulderstand" (September/October 1990); "Urdhva Dhanurasana-Upward-Facing Bow Pose" (November/December 1992); "Sirsasana-Headstand" (July/August 1990).

Lasater, Judith, Ph.D., RPT. "The Subtle Art of Standing Well" (September/October 1985); "Supta Virasana: Lying Down Hero Pose" (March/April 1986); "Uttanasana:

Intense Stretch Pose" (March/April 1988); "Yoga and Your Heart: Interview with Dean Ornish, M.D." (September/October 1989); "Sirsasana: Headstand" (May/June 1991); "How to Relax Deeply" (May/June 1992).

Moyer, Donald. "Baddha Konasana–Bound Angle Pose" (January/February 1987); "Virasana: Hero Pose" (March/April 1989); "Adho Mukha Svanasana: Downward Facing Dog Pose" (November/December 1989); "Ardha Matsyendrasana I: Lord of the Fishes Pose" (May/June 1992).

Myers, Esther. "Awakening the Spine With Vanda Scaravelli" (June 1996).

Sander, Ellen. "Moving Through Menopause With Yoga" (February, 1996).

Schatz, Mary, M.D. "Yoga Relief for Arthritis: A Pathologist and Yoga Teacher Offers Comprehensive Guidelines for Restoring and Maintaining Joint Health" (May/June 1985; reprints available from *Yoga Journal*); "Yoga, Circulation and Imagery" (January/February 1987); "Restorative Asanas for a Healthy Immune System" (July/August 1987); "You Can Have Healthy Bones! Preventing Osteoporosis with Exercise, Diet and Yoga" (March/ April 1988); "Exercises and Yoga Poses for Those at Risk for Osteoporotic Fractures" (March/April 1988); "Yoga and Aging" (May/June 1990); "Relief for Your Aching Back" (May/June 1992).

Steiger, Ruth. "Take-It-Easy Yoga" (November/December 1987).

Thomson, Bill. "Aging with Grace" (May/June 1990).

Wakefield, Dan. "Be Old Now" (September/October 1995).

White, Ganga. " Sting on Yoga" (December 1995).

Iyengar Yoga Institute Review:

Cavanaugh, Carol. "An Interview with Dr. S.V. Karandikar" (February 1984).

_____. "Vera Sida Interview" (November 1986).

Cole, Roger, Ph.D. "Physiology of Yoga" (October 1985).

Schatz, Mary Pullig, M.D. "Stress and Relaxation—Hypertension and Yoga" (March 1984).

Transcription of the video on *"Menopause"* produced by the Institute in Pune, India. Transcript by Kay Parry, with help from Janet and Susan Robertson. Edited by Geeta Iyengar.

The Journal of the International Association of Yoga Therapists:

Arpita, Ph.D., "Physiological and Psychological Effects of Hatha Yoga: A Review of the Literature" volume 1, nos. I & II (1990).

Chandra, F.J. "Medical and Physiological Aspects of Headstand" volume 1, nos. I and II (1990). (Article from booklet series by Dr. Chandra and Ian Rawlinson.)

_____. "Yoga and the Cardiovascular System" volume 2, no. I (1991).

Dreaver, Jim. "The Ultimate Cure: Enlightenment in Daily Life", nos. I and II, (1990).

Hymes, Alan, and Phil Nuernberger, Ph.D. "Breathing Patterns in Heart Attack Patients" 2, no. I (1991). (Article from *Research Bulletin of the Himalayan International Institute,* 2, no. 2, 1980).

Martin, Donna. "Chronic Pain and Yoga Therapy" 1, nos. I and II (1990).

Lyn, Brian. "Reflex (psychophysical yoga)" 1, nos. I and II (1990).

Mayer, Tania. "The Dance of Healing: Multiple Sclerosis and Yoga Therapy" 1, nos. I and II (1990).

Miller, Richard. "The Psychophysiology of Respiration: Eastern and Western Perspectives" 2, no. I (1991). (Article from booklet series published by Cambridge Yoga Publications.)

Miscellaneous Publications

Adolph, Jonathan. "The Wisdom Years: Five Reasons to Look Forward to Old Age." *New Age Journal,* (March/April 1992).

Blakeney, Laurie, Rose Richardson, Sue Salaniuk, and Toni Fuhrman. "Interview with B.K.S. Iyengar." Yoga '93 conference publication.

Clark, Etta. "Growing Old Is Not for Sissies." *The Sun* 196 (1992).

Dunn, Mary. "In Praise of Props: Utilizing the Mundane to Effect the Miraculous." Yoga '87 conference publication.

Eskenazi, Kay and Ruth Steiger. "Backbending Bench Usage Guide," "Eyesbag Usage Guide," "Halasana Bench Usage Guide," "Headstander Usage Guide," "Pelvic Swing Usage Guide, Pranayama Bolster Usage Guide, Wall Ropes Usage Guide." These booklets are highly recommended and may be purchased through Yoga Props. Please see address, page 268.

Hendrix, Paula. "Confessions of a Nursing Home Worker" and "Natural Dying." *In Context* no. 31 (1992).

Johns Hopkins White Papers, 1993: Arthritis, Coronary Heart Disease, Hypertension.

Schatz, Mary Pullig, M.D. "Minimizing Pain: The Principles of Therapeutic Yoga." Yoga '87 conference publication.

Weil, Andrew. "High Blood Pressure: Controlling Hypertension Without Drugs". *East-West Journal* (May/June 1992).

Permissions *(Continued from page iv)*

About the Author

Suza Francina is a certified Iyengar Yoga instructor with more than 20 years of experience in the field of yoga and exercise therapy. Her interest in teaching older people began many years ago while she worked as a home health-care provider for elderly and convalescing people. Her relationship with her clients extended through the last years of their lives. Francina's articles on health and aging have appeared in *Yoga Journal, Women's Health Care—A Guide to Alternatives, The Holistic Health Handbook, American Yoga* and other publications. Her first book, a completely different *Yoga for People Over 50*, was published in 1977.

Suza Francina has lived in Ojai, California for over 40 years. She is director of the Ojai Yoga Center where she specializes in classes and workshops for people over 50. In addition to teaching yoga, Francina has a deep interest in spiritual politics—a new emerging paradigm that recognizes the sacred interconnection of all life. She is a spokesperson for sustainable lifestyles, serves on the Ojai City Council and writes extensively on health and environmental issues.

Index of Poses

AUTHOR'S NOTE: *This index does not list the more advanced postures demonstrated in this book. These are best learned under the guidance of a yoga teacher.*